METHODS IN MOLECULAR BIOLOGY

Series Editor
John M. Walker
School of Life and Medical Sciences
University of Hertfordshire
Hatfield, Hertfordshire, UK

For further volumes:
http://www.springer.com/series/7651

For over 35 years, biological scientists have come to rely on the research protocols and methodologies in the critically acclaimed *Methods in Molecular Biology* series. The series was the first to introduce the step-by-step protocols approach that has become the standard in all biomedical protocol publishing. Each protocol is provided in readily-reproducible step-by-step fashion, opening with an introductory overview, a list of the materials and reagents needed to complete the experiment, and followed by a detailed procedure that is supported with a helpful notes section offering tips and tricks of the trade as well as troubleshooting advice. These hallmark features were introduced by series editor Dr. John Walker and constitute the key ingredient in each and every volume of the *Methods in Molecular Biology* series. Tested and trusted, comprehensive and reliable, all protocols from the series are indexed in PubMed.

Computational Design of Membrane Proteins

Edited by

Irina S. Moreira

Department of Life Sciences, University of Coimbra, Coimbra, Portugal

Miguel Machuqueiro

BioISI, University of Lisboa, Lisboa, Portugal

Joana Mourão

Center for Neuroscience and Cell Biology, University of Coimbra, Coimbra, Portugal

Humana Press

Editors
Irina S. Moreira
Department of Life Sciences
University of Coimbra
Coimbra, Portugal

Miguel Machuqueiro
BioISI
University of Lisboa
Lisboa, Portugal

Joana Mourão
Center for Neuroscience and Cell Biology
University of Coimbra
Coimbra, Portugal

ISSN 1064-3745 ISSN 1940-6029 (electronic)
Methods in Molecular Biology
ISBN 978-1-0716-1467-9 ISBN 978-1-0716-1468-6 (eBook)
https://doi.org/10.1007/978-1-0716-1468-6

This Humana imprint is published by the registered company Springer Science+Business Media, LLC, part of Springer Nature.
The registered company address is: 1 New York Plaza, New York, NY 10004, U.S.A.

Preface

Membrane proteins are targeted by over 50% of all available pharmaceutical drugs, which highlights their central role in the intricate world of biological pathways modulating cellular processes and disease outcomes. However, membrane proteins complex structural machinery is still far from being well understood or characterized due to limitations in the experimental determination of their structure, function, and mechanism of action.

In silico methods have been pushing boundaries for impactful discoveries in biochemical research. Several areas of study addressing membrane proteins have benefited from the intertwinement of both experimental and computational approaches, namely membrane protein design, 3D structure prediction, drug design, dynamic simulations, among others.

We split this volume of *Computational Design of Membrane Proteins* from the book series "Methods in Molecular Biology" into six parts, beginning with an introductory chapter (Chapter 1: Ferreira et al.) to the world of *in silico* portrayal of membrane proteins. An overview of major problems and accomplishments of the latest computational methods developed for these macromolecular systems were listed herein. The second part focuses on emerging tools developed within the field of synthetic biology: a multidisciplinary approach using both experimental work and computational analysis of the incorporation of a membrane transporter into a simple two Droplet Interface Bilayer system (Chapter 2: Findlay et al.); and two protein engineering methods that use point mutations to improve protein stability using the Rosetta software (Chapter 3: Alford et al.) or the FoldX force field (Chapter 4: Junk et al.).

In the third part, there are different methods addressing protein structure prediction in membranes: a homology modeling approach based on the MODELLER software to obtain new G-Protein-Coupled Receptors (GPCR) structures (Chapter 5: Tiss et al.); and two different protocols to calculate the interface prediction (Chapter 6: Nemoto et al.) and docking poses between transmembrane helices (Chapter 7: Hurwitz et al.).

The fourth part focuses on good practices and recommendations to set up membrane molecular dynamics (MD) simulations, either using coarse-grained force fields to increase the timescales of large membrane complexes (Chapter 8: Corey et al., Chapter 9: Nair et al.) or using atomistic simulations of smaller membrane-active peptides (Chapter 10: Frazee et al.).

Part five highlights the efforts from the community to improve the realism of the models used in MD simulations: by calculating pK_a values of key protein residues in membrane environment (Chapter 11: Oliveira et al.); by estimating proper pH-dependent protonation states to use in MD simulations (Chapter 12: Gupta et al.); and by introducing polarizable force fields to better describe these unique heterogeneous environments (Chapter 13: Lemkul).

The final part deals with membrane permeability and passive transport where the presented protocols help to calculate: membrane permeabilities of small molecules (Chapter 14: Ferreira), the potential of these molecules to alter membrane properties and behave as pan assay interference compounds (PAINS) (Chapter 15: Magalhães et al.), and their ability to transport anions using halogen bonds (Chapter 16: Costa).

Altogether, these innovative methods comprise an important contribution to the field of membrane proteins and can be the most desired complement to the available experimental methods which still struggle to obtain information at the molecular level on these systems.

Coimbra, Portugal — *Irina S. Moreira*
Lisboa, Portugal — *Miguel Machuqueiro*
Coimbra, Portugal — *Joana Mourão*

Contents

Contributors

REBECCA F. ALFORD • *Department of Chemical and Biomolecular Engineering, Johns Hopkins University, Baltimore, MD, USA*

PAULA J. BOOTH • *Department of Chemistry, Kings College London, London, UK*

RYM BEN BOUBAKER • *UMR CNRS 6015 – INSERM 1083, Laboratoire MITOVASC, Université d'Angers, Angers, France*

VIOLETA BURNS • *C. Eugene Bennett Department of Chemistry, West Virginia University, Morgantown, WV, USA*

MARIE CHABBERT • *UMR CNRS 6015 – INSERM 1083, Laboratoire MITOVASC, Université d'Angers, Angers, France*

ROBIN A. COREY • *Department of Biochemistry, University of Oxford, Oxford, UK*

PAULO J. COSTA • *University of Lisboa, Faculty of Sciences, BioISI - Biosystems & Integrative Sciences Institute, Lisbon, Portugal*

RICARDO J. FERREIRA • *Red Glead Discovery AB, Lund, Sweden*

HEATHER E. FINDLAY • *Department of Chemistry, Kings College London, London, UK*

NICOLAS FRAZEE • *C. Eugene Bennett Department of Chemistry, West Virginia University, Morgantown, WV, USA*

ALEMAYEHU A. GORFE • *Department of Integrative Biology and Pharmacology, McGovern Medical School, University of Texas Health Science Center, Houston, TX, USA; MD Anderson Cancer Center UTHealth Graduate School of Biomedical Sciences, University of Texas Health Science Center, Houston, TX, USA*

RAQUEL P. GOUVEIA • *Center for Neuroscience and Cell Biology, Center for Innovative Biomedicine and Biotechnology, University of Coimbra, Coimbra, Portugal*

JEFFREY J. GRAY • *Department of Chemical and Biomolecular Engineering, Johns Hopkins University, Baltimore, MD, USA; Program in Molecular Biophysics, Johns Hopkins University, Baltimore, MD, USA*

HAJER GUISSOUMA • *Laboratoire de Génétique, Immunologie et Pathologies Humaines, Dé partement de Biologie, Faculté des Sciences de Tunis, Université de Tunis El Manar, Tunis, Tunisie*

M. R. GUNNER • *Department of Physics, City College of New York, New York, NY, USA; Department of Physics, City University of New York, New York, NY, USA*

CHITRAK GUPTA • *The School of Molecular Sciences, Arizona State University, Tempe, AZ, USA; Biodesign Institute, Arizona State University, Tempe, AZ, USA*

NICOLA J. HARRIS • *Department of Chemistry, Kings College London, London, UK*

DANIEL HENRION • *UMR CNRS 6015 – INSERM 1083, Laboratoire MITOVASC, Université d'Angers, Angers, France*

NAAMA HURWITZ • *Blavatnik School of Computer Science, Tel Aviv University, Tel Aviv, Israel*

PHILIPP JUNK • *Systems Biology Ireland and UCD Charles Institute of Dermatology, School of Medicine, University College Dublin, Dublin 4, Ireland*

UMESH KHANIYA • *Department of Physics, City College of New York, New York, NY, USA; Department of Physics, City University of New York, New York, NY, USA*

CHRISTINA KIEL • *Systems Biology Ireland and UCD Charles Institute of Dermatology, School of Medicine, University College Dublin, Dublin 4, Ireland*

Justin A. Lemkul • *Department of Biochemistry, Virginia Tech, Blacksburg, VA, USA*
Miguel Machuqueiro • *Faculty of Sciences, BioISI – Biosystems and Integrative Sciences Institute, University of Lisboa, Lisboa, Portugal*
Pedro R. Magalhães • *Faculty of Sciences, BioISI – Biosystems and Integrative Sciences Institute, University of Lisboa, Lisbon, Portugal*
Junjun Mao • *Levich Institute, City College of New York, New York, NY, USA*
Catarina Marques-Pereira • *Center for Neuroscience and Cell Biology, Center for Innovative Biomedicine and Biotechnology, University of Coimbra, Coimbra, Portugal; PhD Programme in Experimental Biology and Biomedicine, Institute for Interdisciplinary Research (IIIUC), University of Coimbra, Coimbra, Portugal*
Blake Mertz • *C. Eugene Bennett Department of Chemistry, West Virginia University, Morgantown, WV, USA; WVU Cancer Institute, West Virginia University, Morgantown, WV, USA*
Irina S. Moreira • *Department of Life Sciences, University of Coimbra, Coimbra, Portugal*
Joana Mourão • *Center for Neuroscience and Cell Biology, University of Coimbra, Coimbra, Portugal*
Vinay V. Nair • *Department of Integrative Biology and Pharmacology, McGovern Medical School, University of Texas Health Science Center, Houston, TX, USA; MD Anderson Cancer Center UTHealth Graduate School of Biomedical Sciences, University of Texas Health Science Center, Houston, TX, USA*
Wataru Nemoto • *Division of Life Science, Department of Science and Engineering, School of Science and Engineering, Tokyo Denki University (TDU), Tokyo, Japan; Division of Life Science and Engineering, Graduate School of Science and Engineering, Tokyo Denki University (TDU), Tokyo, Japan*
Nuno F. B. Oliveira • *BioISI – Biosystems and Integrative Sciences Institute, Faculdade de Ciências da Universidade de Lisboa, Lisbon, Portugal*
Pedro B. P. S. Reis • *Faculty of Sciences, BioISI – Biosystems and Integrative Sciences Institute, University of Lisboa, Lisbon, Portugal*
Nícia Rosário-Ferreira • *Coimbra Chemistry Center, Department of Chemistry, University of Coimbra, Coimbra, Portugal; Center for Neuroscience and Cell Biology, Center for Innovative Biomedicine and Biotechnology, University of Coimbra, Coimbra, Portugal*
Akira Saito • *Division of Life Science, Department of Science and Engineering, School of Science and Engineering, Tokyo Denki University (TDU), Tokyo, Japan; Software Service, Inc., Osaka, Japan*
Mark S. P. Sansom • *Department of Biochemistry, University of Oxford, Oxford, UK*
Mrinal Shekhar • *Center for Development of Therapeutics, Broad Institute, Cambridge, MA, USA*
Tomás F. D. Silva • *BioISI – Biosystems and Integrative Sciences Institute, Faculdade de Ciências da Universidade de Lisboa, Lisbon, Portugal*
Abhishek Singharoy • *The School of Molecular Sciences, Arizona State University, Tempe, AZ, USA; Biodesign Institute, Arizona State University, Tempe, AZ, USA*
Phillip J. Stansfeld • *School of Life Sciences and Department of Chemistry, University of Warwick, Coventry, UK*
Asma Tiss • *UMR CNRS 6015 – INSERM 1083, Laboratoire MITOVASC, Université d'Angers, Angers, France; Laboratoire de Génétique, Immunologie et Pathologies Humaines, Département de Biologie, Faculté des Sciences de Tunis, Université de Tunis El Manar, Tunis, Tunisie*

JOHN W. VANT • *The School of Molecular Sciences, Arizona State University, Tempe, AZ, USA; Biodesign Institute, Arizona State University, Tempe, AZ, USA*
BRUNO L. VICTOR • *Faculty of Sciences, BioISI – Biosystems and Integrative Sciences Institute, University of Lisboa, Lisbon, Portugal*
DIOGO VILA-VIÇOSA • *Faculty of Sciences, BioISI – Biosystems and Integrative Sciences Institute, University of Lisboa, Lisbon, Portugal*
HAIM J. WOLFSON • *Blavatnik School of Computer Science, Tel Aviv University, Tel Aviv, Israel*

Part I

Introduction

Chapter 1

Guardians of the Cell: State-of-the-Art of Membrane Proteins from a Computational Point-of-View

Nícia Rosário-Ferreira, Catarina Marques-Pereira, Raquel P. Gouveia, Joana Mourão, and Irina S. Moreira

Abstract

Membrane proteins (MPs) encompass a large family of proteins with distinct cellular functions, and although representing over 50% of existing pharmaceutical drug targets, their structural and functional information is still very scarce. Over the last years, in silico analysis and algorithm development were essential to characterize MPs and overcome some limitations of experimental approaches. The optimization and improvement of these methods remain an ongoing process, with key advances in MPs' structure, folding, and interface prediction being continuously tackled. Herein, we discuss the latest trends in computational methods toward a deeper understanding of the atomistic and mechanistic details of MPs.

Key words Membrane proteins, Computational methods, Interfaces, Molecular dynamics, Machine learning, Sequence–structure–function Prediction

1 Key Concepts

Biological membranes split the extra- and intracellular compartments of cells and balance the in- and outward traffic. Their macromolecular composition and ability to create polar and nonpolar environments outside and inside the membrane modulate complex biological pathways via several different proteins [1].

Membrane proteins (MPs) are a diverse group of proteins that can be a part of or interact with biological membranes, representing around a third of the proteome and over half of all existing drug targets [2]. MPs are usually classified as peripheral or transient and integral or permanent according to the type of interaction with the lipid bilayer. According to their structural variability, MPs encompass a wide range of cellular functions as they can be enzymes, transporters, receptors, ion channels, serve as cellular junctions, or as points of anchorage to other macromolecules [3]. MPs are indeed crucial for the modulation of cellular processes, which emphasizes the importance of understanding their structure–

Irina S. Moreira et al. (eds.), *Computational Design of Membrane Proteins*, Methods in Molecular Biology, vol. 2315, https://doi.org/10.1007/978-1-0716-1468-6_1,

function paradigm [4]—postulating that protein 3D structure is directly related to the function it exerts. The characterization of the route from protein sequence, to structure, and their specific function, as well as the dynamics between different structural conformations leading to an array of cellular functions, is of utmost importance for rational drug design and for the mechanistic understanding of existent ones [5]. MPs are peripheral (PMPs) when they establish reversible interactions with the lipids or other integral proteins in the membrane [6]. PMPs have different roles since the transient nature of their connection promotes the reshuffling of membrane lipids, which will, in turn, impact and modulate several cellular functions as a result of changes in the membrane dynamics [7]. As a consequence of both the composition and the nature of membrane protein–lipid interactions, several microenvironments can arise, impacting the membrane organization and communication between the intra and extracellular environments [7]. Integral MPs (IMPs) are a much more diverse group encompassing anchored and transmembrane proteins. Anchored MPs are classified as monotypic proteins since their position is fixed on one side of the biological membrane [3]. Contrarily, transmembrane proteins (TMPs) span the entire phospholipid bilayer [3] and can be of type I, type II, type III, and type IV according to the orientation and anchoring of the N- and C-termini concerning the membrane [8]. TMPs can also be categorized as α-helical or β-barrels according to their structure. The α-helical TMPs present different topologies and are classified according to the total helices number, boundaries, and membrane orientation [9]. Thereby, α-helical TMPs can be bitopic, if both N- and C-termini domains are anchored, and the protein only crosses the membrane once or polytopic, if the protein crosses the membrane multiple times, which can also happen in β-barrel TMPs [3].

2 Background

The several limitations found in the experimental determination of tridimensional (3D) structures of MPs have hindered the number of available structures in public databases compared to soluble proteins [10]. These difficulties can arise in initial phases, such as in the biosynthesis of MPs, leading to lower yield and functionality levels [11] or in the experimental determination of MPs' structure by commonly used approaches such as nuclear magnetic resonance (NMR), X-ray crystallography, and cryo-electron microscopy (Cryo-EM) [12]. One main limitation of both X-ray crystallography and Cryo-EM is the extremely challenging crystallization conditions [13] due to the presence of MPs hydrophobic transmembrane motifs [12]. Even when crystallization is successful, MPs often acquire incorrect folds resulting from a misleading conformation that mismatch the native one, thus hindering the retrieval of their structural data [12, 14]. Furthermore, whereas Cryo-EM and X-ray crystallography provide us with static portrayals of systems, NMR can describe their dynamics, but frequently, MP's exceed NMR's size limitations hampering its use for their structural analysis [15].

In recent years, the next-generation sequencing (NGS) revolution provided thousands of genome sequences and, subsequently, numerous proteins to be structurally characterized. However, these protein sequences fluctuate regarding amino acids number and composition, which can jeopardize, in terms of time, the experimental determination of MPs, raising even more the importance of computational methods to assess structural and functional differences. Non-redundant annotated sequences from the Universal Protein Resource (UniProt; https://www.uniprot.org) [16] are 1000× more than the number of structures available on the Protein Data Bank (PDB; https://www.rcsb.org) [17, 18] (as of September 14, 2020), including redundant structures. Furthermore, this difference is even more prominent if data on MPs are compared, since MP sequences in UniProt are 7000× over MP structures in PDB, highlighting once more the necessity to develop methods able to attain their 3D structure (Fig. 1). As such, despite the efforts made, only an underwhelming 3% of all the proteins deposited in the PDB [17, 18] are MPs.

MP-dedicated databases contain valuable information to be used in distinct research approaches [19–31]. Shimizu and coworkers [17] compared several databases regarding their scope, entries overlap, and consistency regarding annotations. They included some of the most important MP-dedicated databases such as the Membrane Proteins of Known Structure (mpstruct; https://blanco.biomol.uci.edu/mpstruc/) [24], Orientation of Proteins in Membranes (OPM; https://opm.phar.umich.edu) [25, 27], Protein Data Bank of Transmembrane Proteins (PDBTM; http://pdbtm.enzim.hu) [27], the extended mpstruct dataset, entries from PDB [17, 18] and entries from Structural Classification of Proteins (SCOP; http://scop.mrc-lmb.cam.ac.uk) [32, 33]. The authors' findings highlighted the need for uniform protein classification across platforms and exposed disparities in nomenclature hindering the use of such databases [17].

As mentioned, although experimental methods are still crucial for the determination of MP structure, folding, and interface prediction, they present severe limitations that can hinder the study of this complex and relevant group of proteins. The standard approach used for sequence–structure–function determination is not suitable for MPs since differences in the membranes' constitution affect their folding pre- and post-translationally leading to differences in topology, important to the function they exert [4]. Thus, computational methods present themselves as valuable

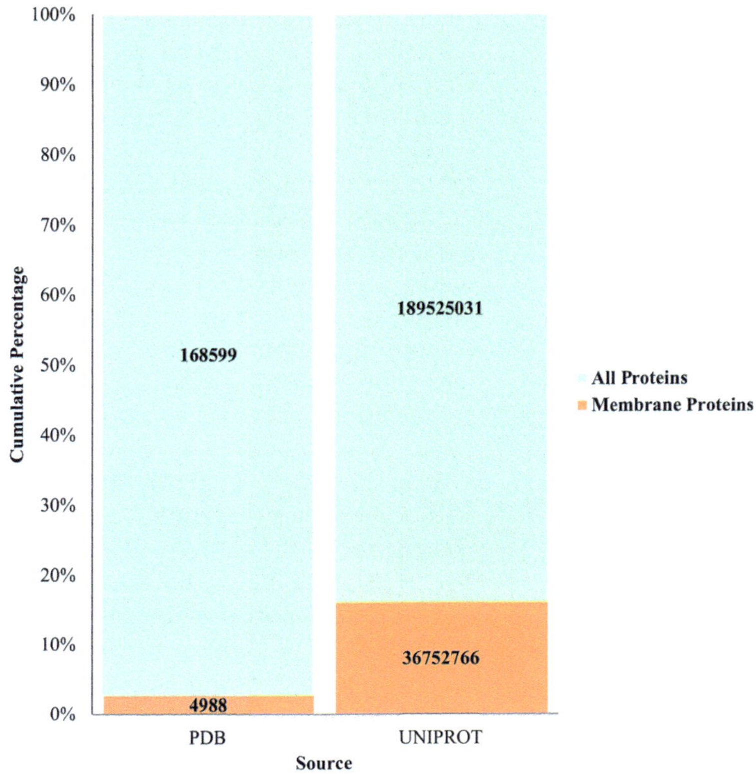

Fig. 1 A bar chart representing the percentage of PDB structures (left bar) and Uniprot annotated sequences (right bar). The total entries of all proteins are represented in blue and MP entries in orange (as of June 14, 2020). A significant discrepancy is observed not only in the number of sequences and structures but also in a data imbalance between all proteins and membrane proteins

tools to bridge the knowledge gap left by experimental approaches driving toward new insights about MP structure and function. From the structure–function paradigm, another realm of action is protein engineering, where, generally from mutations in the amino acid sequence, new or directed functions are achieved through the modulation of protein structure [34]. This approach is a stepping stone and an admitted holy grail for de novo design, where MP functions are shaped to manipulate biological pathways according to the need in question [35].

This chapter focuses solely on TMPs (ergo, for simplification, MPs) since they are crucial for most cell signaling processes, increasing their biological relevance [3]. Herein, we provide an overview of the current state of the art of computational methods to study MP sequence–structure–function (Fig. 2). We also encourage readers to look for other comprehensive literature reviews to attain a broader perspective [3, 12, 36, 37].

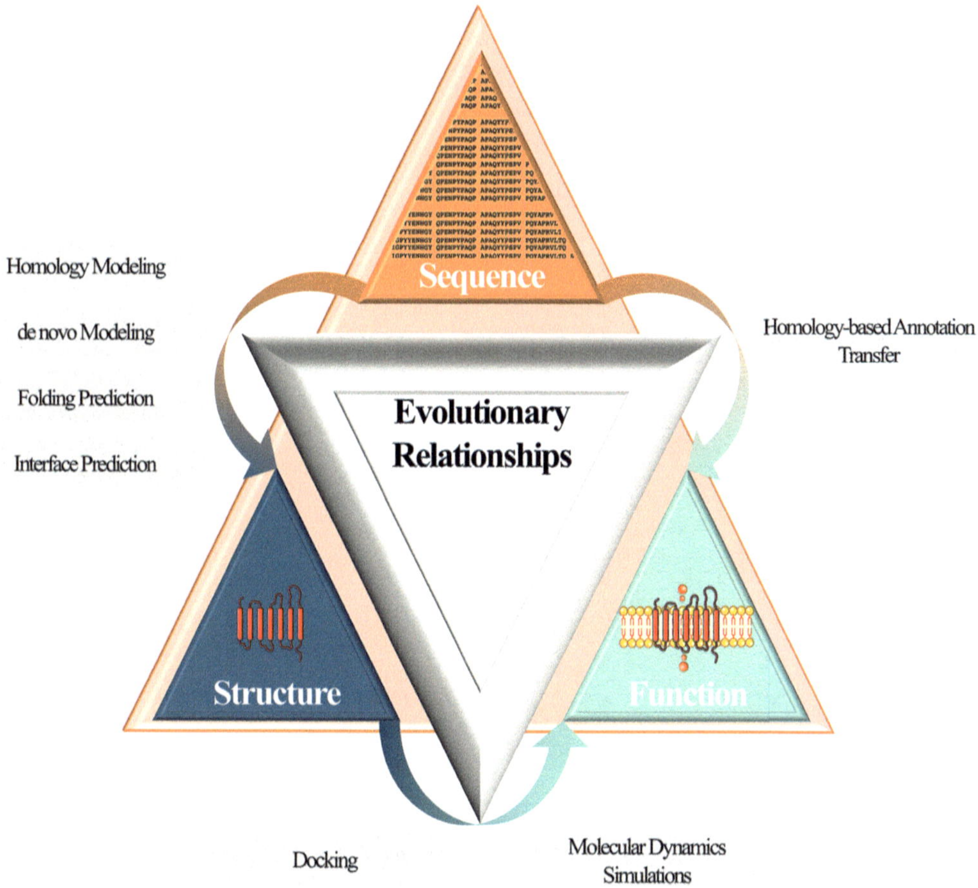

Fig. 2 Sequence–structure–function paradigm. Schematic figure depicting the prediction methods for structural-related features attainable from sequence data (orange to dark blue curved arrow), the methods available for understanding MP interactions and dynamics for function prediction from the structure (dark blue to the light blue curved arrow), and function prediction directly from the sequence (orange to the light blue curved arrow). The basis for this paradigm is that the relationship between sequence, structure, and function are evolutionary driven

3 Solving the Sequence–Structure–Function Paradigm

The sequence–structure–function paradigm highlights that different sequences can potentially lead to different structures, and, in turn, different structural conformations can potentially lead to different cellular functions [4]. Using protein sequences as a starting point and through the application of diverse in silico tools, it is possible to predict MP-related features. The increasing number of computational studies in PubMed used toward MP study reflect its relevance (Fig. 3a). Computational methods such as homology or de novo modeling, models used for interface and folding prediction, as well as docking and molecular dynamics (MD) simulations

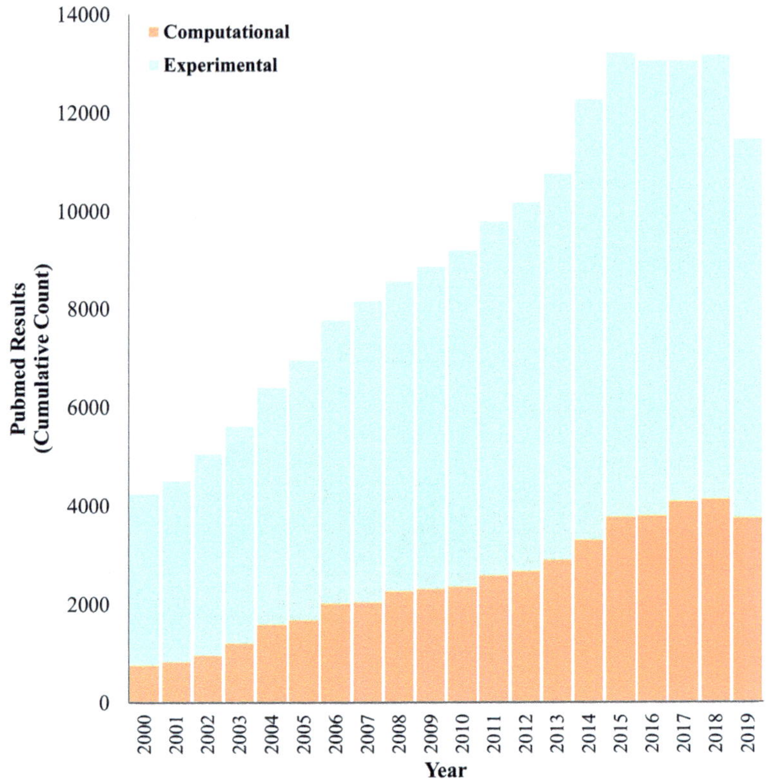

Fig. 3 Scientific publications retrieved from PubMed on September 14, 2020. (**a**) A bar chart representing the yearly number of papers of computational and experimental studies in membrane proteins between 2000 and 2019. PubMed criteria: (membrane protein) AND (experimental) and (membrane protein) AND (computational). (**b**) A bar chart representing the yearly number of in silico studies related to membrane proteins between 2000 and 2019. PubMed criteria: (membrane protein) AND (machine learning) and (membrane protein) AND (deep learning) and (membrane protein) AND (docking) and (membrane protein) AND (molecular dynamics) and (membrane protein) AND (homology modeling) and (membrane protein) AND (de novo modeling)

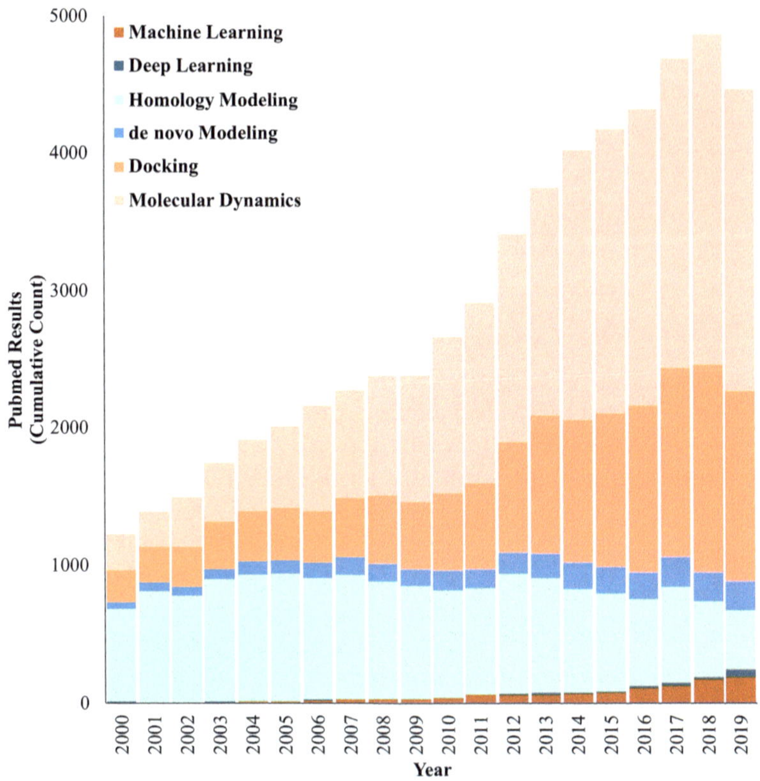

Fig. 3 (continued)

are increasingly used (Fig. 3b) to overcome some of the experimental limitations regarding MP structure and function prediction [38].

3.1 Structure Prediction

3.2 Homology Modeling

The main concept behind homology modeling is that proteins that share common ancestors display similarities, which, in turn, enables the comparison and prediction of their structure and function from sequence [39]. An initial high-quality Multiple Sequence Alignment (MSA) between the target and one or more protein templates is pivotal [36]. It can be performed using specific alignment algorithms that use substitution matrices to predict the likelihood of evolutionary amino acids substitutions. Specifically, in MP modeling, the high prevalence of loops and transmembrane (TM) regions must be directly addressed [40]. Different alignment methods, such as SLIM (Scorematrix Leading to Intra-Membrane domains) [41] and PHAT (Predicted Hydrophobic And Transmembrane) [42], with substitution matrices suited to the specific membrane environment of MPs, and that takes into consideration their evolutionary divergence were proposed [42]. Although homology modeling can predict non-MPs accurately, for MPs if the similarity between target and template sequences is beneath 30%, their structure prediction is primarily limited by the scant number of experimental 3D structures [43]. Homology modeling tools, like MEDELLER [44] and MEMOIR (MEMbrane prOtein modelleR) [45] software, are specific for MPs and can predict their structure through template–target sequence alignments. However, tools for 3D protein structure prediction, but nonspecific of MPs, like MODELLER [46], are also successfully used for MP structure prediction [47]. Comparative testing of MEDELLER versus MODELLER showed that MEDELLER outperformed MODELLER in the core model group with lower root mean square deviation (RMSD) values. Authors state that before FREAD (a loop-modeling prediction step), MEDELLER can predict the MPs' backbone structure more accurately. However, when it comes to higher coverage levels and lower sequence identities percentages, MEDELLER tends to fall behind. It seems to be mainly caused by the lower core modeling that leads to erroneous loop modeling, hence decreasing the overall quality and, simultaneously, leading to a lower confidence homology and incomplete models, which MODELLER counteracts by encompassing a refinement step [44]. More recently, MEMOIR's pipeline allowed it to generate 109 complete MP loop models out of the 150 tested [45]. The generated loop models were more precise than MODELLER's ab initio models [45] mainly due to the integration of MEDELLER to predict core coordinates and FREAD [48]. A more specific method, GPCRM, was developed to accurately predict G-protein-coupled receptors (GPCR) structures also using homology modeling, which encompasses MODELLER as a loop refinement tool [48, 49].

3.3 De Novo Modeling

When lacking a template 3D structure to use for structure prediction, *de novo* methods are crucial to overcoming this limitation. Despite the existence of several methods, a careful comparison of their performance is periodically assessed and reported in the last Critical Assessment of Protein Structure Prediction (CASP13)—a blind assessment of the current state of the field—where authors emphasized the progress of the latest de novo modeling tools [50]. The improvement of the de novo modeling tools observed at CASP13 was linked to the implementation of Deep Learning (DL) algorithms to residue distance prediction, such as RaptorX [51] used in PredMP software. PredMP [43] is a sequence-based software integrated into a web server that uses Deep Transfer Learning (DTL) [52] for de novo prediction of 3D structures of MPs. After an initial step of MP contact prediction from non-MP information, it uses this data as distance restraints toward the final prediction of the 3D structure, including a final step of embedding the structure in the lipid bilayer of the membrane. Differently from

PredMP that predicts distances between MPs residues [13], MemBrain, a webserver for MP structure prediction, applies machine-learning (ML) algorithms to predict MP residue contacts accurately [13]. MemBrain [13] comprises three prediction modules for α-helical MPs, namely MemBrain-TMH for structure, MemBrain-Contact for residue–residue contact maps, and MemBrain-Rasa for relative accessible surface area. Initially, all three modules used ML algorithms, but, recently, the second predictor, MemBrain-Contact [53] was updated. The authors demonstrated improved results when using a more powerful DL algorithm to analyze latent structural features of contact patterns in the full chain instead of only using the inter-helix contacts. Although MemBrain exclusively supports predictions regarding α-helical MPs, the developers want to adjust the method also to manage the prediction of β-barrel MPs [53].

AlphaFold [54], was the method with the highest accuracy in CASP13, particularly in de novo modeling and a hybrid approach that combined de novo with homology modeling (referred to as Free Modeling and Template-Based Modeling—FM/TBM—in the competition). After performing MSA, AlphaFold uses DLs to, first, predict the distances between pairs of amino acids and the angles between chemical bonds of those pairs of amino acids as constraints, and second, to predict possible structures. The information gathered is then combined into a score, which, after a gradient descent, will predict the lowest potential transcribed to the 3D structure of a protein [54]. Despite being developed initially to proteins in general, specific performance regarding MPs was unknown. AlphaFold system was recently used to predict the SARS-CoV2 protein 3D structures that comprises one MP. The results of this MP modeling agree with the experimental data demonstrating the potential of this method to MPs too (https://deepmind.com/research/open-source/computational-predictions-of-protein-structures-associated-with-COVID-19). Recently, an end-to-end (E2E) model was developed by AlQuraishi for structure prediction through a single Recurrent Geometric Network (RGN) model created in a learnable system and using Position-Specific Scoring Matrices (PSSMs) as input instead of co-evolutionary data [55]. RGNs' main parameterizations were torsional angles, cartesian descriptors, and a matrix of interatomic distances, making them suitable to complement the preexisting methods that incorporate coevolution data. This approach strengthens the ability of RGNs to predict folding over the secondary structure. However, PSSMs as an input introduce some limitations to the system since they are less robust and provide less information than MSAs. Nevertheless, one of the future E2E model improvements to make RGNs more powerful could be the exclusive use of raw sequences as input instead of PSSMs [55].

3.4 Folding Prediction

When only low similarity templates (generally <30% homology) are available, instead of focusing on the overall picture of MP structure, some methods focus on specific domains that characterize them such as folding prediction [56]. The abovementioned DTL method (used in de novo modeling by PredMP) was applied beforehand to predict MPs folding [52]. After an MSA of homologous proteins, sequential and pairwise features were extracted using RaptorX-Property [57, 58] and CCMpred [59], respectively. The DTL method uses two deep residual neural networks (ResNet) to predict MPs contact occurrence patterns and supply them as contact restraints for Crystallography and NMR System (CNS) [60]. To overcome the limited training data related to the scarcity of MP 3D solved structures, the authors trained this DL model with non-MP structures. Hence, the name deep transfer since they extrapolate from a non-MP trained model to MP 3D structure prediction. This method's success relies on the importance of pairwise- over sequence-based features since the results showed that contact maps for both non-MPs and MPs share similar occurrence frequencies. Ultimately, this method unveiled that by training the model using only the small MP experimental data available, not all the contact patterns could be predicted, which was then improved by covering non-MPs. RaptorX-Contact, a model that participated in the Continuous Automated Model EvaluatiOn (CAMEO) project and that was based on the previous DTL model, was the only webserver able to predict two MP folds from sequence data correctly. It benefits from distance- rather than contact-based folding calculations improved by DL algorithms such as ResNet trained with a distance matrix. Recently, Sibe [61], a sequence-based software applied ML and performed statistical analysis on MSA and amino acid evolutionary coupling features to predict protein folding.

3.5 Interface Prediction

MP interactions are challenging to analyze due to membrane fluidity which hampers interface characterization and prediction [62]. Membrane protein Binding residues Prediction (MBPred) [63], a sequence-based method, uses residue coevolution as a feature to accurately predict TMP interface amino acids. This method achieved better results than others specifically trained on globular proteins (e.g., Protein–protein interaction SItes prediction seVER (PSIVER) [64]), justifying the need for MP-dedicated tools. MBPred uses an ensemble of four random forest ML classifiers that differ in the region involved in the interactions (transmembrane, cytoplasmic, extracellular segments, and whole amino acid sequence). Another sequence-based predictor, MP-Ligand binding sites Predictor (MPLs-Pred) [65], uses a random forest classifier trained with features such as evolutionary profiles, topology structure, physicochemical properties, and primary sequence segment descriptors to predict and discriminate ligand-binding residues into drug-like metal or biomacromolecule categories. More recently, tools dedicated to GPCR oligomers, such as G-protein-coupled Receptors Interaction Partners (GRIP) [66] and GPCR–GPCR Interaction pair Predictor (GGIP) [67], were also developed. GRIP, a recently updated structure-based method, aims to identify specific subtypes of GPCR oligomer interfaces by calculating conserved residues among transmembrane helices [68]. GGIP, a Support Vector Machine (SVM)-ML hybrid predictor method trained with structure- and sequence-based features, was successfully applied for protein interface prediction, especially for MP pair prediction [67]. Computational models for interface prediction are beginning to accommodate interface changes over time, which provide valuable insights regarding MP interactions when combined with experimental results.

3.6 Characterization of MPs

3.6.1 Docking

Docking approaches aim to provide structural insights (mainly molecule orientation upon interaction) on protein-protein complexes and protein–ligand interactions, critical for successful rational drug design [69]. MP docking predictions are rapidly evolving despite the lagging of available methods when compared to soluble proteins, and, at the same time, non-MP approaches are being optimized for MPs docking altogether [69]. A falsely perceived advantage would be that MPs have reduced conformational space for interactions mainly due to transmembrane motifs that highly restrain movement. However, it is well known that MP interactions within the membrane play an essential role in their structure and conformation, strengthening the fact that they must be accurately predicted [69].

Docking algorithms can be split into rigid, flexible, and information-driven docking [70]. Rigid docking generates numerous conformations, ranked according to energy scores, with promising surface complementarity [71]. It can be accomplished through Fast Fourier Transform (FFT)-based methods [69, 71]. As MP side chain malleability regulates ligand interaction, it is important to consider protein flexibility [71]; hence, the use of flexible docking is typically advantageous. It can be performed by using Monte Carlo-based methods as they are capable of accepting or rejecting changes in thermodynamic states [71]. Manuel Pastor et al. [72] evaluated recently seven different docking approaches (ZDOCK [73], ClusPro [74], Hex [75], Global RAnge Molecular Matching (GRAMM)-X [76], PatchDock [77], SymmDock [78], and High-Ambiguity Driven DOCKing (HADDOCK) [79]) for experimentally known MPs using a blind docking approach with default parameters. The authors considered GRAMM-X as the best approach for MPs as it was able to generate a higher number of structures and accurate models. Since this docking software includes evolutionary conservation ponderation, it seems to be, according to authors, more effective in areas where interface residues are conserved [69, 72]. Manuel Pastor et al. mentioned the water solvation method as unfit for MP application [72].

Some docking approaches optimized for MPs that can be found in the literature are RosettaMP [80], MemDock [69], and HADDOCK [81]. RosettaMP is a MP-adapted framework, part of the Rosetta3 platform [82] encompassing different applications for different purposes. For example, MPdock, through a combination of RosettaMP and RosettaDock algorithm, predicts the ensemble structure of protein–protein complexes inserted in the biological membrane. The protein–protein complex undergoes a packing step before being placed into the biological membrane according to a combination of general docking with membrane-specific scoring functions. Furthermore, MPdock retrieved more than 50% of native contacts fulfilling the Critical Assessment of PRedicted Interactions (CAPRI) criteria for high-quality predictions [83]. In addition to MPdock, RosettaMP also encompasses an application specific for symmetric MP complexes (MPsymdock) that merges Rosetta symmetric protocol [84] onto RosettaMP. RosettaMP can have a broad scope of application going from their use alone or as an approach to complement other computational methods. Recently, RosettaMP has been updated with a new algorithm, mp_lipic_acc, which can predict MP residues exposed to lipids with 90% accuracy and speed, being suitable for different protein shapes. This algorithm can classify residues with lipid exposure by providing membrane coordinates [80]. MemDock [69] is an integrated docking algorithm that adapted the existing methods toward MPs. In the first step of the rigid docking, an adaptation of PatchDock [77, 85], Mem-PatchDock, is used, and for the refinement step, Mem-FiberDock based on FiberDock is applied [86, 87]. Mem-PatchDock and Mem-FiberDock consider the lipid membrane by implementing membrane-consistent orientation in docking and refinement steps [69]. All the steps of MemDock fit to MPs through the implementation of suited sampling and MemScore scoring functions. This approach showed higher docking accuracy and speed when compared to standard docking algorithms, including GRAMM-X [69]. HADDOCK [88], an information-driven docking method, was recently benchmarked using the mpstruct database. Despite the 48% success rate, the low number of acceptable models after the rigid docking refinement leaves the ground for substantial improvements in the near future [81]. HADDOCK uses user-provided information from both experimental and/or computational methods to define Ambiguous Interaction Restraints (AIRs) for interfacial residues allowing for further calculations on all distances between molecules to be docked. After considering all distances and orientations, HADDOCK calculates the final, binding score. Three steps of gradient-driven energy minimization and short MD simulations are required to make these predictions: it0, it1, and itw. In the first step it0, rigid docking is performed backed by AIRs; then, it1 introduces flexibility and increases the temperature whilst promoting the interface packing. Finally, during the itw step, the solvate is added, and the system undergoes short MD simulations. After a clustering analysis, the structures are ranked according to the HADDOCK score, a combined energy scoring function.

3.6.2 Molecular Dynamics Simulations

Molecular dynamics (MD) simulation is used to study macromolecules' dynamic behaviour from a privileged point of view. As a result, it is a valuable tool to deepen our understanding of macromolecular structure–function relationships [89, 90]. MD simulations are crucial for a multitude of uses spanning, but not limited to: protein docking [91], folding and insertion onto the membrane [92], protein aggregation [93], large conformational protein changes [94], protein–ligand binding [95], protein crowding [96], channel gating [97], and protein–lipid interactions [93, 98]. Adding to this list, MD is also useful for characterizing the interactions between the protein and other molecules, a pivotal step in the drug development pipeline [99–104]. Regarding MPs in particular, these simulations are useful in the prediction of the uptake of small molecules [105], identification of protein activity effects caused by interactions with lipids [106], and protein structure refinement [107]. Before setting an MD simulation, system size and timestep should be defined a priori to establish a feasible calculation span. Advances in both hardware and mathematical algorithms used in MD simulations have increased their performance, allowing for a longer run time of MD simulations on increasingly larger systems [108]. Sampling in MD simulations can be restrictive; however, enhanced sampling methods such as umbrella sampling [109] or metadynamics [110] were successfully used to overcome this limitation [111, 112]. Another shortcoming of MD simulations is the accuracy of the Force Fields (FFs)—mathematical expressions that can describe the dependence of the potential energy of a system on the coordinates of its atoms [104]—i.e., how accurate the simulation is compared to experimental data [113]. FFs are in continuous improvement [114–116].

To be able to run an MD simulation, the 3D structure of MPs must be available in a public database (e.g., PDB) [18, 113, 117] or previously modeled as described in Subheading 3.1 [18, 113, 117]. When the protein structure is unavailable, and there are no similar proteins that can be obtained for homology modeling, MD can be used as a physics-based ab initio modeling tool, by relying on its ability to predict protein folding. Since experimental methods to attain a protein in its native conformation are laborious, computational methods to accelerate MD simulations were already developed [118]. Recently, a Cryo-EM iterative Rosetta–MD protocol [119] was extended to MPs. In this protocol, an initial protein structure predicted via de novo modeling with a RMSD value under 5 Å is embedded in the membrane in its correct orientation by using adequate coordinates. Next, a density map is generated using the previous coordinates, and a new membrane embedding is done followed by minimization and equilibration steps with a refinement of the system via a density-guided MD simulation step. The last frame from the MD simulation is used as input for Rosetta1 density-guided local building and refinement [119]. This iterative step stops once the top five scoring models achieve a 6.9 Å for membrane refinement. When the protein is ready, it needs to be embedded in a lipid membrane, and to simulate better the biological environment, solvate and ions should be added to the system [120].

The lipid composition of biological membranes varies according to different conditions such as cell type, cell cycle phase, and environmental conditions. Since the interaction with specific lipids influences MP functions, protein behavior after such interactions must be assessed to avoid acquiring non-native conformations [121]. The lipids' location in the inner or outer leaflet of the membrane can also influence MP function; thus, it is important to create asymmetric bilayer membranes, with 0 leaflet tension [122]. Recently, researchers concluded that asymmetric bilayers are less likely to produce artifacts and created a method where the number of lipids in one leaflet of a preexisting membrane is adjusted to satisfy the 0 leaflet tension condition [122]. To automatically build the system, the CHARMM-GUI [123] tool can generate several input files that can be used to start the simulation and avoid an additional step of solvation.

Different approaches can be taken to insert the protein of interest into the membrane as inflategro or -membed (flag applied to the command gmx mdrun from the GROMACS software). The first, inflategro, starts by inflating the membrane before removing overlapping lipids, followed by deflation. The second, -membed, starts by contracting the protein width and removing overlapping molecules, followed by a gradual protein decompression until it reaches its full size in a short MD simulation. While these methods decrease the overall MD simulation time by reducing the time needed to perform equilibration, they jeopardize high-throughput processing since they require several cycles [124, 125]. Different multiscale methods can be used to achieve simultaneously or sequentially levels of resolution and overcome this drawback. Insertion and equilibration can be done using a Coarse-Grained (CG) approach and afterward converted to an All-Atom (AA) resolution before the production phase using methods like INSANE [124] and Backward [126]. The database MemProtMD [127] was recently developed to automatically identify new integral membrane proteins from PDB, insert them into a membrane, run a

CG simulation for 1μs, and transform the system to an atomic resolution to perform membrane–protein interaction analysis [128].

A mixture of implicit and explicit methods can be applied to represent the whole MP MD simulation system, comprising proteins, membrane, water, and ions [129–134]. An implicit solvent model is an approach that enhances the speed of MD simulations by treating the solvent (water) as a simplified continuum media while doing an atomistic representation of the solute (ions, MP, and ligand when applicable) [135]. Some methods were developed to attain an implicit solvent model, either based on electrostatics and the Poisson–Boltzmann equation, or semi-empirical, where experimental knowledge and theoretical concepts are combined. An example of the last one is like in EEF1, where experimental solvation free energy and enthalpy data are combined with theoretical knowledge to assess the effect of neighboring atoms solvation free energy [37]. In this EEF1 method, the membrane, yet not accurately portrayed, is defined as an infinite surface with distinct thickness and features from the water. The promising expertise toward modeling of protein–protein interactions through the ability to correctly predict interfaces, and protein features allow for multiple MD independent runs [136].

Implicit methods are useful to study large-scale phenomena of proteins [137], ligand docking [138], protein–protein interactions [139], creating free energy landscapes [140], and folding of proteins [141]. Also, MP stability fluctuates with pH; hence, pH oscillations can cause changes in protein folding and residue charges, disrupting local conformation and, subsequently, protein interactions [142, 143]. Some approaches attempt to describe biomolecular electrostatics and calculate pKa to study MPs through MD simulations accurately [144]. PDB2PQR-, PDB-format to PQR-format, and PBEQ are methods to solve the Poisson–Boltzmann equation and calculate continuum electrostatic potentials [145, 146]. However, when solvent-mediated interactions are important, or when the objective is to simulate membrane fluctuations caused by proteins, the implicit solvent method is not recommended due to the solvent's oversimplification [37, 147].

Explicit representations of the system, where the solvent molecules are defined, can be split into All-Atom or Atomistic Molecular Dynamics approach (AAMD) and Coarse-Grained Molecular Dynamics approach (CGMD) [37, 89]. While AAMD is computationally more demanding, hindering longer time scales beyond millisecond in large systems, it can offer more detailed information about lipid and protein local structural properties. On the other hand, CGMD, with a less accurate representation of the number of atoms in proteins and lipids, enables the system's MD simulations to reach longer time scales compared with the AAMD [37, 89,

104]. CGMD can be further divided according to how non-bonded interactions, interactions between side chains, or interactions between side chains and backbone are treated [148]. Models whose interactions rely on the initial protein structure, like Elastic Network Models (ENM) [149], and models whose interactions are defined like molecular mechanics FFs [37]. With their parameterization and low computational costs, are more convenient at making analyses like characterizing mechanisms of large allosteric transitions and Brownian dynamics [150]. Differently, interactions similarly defined to molecular mechanics FFs benefit from being less dependent on the initial structure of the protein, unlike ENM [37].

Depending on how the system is represented, different FFs are available. The FFs used for MPs need to have the proper parameterization, not only for the protein in study but also for the lipids in the membrane to ensure result reliability. FFs can be categorized according to the environment's ability to influence atoms' partial charge as addictive or non-polarizable and non-addictive or polarizable. Despite the apparent advantage of non-additive FFs, their high cost and effort to develop hindered is progression and usage [114], making addictive FFs the most commonly applied in MPs (some recent studies are listed in Table 1) [169]. These FFs are integrated into a wide diversity of software used to run MD

Table 1
Most commonly used addictive FFs in MD simulations and examples of their application to MP-related studies

			Applications	
Method	**Force field**	**Main features**	**General**	**MPs**
AAMD	AMBER	• High accuracy at reproducing experimental data • Good analytic tools provided for the MD analysis • Include a limited number of lipids	[108, 151, 152]	[153–155]
	CHARMM	• Good accuracy at reproducing experimental data • Include a large number of lipids • Head groups of lipids need additional parameterization	[108, 123, 156]	[157–164]
	GROMOS	• Good accuracy at reproducing experimental data • Include a reasonable number of lipids • Represent each nonpolar group of hydrocarbons as a single particle	[108]	[165]
CGMD	MARTINI	• Include a large number of lipids • Acceptable accuracy at reproducing experimental data • Normally represent four atoms with one CG bead • Lost atomistic resolution and information about properties derived from the individual atomic constitution	[108, 166]	[167, 168]

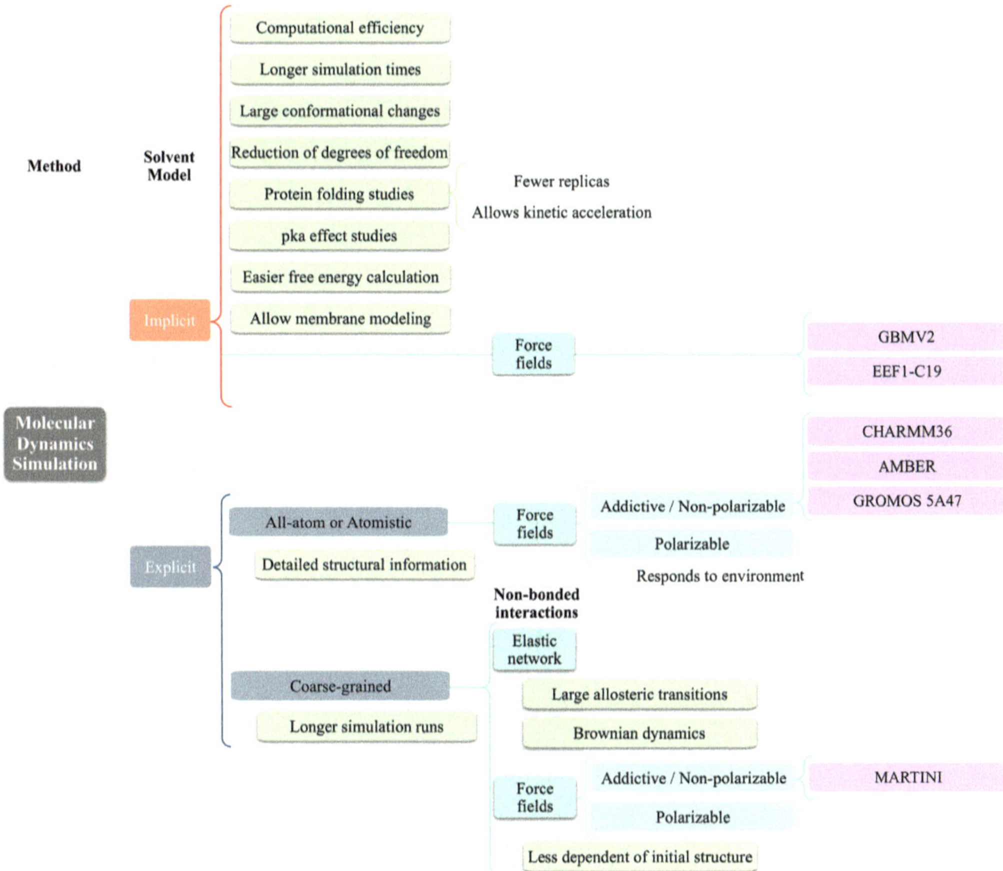

Fig. 4 Schematic diagram representing the main features of MD simulations according to the type of solvent model. The solvent model could be implicit when the solvent is represented as a continuum (focusing on solvent conditions) and where some properties are more easily studied (orange box), or explicit when the solvent is represented atomistically (focusing on incorporating solute molecules) and where analysis more detailed analysis that typically depends on the environment, such as regarding hydrogen bonds, salt bridges, and ion distribution are possible (dark gray box). The main advantages of using each model or approach are represented in light green boxes. The light pink boxes represent the main FF applicable to AA or CG MD simulations mentioned in this section

simulations, with the most commonly used being GROningen MAchine for Chemical Simulation (GROMACS) [170, 171], NAnoscale Molecular Dynamics (NAMD) [172], Chemistry at HARvard Molecular Mechanics (CHARMM) [173], and Assisted Model Building with Energy Refinement (AMBER) [174]. Figure 4 provides a graphical summary of the main characteristics of various types of MD simulation protocols.

In terms of MD simulation analysis, a wide range of features can be retrieved depending on the primary goal. Examples of quality and stability evaluation are the radius of gyration, RMSDs, and minimal distance evaluation [175]. The analysis of these features is essential to guarantee that the system equilibration was successfully attained and provided the conformation change throughout the simulation. It is well known that membrane energetics and K_A (area compressibility) values affect the MP properties. To understand and predict the membrane's mechanical properties, Weinstein and colleagues defined a new method to evaluate the membrane's compressibility in which successive constrained area simulations are performed. The study results allowed the authors to conclude that the obtained area compressibility is in good agreement with the available experimental data [176]. Future applications of this methodology include its application to the fluctuation of leaflet compressibility along regions of MPs. QwikMD [177], a relatively new VMD [178] plugin, can not only help set up the system to run MD simulations but also be used to make basic and advanced analyses of results. This year, the GPCRmd platform was created to enable users to visualize, analyze, and share MD data from GPCRs. By allowing the comparative analysis of several receptors, this platform was able to identify critical structural features that affect these receptor function [179]. More recently, new ML/DL approaches were developed to analyze MDs simulations, mainly for function prediction. Trilce Estrada group addressed the problem of predicting function from folding and conformational changes arising from MD simulations' trajectories using GEM-net [180]. This convolutional neural architecture retrieves information from images depicting structural and conformational changes, named graphic encodings. Weinstein's group [181] also applied ML algorithms to understand GPCR responses to different ligands by exploring their MD simulation differences. They used MD trajectories of GPCRs bound to ligands to extract information regardless of position and orientation data and feed it forward to train a Deep Neural Network (DNN) in the form of a graphical depiction. After validation and accuracy testing, a sensitivity protocol fetches important structural features to the interaction [181].

4 Future Perspectives and Conclusion

Herein, we provide an update of some computational methods used to study MPs structure, folding, interface, interactions, and dynamics. We observe high demand for adapting tools, typically used for non-MPs, toward MP usage or to create new, optimized ones. To the best of our knowledge, we still miss detailed studies comparing the available prediction models, useful to highlight their strengths, limitations, and other aspects with room for improvement. Concerning all the available models and software, there is not a universal one that stands out and outperforms the competition in all parameters; hence, the user must decide accordingly to the problem at hand.

With the NGS advances and increased available data, new and better databases can be created, which is highly beneficial to train and develop future in silico models. Whilst MD simulations can provide extra data and increasing sampling, big data meets improved DL algorithms with exciting outcomes, as listed throughout this chapter. The constant improvement of computational methods for MP study can fill some limitations of current experimental approaches and be a valuable tool to provide insights into MP structure and function as well as to improve the structural knowledge of membrane-protein interactions.

Acknowledgments

This work was funded by COMPETE 2020—Operational Programme for Competitiveness and Internationalisation and Portuguese national funds via FCT—Fundação para a Ciência e a Tecnologia, under projects DSAIPA/DS/0118/2020, POCI-01-0145-FEDER-031356, PTDC/QUI-OUT/32243/2017, UIDB/04539/2020 and LA/P/0058/2020. The authors would also like to acknowledge ERNEST—European Research Network on Signal Transduction, CA18133, and STRATAGEM —New diagnostic and therapeutic tools against multidrug-resistant tumors, CA17104. Nícia Rosário-Ferreira and Catarina Marques-Pereira were also supported by FCT, through Ph.D. scholarships PD/BD/135179/2017 and 2020.07766.BD (DOCTORATES 4 COVID-19), respectively. Raquel Pina Gouveia was also supported by FCT and DGES—Direção Geral do Ensino Superior, through fellowship for Curso de Verão in "Metodologias de Investigação Científica"—Módulo de I&D "Metodologias Avançadas para o Estudo do Cérebro", under the program "Verão com Ciência."

References

1. Luckey M (2016) Chapter 1. Introduction to the structural biology of membrane proteins. In: Domene C (ed) Computational biophysics of membrane proteins. Royal Society of Chemistry, Cambridge, pp 1–18
2. Gong J, Chen Y, Pu F et al (2019) Understanding membrane protein drug targets in computational perspective. Curr Drug Targets 20:551–564
3. Nugent T, Jones D, Hayat S (2017) Advances in computational methods for Transmembrane protein structure prediction. In: Rigden DJ (ed) From protein structure to function with bioinformatics. Springer Netherlands, Dordrecht, pp 135–165
4. Nwamba OC (2020) Membranes as the third genetic code. Mol Biol Rep 47:4093–4097
5. Fuxreiter M (2018) Towards a stochastic paradigm: from fuzzy ensembles to cellular functions. Molecules 23:3008
6. Monje-Galvan V, Klauda JB (2016) Peripheral membrane proteins: tying the knot between experiment and computation. Biochim Biophys Acta 1858:1584–1593
7. Whited AM, Johs A (2015) The interactions of peripheral membrane proteins with biological membranes. Chem Phys Lipids 192:51–59
8. Raj NN, Mahalekshmi T (2018) Multilabel classification of membrane protein in human

by decision tree (DT) approach. Biomed Pharmacol J 11:113–121
9. Nugent T, Jones DT (2009) Transmembrane protein topology prediction using support vector machines. BMC Bioinformatics 10:1–11
10. Yin H, Flynn AD (2016) Drugging membrane protein interactions. Annu Rev Biomed Eng 18:51–76
11. Pedro AQ, Queiroz JA, Passarinha LA (2019) Smoothing membrane protein structure determination by initial upstream stage improvements. Appl Microbiol Biotechnol 103:5483–5500
12. Almeida JG, Preto AJ, Koukos PI et al (2017) Membrane proteins structures: a review on computational modeling tools. Biochim Biophys Acta Biomembr 1859:2021–2039
13. Yin X, Yang J, Xiao F et al (2018) MemBrain: an easy-to-use online webserver for Transmembrane protein structure prediction. Nanomicro Lett 10:2
14. Liang B, Tamm LK (2016) NMR as a tool to investigate the structure, dynamics and function of membrane proteins. Nat Struct Mol Biol 23:468–474
15. Seeger MA (2018) Membrane transporter research in times of countless structures. Biochim Biophys Acta Biomembr 1860:804–808
16. UniProt Consortium (2019) UniProt: a worldwide hub of protein knowledge. Nucleic Acids Res 47:D506–D515
17. Shimizu K, Cao W, Saad G et al (2018) Comparative analysis of membrane protein structure databases. Biochim Biophys Acta Biomembr 1860:1077–1091
18. Berman HM, Westbrook J, Feng Z et al (2000) The Protein Data Bank. Nucleic Acids Res 28:235–242
19. Kim M-S, Yi G-S (2013) HMPAS: human membrane protein analysis system. Proteome Sci 11:S7
20. Pándy-Szekeres G, Munk C, Tsonkov TM et al (2018) GPCRdb in 2018: adding GPCR structure models and ligands. Nucleic Acids Res 46:D440–D446
21. Tsirigos KD, Bagos PG, Hamodrakas SJ (2011) OMPdb: a database of {beta}-barrel outer membrane proteins from Gram-negative bacteria. Nucleic Acids Res 39: D324–D331
22. Gromiha MM, Ou Y-Y (2014) Bioinformatics approaches for functional annotation of membrane proteins. Brief Bioinform 15:155–168
23. Elbourne LDH, Tetu SG, Hassan KA, Paulsen IT (2017) TransportDB 2.0: a database for exploring membrane transporters in sequenced genomes from all domains of life. Nucleic Acids Res 45:D320–D324
24. White SH (2009) Biophysical dissection of membrane proteins. Nature 459:344–346
25. Lomize MA, Lomize AL, Pogozheva ID, Mosberg HI (2006) OPM: orientations of proteins in membranes database. Bioinformatics 22:623–625
26. Tusnady GE, Dosztanyi Z, Simon I (2004) Transmembrane proteins in the Protein Data Bank: identification and classification. Bioinformatics 20:2964–2972
27. Kozma D, Simon I, Tusnády GE (2012) PDBTM: Protein Data Bank of transmembrane proteins after 8 years. Nucleic Acids Res 41:D524–D529
28. Fox NK, Brenner SE, Chandonia J-M (2014) SCOPe: structural classification of proteins—extended, integrating SCOP and ASTRAL data and classification of new structures. Nucleic Acids Res 42:D304–D309
29. Nastou KC, Tsaousis GN, Iconomidou VA (2020) PerMemDB: a database for eukaryotic peripheral membrane proteins. Biochim Biophys Acta Biomembr 1862:183076
30. Lomize AL, Hage JM, Pogozheva ID (2018) Membranome 2.0: database for proteome-wide profiling of bitopic proteins and their dimers. Bioinformatics 34:1061–1062
31. Saier MH Jr, Reddy VS, Tsu BV et al (2016) The transporter classification database (TCDB): recent advances. Nucleic Acids Res 44:D372–D379
32. Andreeva A, Howorth D, Chothia C et al (2014) SCOP2 prototype: a new approach to protein structure mining. Nucleic Acids Res 42:D310–D314
33. Andreeva A, Kulesha E, Gough J, Murzin AG (2020) The SCOP database in 2020: expanded classification of representative family and superfamily domains of known protein structures. Nucleic Acids Res 48:D376–D382
34. Poluri KM, Gulati K (2017) World of proteins: structure-function relationships and engineering techniques. In: Poluri KM, Gulati K (eds) Protein engineering techniques. Springer Singapore, Singapore, pp 1–25
35. Huang P-S, Boyken SE, Baker D (2016) The coming of age of de novo protein design. Nature 537:320–327
36. Koehler Leman J, Ulmschneider MB, Gray JJ (2015) Computational modeling of membrane proteins. Proteins 83:1–24
37. Latek D, Trzaskowski B, Niewieczerzał S et al (2019) Modeling of membrane proteins:

from bioinformatics to molecular quantum mechanics. In: Liwo A (ed) Computational methods to study the structure and dynamics of biomolecules and biomolecular processes. Springer International Publishing, Cham, pp 371–451
38. Moult J, Fidelis K, Kryshtafovych A et al (2014) Critical assessment of methods of protein structure prediction (CASP)--round x. Proteins 82(Suppl 2):1–6
39. Muhammed MT, Aki-Yalcin E (2019) Homology modeling in drug discovery: overview, current applications, and future perspectives. Chem Biol Drug Des 93:12–20
40. Kaczanowski S, Zielenkiewicz P (2010) Why similar protein sequences encode similar three-dimensional structures? Theor Chem Accounts 125:643–650
41. Müller T, Rahmann S, Rehmsmeier M (2001) Non-symmetric score matrices and the detection of homologous transmembrane proteins. Bioinformatics 17(Suppl 1):S182–S189
42. Ng PC, Henikoff JG, Henikoff S (2000) PHAT: a transmembrane-specific substitution matrix. Predicted hydrophobic and transmembrane. Bioinformatics 16:760–766
43. Wang S, Fei S, Wang Z et al (2019) PredMP: a web server for de novo prediction and visualization of membrane proteins. Bioinformatics 35:691–693
44. Kelm S, Shi J, Deane CM (2010) MEDELLER: homology-based coordinate generation for membrane proteins. Bioinformatics 26:2833–2840
45. Ebejer J-P, Hill JR, Kelm S et al (2013) Memoir: template-based structure prediction for membrane proteins. Nucleic Acids Res 41: W379–W383
46. Webb B, Sali A (2016) Comparative protein structure modeling using MODELLER. Curr Protoc Bioinformatics 54:5.6.1–5.6.37
47. Tabassum A, Rajeshwari T, Soni N et al (2014) Structural characterization and mutational assessment of podocin - a novel drug target to nephrotic syndrome - an in silico approach. Interdiscip Sci 6:32–39
48. Choi Y, Deane CM (2010) FREAD revisited: accurate loop structure prediction using a database search algorithm. Proteins 78:1431–1440
49. Miszta P, Pasznik P, Jakowiecki J et al (2018) GPCRM: a homology modeling web service with triple membrane-fitted quality assessment of GPCR models. Nucleic Acids Res 46:W387–W395
50. Kryshtafovych A, Schwede T, Topf M et al (2019) Critical assessment of methods of protein structure prediction (CASP)-round XIII. Proteins 87:1011–1020
51. Xu J, Wang S (2019) Analysis of distance-based protein structure prediction by deep learning in CASP13. Proteins 87:1069–1081
52. Wang S, Li Z, Yu Y, Xu J (2017) Folding membrane proteins by deep transfer learning. Cell Syst 5:202–211.e3
53. Yang J, Shen H-B (2018) MemBrain-contact 2.0: a new two-stage machine learning model for the prediction enhancement of transmembrane protein residue contacts in the full chain. Bioinformatics 34:230–238
54. Senior AW, Evans R, Jumper J et al (2020) Improved protein structure prediction using potentials from deep learning. Nature 577:706–710
55. AlQuraishi M (2019) End-to-End Differentiable Learning of Protein Structure. Cell Syst 8:292–301.e3
56. Pellowe GA, Booth PJ (2020) Structural insight into co-translational membrane protein folding. Biochim Biophys Acta Biomembr 1862:183019
57. Wang S, Li W, Liu S, Xu J (2016) RaptorX-property: a web server for protein structure property prediction. Nucleic Acids Res 44: W430–W435
58. Wang S, Peng J, Ma J, Xu J (2016) Protein secondary structure prediction using deep convolutional neural fields. Sci Rep 6:18962
59. Seemayer S, Gruber M, Söding J (2014) CCMpred--fast and precise prediction of protein residue-residue contacts from correlated mutations. Bioinformatics 30:3128–3130
60. Brünger AT, Adams PD, Clore GM et al (1998) Crystallography & NMR system: a new software suite for macromolecular structure determination. Acta Crystallogr D Biol Crystallogr 54:905–921
61. Cheung NJ, Yu W (2019) Sibe: a computation tool to apply protein sequence statistics to predict folding and design in silico. BMC Bioinformatics 20:455
62. Kharche SA, Sengupta D (2020) Dynamic protein interfaces and conformational landscapes of membrane protein complexes. Curr Opin Struct Biol 61:191–197
63. Zeng B, Hönigschmid P, Frishman D (2019) Residue co-evolution helps predict interaction sites in α-helical membrane proteins. J Struct Biol 206:156–169
64. Murakami Y, Mizuguchi K (2010) Applying the Naïve Bayes classifier with kernel density estimation to the prediction of protein-protein interaction sites. Bioinformatics 26:1841–1848

65. Lu C, Liu Z, Zhang E et al (2019) MPLs-Pred: predicting membrane protein-ligand binding sites using hybrid sequence-based features and ligand-specific models. Int J Mol Sci 20:3120
66. Nemoto W, Fukui K, Toh H (2009) GRIP: a server for predicting interfaces for GPCR oligomerization. J Recept Signal Transduct Res 29:312–317
67. Nemoto W, Yamanishi Y, Limviphuvadh V et al (2016) GGIP: structure and sequence-based GPCR-GPCR interaction pair predictor. Proteins 84:1224–1233
68. Saito A, Tsuchiya D, Sato S et al (2020) Update of the GRIP web service. J Recept Signal Transduct Res 40:348–356
69. Hurwitz N, Schneidman-Duhovny D, Wolfson HJ (2016) Memdock: an α-helical membrane protein docking algorithm. Bioinformatics 32:2444–2450
70. Halperin I, Ma B, Wolfson H, Nussinov R (2002) Principles of docking: an overview of search algorithms and a guide to scoring functions. Proteins 47:409–443
71. Pagadala NS, Syed K, Tuszynski J (2017) Software for molecular docking: a review. Biophys Rev 9:91–102
72. Kaczor AA, Selent J, Sanz F, Pastor M (2013) Modeling complexes of Transmembrane proteins: systematic analysis of protein–protein docking tools. Mol Inform 32:717–733
73. Chen R, Li L, Weng Z (2003) ZDOCK: an initial-stage protein-docking algorithm. Proteins 52:80–87
74. Alekseenko A, Ignatov M, Jones G et al (2020) Protein-protein and protein-peptide docking with ClusPro server. Methods Mol Biol 2165:157–174
75. Ghoorah AW, Devignes M-D, Smaïl-Tabbone M, Ritchie DW (2013) Protein docking using case-based reasoning. Proteins 81:2150–2158
76. Tovchigrechko A, Vakser IA (2005) Development and testing of an automated approach to protein docking. Proteins 60:296–301
77. Duhovny D, Nussinov R, Wolfson HJ (2002) Efficient unbound docking of rigid molecules. In: Guigó R, Gusfield D (eds) Algorithms in bioinformatics. Springer, Berlin, pp 185–200
78. Schneidman-Duhovny D, Inbar Y, Nussinov R, Wolfson HJ (2005) Geometry-based flexible and symmetric protein docking. Proteins 60:224–231
79. Dominguez C, Boelens R, Bonvin AMJJ (2003) HADDOCK: a protein-protein docking approach based on biochemical or biophysical information. J Am Chem Soc 125:1731–1737
80. Koehler Leman J, Lyskov S, Bonneau R (2017) Computing structure-based lipid accessibility of membrane proteins with mp_lipid_acc in RosettaMP. BMC Bioinformatics 18:115
81. Koukos PI, Faro I, van Noort CW, Bonvin AMJJ (2018) A membrane protein complex docking benchmark. J Mol Biol 430:5246–5256
82. Alford RF, Koehler Leman J, Weitzner BD et al (2015) An integrated framework advancing membrane protein modeling and design. PLoS Comput Biol 11:e1004398
83. Lensink MF, Wodak SJ (2014) Score_set: a CAPRI benchmark for scoring protein complexes. Proteins 82:3163–3169
84. André I, Bradley P, Wang C, Baker D (2007) Prediction of the structure of symmetrical protein assemblies. Proc Natl Acad Sci U S A 104:17656–17661
85. Schneidman-Duhovny D, Inbar Y, Nussinov R, Wolfson HJ (2005) PatchDock and SymmDock: servers for rigid and symmetric docking. Nucleic Acids Res 33: W363–W367
86. Mashiach E, Nussinov R, Wolfson HJ (2010) FiberDock: a web server for flexible induced-fit backbone refinement in molecular docking. Nucleic Acids Res 38:W457–W461
87. Mashiach E, Nussinov R, Wolfson HJ (2010) FiberDock: flexible induced-fit backbone refinement in molecular docking. Proteins 78:1503–1519
88. van Zundert GCP, Rodrigues JPGLM, Trellet M et al (2016) The HADDOCK2.2 web server: user-friendly integrative modeling of biomolecular complexes. J Mol Biol 428:720–725
89. Goossens K, De Winter H (2018) Molecular dynamics simulations of membrane proteins: an overview. J Chem Inf Model 58:2193–2202
90. Hospital A, Goñi R, Orozco M, Gelpi J (2015) Molecular dynamics simulations: advances and applications. Adv Appl Bioinform Chem 8:37–47
91. Alford RF, Smolin N, Young HS et al (2020) Protein docking and steered molecular dynamics suggest alternative phospholamban-binding sites on the SERCA calcium transporter. J Biol Chem 295:11262–11274
92. Im W, Brooks CL 3rd (2005) Interfacial folding and membrane insertion of designed peptides studied by molecular dynamics simulations. Proc Natl Acad Sci U S A 102:6771–6776

93. Kargar F, Emadi S, Fazli H (2020) Dimerization of Aβ40 inside dipalmitoylphosphatidylcholine bilayer and its effect on bilayer integrity: atomistic simulation at three temperatures. Proteins 88(11):1540–1552
94. Dhusia K, Su Z, Wu Y (2020) Understanding the impacts of conformational dynamics on the regulation of protein-protein association by a multiscale simulation method. J Chem Theory Comput 16:5323–5333
95. Radwan A, Mahrous GM (2020) Docking studies and molecular dynamics simulations of the binding characteristics of waldiomycin and its methyl ester analog to Staphylococcus aureus histidine kinase. PLoS One 15:e0234215
96. Goose JE, Sansom MSP (2013) Reduced lateral mobility of lipids and proteins in crowded membranes. PLoS Comput Biol 9:e1003033
97. Fagnen C, Bannwarth L, Oubella I et al (2020) New structural insights into Kir channel gating from molecular simulations, HDX-MS and functional studies. Sci Rep 10:8392
98. Jenkins K, Mateeva T, Szabó I et al (2020) Combining data integration and molecular dynamics for target identification in α-Synuclein-aggregating neurodegenerative diseases: structural insights on Synaptojanin-1 (Synj1). Comput Struct Biotechnol J 18:1032–1042
99. Adam Hospital, Goñi JR, Orozco M, Gelpí JL (2015) Molecular dynamics simulations: advances and applications. Adv Appl Bioinform Chem 8:37
100. Niesen MJM, Zimmer MH, Miller TF 3rd (2020) Dynamics of co-translational membrane protein integration and translocation via the sec translocon. J Am Chem Soc 142:5449–5460
101. Lee Y, Lazim R, Macalino SJY, Choi S (2019) Importance of protein dynamics in the structure-based drug discovery of class A G protein-coupled receptors (GPCRs). Curr Opin Struct Biol 55:147–153
102. Venko K, Roy Choudhury A, Novič M (2017) Computational approaches for revealing the structure of membrane transporters: case study on bilitranslocase. Comput Struct Biotechnol J 15:232–242
103. Harpole TJ, Delemotte L (2018) Conformational landscapes of membrane proteins delineated by enhanced sampling molecular dynamics simulations. Biochim Biophys Acta Biomembr 1860:909–926
104. Zheng L, Alhossary AA, Kwoh C-K, Mu Y (2019) Molecular dynamics and simulation. In: Encyclopedia of bioinformatics and computational biology. Elsevier, Amsterdam, pp 550–566
105. Ferreira RJ, Kasson PM (2019) Antibiotic uptake across gram-negative outer membranes: better predictions towards better antibiotics. ACS Infect Dis 5:2096–2104
106. Corey RA, Stansfeld PJ, Sansom MSP (2020) The energetics of protein–lipid interactions as viewed by molecular simulations. Biochem Soc Trans 48:25
107. Dutagaci B, Heo L, Feig M (2018) Structure refinement of membrane proteins via molecular dynamics simulations. Proteins 86:738–750
108. Loschwitz J, Olubiyi OO, Hub JS et al (2020) Computer simulations of protein–membrane systems. Prog Mol Biol Transl Sci 170:273
109. Kästner J (2011) Umbrella sampling. WIREs Comput Mol Sci 1:932–942
110. Barducci A, Bonomi M, Parrinello M (2011) Metadynamics. WIREs Comput Mol Sci 1:826–843
111. Ibrahim P, Clark T (2019) Metadynamics simulations of ligand binding to GPCRs. Curr Opin Struct Biol 55:129–137
112. Domański J, Hedger G, Best RB et al (2017) Convergence and sampling in determining free energy landscapes for membrane protein association. J Phys Chem B 121:3364–3375
113. Shiref H, Bergman S, Clivio S, Sahai MA (2020) The fine art of preparing membrane transport proteins for biomolecular simulations: concepts and practical considerations. Methods S1046-2023:30215-4
114. Marrink SJ, Corradi V, Souza PCT et al (2019) Computational modeling of realistic cell membranes. Chem Rev 119:6184–6226
115. Sandoval-Perez A, Pluhackova K, Böckmann RA (2017) Critical comparison of biomembrane force fields: protein-lipid interactions at the membrane interface. J Chem Theory Comput 13:2310–2321
116. Mustafa G, Nandekar PP, Mukherjee G et al (2020) The effect of force-field parameters on cytochrome P450-membrane interactions: structure and dynamics. Sci Rep 10:7284
117. Srivastava A, Nagai T, Srivastava A et al (2018) Role of computational methods in going beyond X-ray crystallography to explore protein structure and dynamics. Int J Mol Sci 19:3401
118. Dhingra S, Sowdhamini R, Cadet F, Offmann B (2020) A glance into the evolution of template-free protein structure prediction methodologies. Biochimie 175:85–92

119. Leelananda SP, Lindert S (2017) Iterative molecular dynamics-Rosetta membrane protein structure refinement guided by Cryo-EM densities. J Chem Theory Comput 13:5131–5145
120. Oakes V, Domene C (2019) Combining structural data with computational methodologies to investigate structure-function relationships in TRP channels. Methods Mol Biol 1987:65–82
121. Sanders CR, Mittendorf KF (2011) Tolerance to changes in membrane lipid composition as a selected trait of membrane proteins. Biochemistry 50:7858–7867
122. Doktorova M, Weinstein H (2018) Accurate in silico modeling of asymmetric bilayers based on biophysical principles. Biophys J 115:1638–1643
123. Jo S, Kim T, Iyer VG, Im W (2008) CHARMM-GUI: a web-based graphical user interface for CHARMM. J Comput Chem 29:1859–1865
124. Wassenaar TA, Ingólfsson HI, Böckmann RA et al (2015) Computational lipidomics with insane: a versatile tool for generating custom membranes for molecular simulations. J Chem Theory Comput 11:2144–2155
125. Wolf MG, Hoefling M, Aponte-Santamaría C et al (2010) g_membed: efficient insertion of a membrane protein into an equilibrated lipid bilayer with minimal perturbation. J Comput Chem 31:2169–2174
126. Wassenaar TA, Pluhackova K, Böckmann RA et al (2014) Going backward: a flexible geometric approach to reverse transformation from coarse grained to atomistic models. J Chem Theory Comput 10:676–690
127. Newport TD, Sansom MSP, Stansfeld PJ (2019) The MemProtMD database: a resource for membrane-embedded protein structures and their lipid interactions. Nucleic Acids Res 47:D390–D397
128. Stansfeld PJ, Goose JE, Caffrey M et al (2015) MemProtMD: automated insertion of membrane protein structures into explicit lipid membranes. Structure 23:1350–1361
129. Chitrak Gupta BM (2017) Protonation enhances the inherent helix-forming propensity of pHLIP. ACS Omega 2:8536
130. Pallante L, Rocca A, Klejborowska G et al (2020) In silico investigations of the mode of action of novel colchicine derivatives targeting β-tubulin Isotypes: a search for a selective and specific β-III tubulin ligand. Front Chem 8:108
131. Im W, Roux B (2002) Ion permeation and selectivity of OmpF porin: a theoretical study based on molecular dynamics, brownian dynamics, and continuum electrodiffusion theory. J Mol Biol 322:851–869
132. Gupta C, Ren Y, Mertz B (2018) Cooperative nonbonded forces control membrane binding of the pH-low insertion peptide pHLIP. Biophys J 115:2403
133. Mertz B, Feng J, Corcoran C, Neeley B (2015) Explaining the mobility of retinal in activated rhodopsin and opsin. Photochem Photobiol Sci 14:1952–1964
134. Faramarzi S, Feng J, Mertz B (2018) Allosteric effects of the proton donor on the microbial proton pump proteorhodopsin. Biophys J 115:1240
135. Liu Y, Haddadian E, Sosnick TR et al (2013) A novel implicit solvent model for simulating the molecular dynamics of RNA. Biophys J 105:1248–1257
136. Lazaridis T (2003) Effective energy function for proteins in lipid membranes. Proteins 52:176–192
137. Laradji M, Sunil Kumar PB, Spangler EJ (2016) Exploring large-scale phenomena in composite membranes through an efficient implicit-solvent model. J Phys D Appl Phys 49:293001
138. Kellici TF, Ntountaniotis D, Liapakis G et al (2019) The dynamic properties of angiotensin II type 1 receptor inverse agonists in solution and in the receptor site. Arab J Chem 12:5062–5078
139. Mizuhara Y, Parkin D, Umezawa K et al (2017) Over-destabilization of protein-protein interaction in generalized born model and utility of energy density integration cutoff. J Phys Chem B 121:4669–4677
140. Ulmschneider MB, Ulmschneider JP, Sansom MSP, Di Nola A (2007) A generalized born implicit-membrane representation compared to experimental insertion free energies. Biophys J 92:2338–2349
141. Ulmschneider JP, Ulmschneider MB (2007) Folding simulations of the transmembrane helix of virus protein U in an implicit membrane model. J Chem Theory Comput 3:2335–2346
142. Tanford C, Roxby R (1972) Interpretation of protein titration curves. Application to lysozyme. Biochemistry 11:2192–2198
143. Singharoy A, Barragan AM, Thangapandian S et al (2016) Binding site recognition and docking dynamics of a single electron transport protein: cytochrome c2. J Am Chem Soc 138:12077–12089
144. Chen J, Brooks CL 3rd, Khandogin J (2008) Recent advances in implicit solvent-based

methods for biomolecular simulations. Curr Opin Struct Biol 18:140–148

145. Jo S, Vargyas M, Vasko-Szedlar J et al (2008) PBEQ-solver for online visualization of electrostatic potential of biomolecules. Nucleic Acids Res 36:W270–W275

146. Dolinsky TJ, Nielsen JE, McCammon JA, Baker NA (2004) PDB2PQR: an automated pipeline for the setup of Poisson-Boltzmann electrostatics calculations. Nucleic Acids Res 32:W665–W667

147. Mori T, Miyashita N, Im W et al (2016) Molecular dynamics simulations of biological membranes and membrane proteins using enhanced conformational sampling algorithms. Biochim Biophys Acta 1858:1635

148. Zacharias M (2013) Combining coarse-grained nonbonded and atomistic bonded interactions for protein modeling. Proteins 81:81–92

149. Togashi Y, Flechsig H (2018) Coarse-grained protein dynamics studies using elastic network models. Int J Mol Sci 19:3899

150. Fuglebakk E, Reuter N, Hinsen K (2013) Evaluation of protein elastic network models based on an analysis of collective motions. J Chem Theory Comput 9:5618–5628

151. Dickson CJ, Madej BD, Skjevik AA et al (2014) Lipid14: the Amber lipid force field. J Chem Theory Comput 10:865–879

152. AMBAR 2020 Reference Manual. https://ambermd.org/doc12/Amber20.pdf. Accessed 10 July 2020

153. Pieńko T, Trylska J (2020) Extracellular loops of BtuB facilitate transport of vitamin B12 through the outer membrane of E. coli. PLoS Comput Biol 16:e1008024

154. Perona A, Ros MP, Mills A et al (2020) Distinct binding of cetirizine enantiomers to human serum albumin and the human histamine receptor H. J Comput Aided Mol Des 34:1045–1106

155. Guardia CM, Tan X-F, Lian T et al (2020) Structure of human ATG9A, the only transmembrane protein of the core autophagy machinery. Cell Rep 31:107837

156. CHARMM-GUI. http://www.charmm-gui.org/?doc=archive&lib=lipid. Accessed 10 July 2020

157. Loney RW, Panzuela S, Chen J et al (2020) Location of the hydrophobic surfactant proteins, SP-B and SP-C, in fluid-phase bilayers. J Phys Chem B 124:6763–6774

158. Jahan Sajib MS, Wei Y, Mishra A et al (2020) Atomistic simulations of biofouling and molecular transfer of a cross-linked aromatic polyamide membrane for desalination. Langmuir 36:7658–7668

159. Walczewska-Szewc K, Nowak W (2020) Structural determinants of insulin release: disordered N-terminal tail of Kir6.2 affects potassium channel dynamics through interactions with sulfonylurea binding region in a SUR1 partner. J Phys Chem B 124:6198–6211

160. Dos Santos EG, Faria RX, Rodrigues CR, Bello ML (2020) Molecular dynamic simulations of full-length human purinergic receptor subtype P2X7 bonded to potent inhibitors. Eur J Pharm Sci 152:105454

161. Neumann LSM, Dias AHS, Skaf MS (2020) Molecular modeling of aquaporins from. J Phys Chem B 124:5825–5836

162. Kwon B, Mandal T, Elkins MR et al (2020) Cholesterol interaction with the trimeric HIV fusion protein gp41 in lipid bilayers investigated by solid-state NMR spectroscopy and molecular dynamics simulations. J Mol Biol 432:4705–4721

163. Rojas G, Orellana I, Rosales-Rojas R et al (2020) Structural determinants of the dopamine transporter regulation mediated by G proteins. J Chem Inf Model 60:3577–3586

164. Woo H, Park S-J, Choi YK et al (2020) Developing a fully glycosylated full-length SARS-CoV-2 spike protein model in a viral membrane. J Phys Chem B 124:7128–7137

165. Ruiz FM, Lopez J, Ferrara CG et al (2020) Structural characterization of TssL from:a key component of the type VI secretion system. J Bacteriol 202:e00210–e00220

166. Li H, Gorfe AA (2013) Aggregation of lipid-anchored full-length H-Ras in lipid bilayers: simulations with the MARTINI force field. PLoS One 8:e71018

167. Gopal SM, Pawar AB, Wassenaar TA, Sengupta D (2020) Lipid-dependent conformational landscape of the ErbB2 growth factor receptor dimers. Chem Phys Lipids 230:104911

168. Prasanna X, Mohole M, Chattopadhyay A, Sengupta D (2020) Role of cholesterol-mediated effects in GPCR heterodimers. Chem Phys Lipids 227:104852

169. Inakollu VS, Geerke DP, Rowley CN, Yu H (2020) Polarisable force fields: what do they add in biomolecular simulations? Curr Opin Struct Biol 61:182–190

170. Abraham MJ, Murtola T, Schulz R et al (2015) GROMACS: high performance molecular simulations through multi-level parallelism from laptops to supercomputers. SoftwareX 1-2:19–25

171. Páll S, Abraham MJ, Kutzner C et al (2015) Tackling Exascale software challenges in molecular dynamics simulations with GROMACS. In: Markidis S, Laure E (eds) Solving software challenges for exascale. Springer International Publishing, Cham, pp 3–27
172. Phillips JC, Hardy DJ, Maia JDC et al (2020) Scalable molecular dynamics on CPU and GPU architectures with NAMD. J Chem Phys 153:044130
173. Brooks BR, Brooks CL 3rd, Mackerell AD Jr et al (2009) CHARMM: the biomolecular simulation program. J Comput Chem 30:1545–1614
174. Case DA, Belfon K, Ben-Shalom IY et al (2020) AMBER 2020. University of California, San Francisco
175. Patodia S (2014) Molecular dynamics simulation of proteins: a brief overview. J Phys Chem Biophys 4:6
176. Doktorova M, LeVine MV, Khelashvili G, Weinstein H (2019) A new computational method for membrane compressibility: bilayer mechanical thickness revisited. Biophys J 116:487–502
177. Ribeiro JV, Bernardi RC, Rudack T et al (2016) QwikMD - integrative molecular dynamics toolkit for novices and experts. Sci Rep 6:26536
178. Humphrey W, Dalke A, Schulten K (1996) VMD: visual molecular dynamics. J Mol Graph 14:33–38
179. Rodríguez-Espigares I, Torrens-Fontanals M, Tiemann JKS et al (2020) GPCRmd uncovers the dynamics of the 3D-GPCRome. Nat Methods 17:777–787
180. Carrillo-Cabada H, Benson J, Razavi A, et al (2019) A graphic encoding method for quantitative classification of protein structure and representation of conformational changes. IEEE/ACM Trans Comput Biol Bioinform, In Press
181. Plante A, Shore DM, Morra G et al (2019) A machine learning approach for the discovery of ligand-specific functional mechanisms of GPCRs. Molecules 24(11):2097

Part II

Methods in Synthetic Biology

Chapter 2

Integrating Membrane Transporter Proteins into Droplet Interface Bilayers

Heather E. Findlay, Nicola J. Harris, and Paula J. Booth

Abstract

Droplet interface bilayers (DIBs) are an emerging tool within synthetic biology that aims to recreate biological processes in artificial cells. A critical component for the utility of these bilayers is controlled flow between compartments and, notably, uphill transport against a substrate concentration gradient. A versatile method to achieve the desired flow is to exploit the specificity of membrane proteins that regulate the movement of ions and transport of specific metabolic compounds. Methods have been in existence for some time to synthesize proteins within a droplet as well as incorporate membrane proteins into DIBS; however, there have been few reports combining synthesis and DIB incorporation for membrane transporters that demonstrate specific, uphill transport. This chapter presents two methods for the incorporation of a membrane transporter into a simple two-droplet DIB system, with the downhill and uphill transport reaction readily monitored by fluorescence microscopy.

Key words Droplet interface bilayer, Membrane protein, Transporter, Fluorescence microscopy, Cell-free expression

1 Introduction

Biological membranes perform a variety of roles crucial to survival, from osmotic homeostasis and cell compartmentalization to photosynthesis and respiration. Membrane proteins are vital components of many of these processes, including signaling, enzymatic activity, and crucially controlling the import and export of important metabolites and waste products. A variety of synthetic biology approaches seek to create artificial cells using systems such as giant vesicles [1] or polymersomes [2] that can replicate biological functions. Control over the contents and communication between the compartments in these potential protocells are essential, and functionalizing with membrane proteins is an effective way to achieve this.

Irina S. Moreira et al. (eds.), *Computational Design of Membrane Proteins*, Methods in Molecular Biology, vol. 2315,
https://doi.org/10.1007/978-1-0716-1468-6_2,

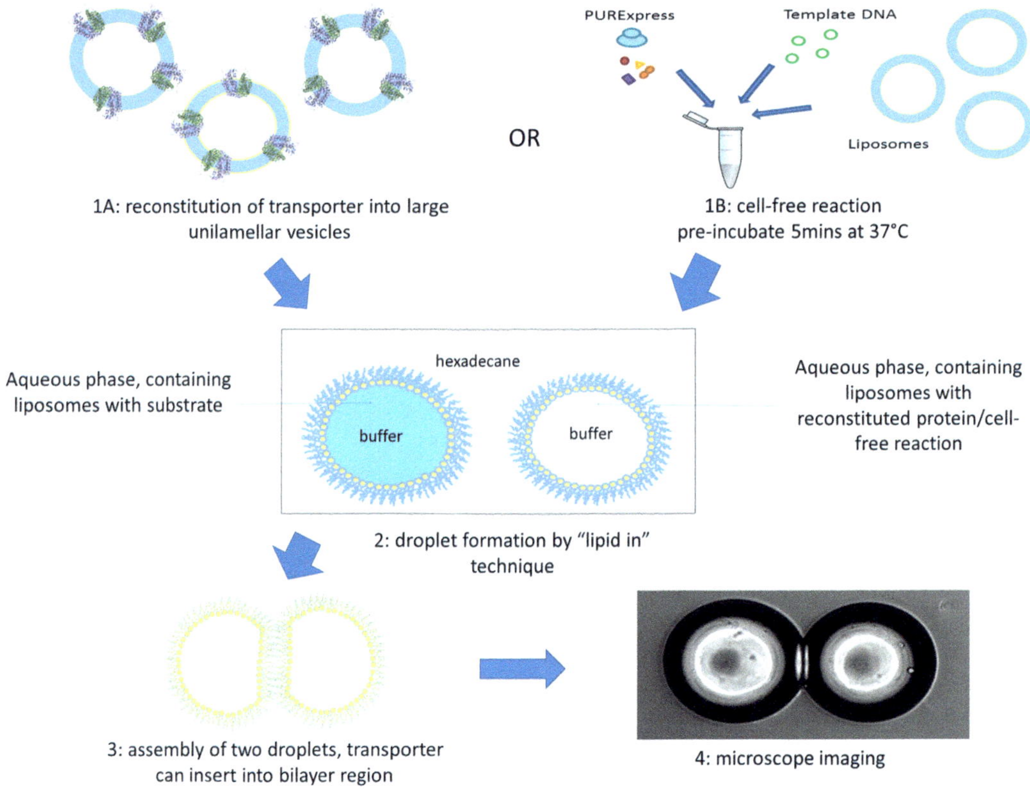

Fig. 1 A simple droplet interface bilayer system for membrane transporter proteins. Step 1: The aqueous solutions are prepared. One solution contains liposomes to provide the lipids for a lipid-in DIB assembly and the fluorescent substrate. The second solution contains reconstituted proteoliposomes, or a cell-free expression system including template DNA and empty liposomes. Step 2: Droplets from each solution are pipetted carefully into the solvent bath and equilibrated. Step 3: The two droplets are pushed together to form a bilayer interface. Step 4: A phase contrast microscopy image of the DIB. Fluorescence microscopy can be used to monitor transport activity. (Recreated from Findlay et al. (2016) [5])

One such system is the droplet interface bilayer (DIB) [3, 4]. This consists of nano- to micro-scale aqueous droplets in an oil-based solvent background, where the edge of the droplet is surrounded by a lipid monolayer with the polar headgroup facing the interior aqueous phase and the acyl tails in the solvent. Where two droplets are brought into contact, a bilayer is formed at the connection (Fig. 1). Many droplets can be combined using microfluidics or 3D printing techniques [6, 7], demonstrating promise as platforms for more complex biological applications such as the development of synthetic tissues. A variety of membrane proteins have been incorporated into the bilayer interface demonstrating a range of functions, such as the mechanosensitive channels MscL [8], the light-driven proton pump bacteriorhodopsin [9], and the multidrug transporter EmrE [10].

Two principal methods are used to create DIBs, the lipid-out and the lipid-in methods [3] The lipid-out method involves aqueous droplets being submerged in oil that contains dissolved lipids. The monolayer forms from the lipids in the solvent, typically over the course of around 30 min [11]. In the lipid-in technique, the lipids are instead donated by liposomes suspended within the aqueous droplet itself and the monolayer forms in a few minutes. Proteins can be added directly by dilution into the aqueous droplet, a method particularly suitable for toxins or membrane-active peptides that have a soluble form such as α-hemolysin and gramicidin [3]. For integral membrane proteins, other techniques have been used, including dilution from detergent micelles [12] or by pre-reconstituting protein into liposomes [8, 10]. The reconstitution method is particularly useful for helical membrane proteins as it protects them from aggregation, and the proteoliposomes can be used as the lipid source for the lipid-in method of DIB assembly. During assembly, some portion of the reconstituted protein integrates with the interface. However, this process of the exchange of protein from the internal proteoliposomes to the bilayer interface is not well understood, and quantifying the amount of protein incorporated into the interface itself is difficult.

Another method of incorporating protein into the bilayer interface is by expressing protein directly within the droplets using a cell-free in vitro protein expression system. This is a suitable option for DIBs due to the small volume of sample and low quantities of protein required. Commercial kits are available either based on cell extracts or using purified recombinant transcription/translation components, and both have been used successfully in DIBs [13]. Protein expression can occur before or during droplet assembly [5] or has also been triggered post-assembly using a light-sensitive DNA promoter [14].

In this protocol, we describe an example of the incorporation and functional measurement of the membrane protein LacY into a simple two-droplet bilayer interface system, firstly by the reconstitution of purified protein followed by assembly of a DIB by the lipid-in method, and secondly by the in situ translation of the gene in a droplet-based cell-free expression system (Fig. 2) [5]. LacY is a secondary transporter that utilizes an inward proton gradient to transport β-galactosidases such as lactose across the membrane up a concentration gradient by the alternating access method [15]. It is a key component of the lac operon, commonly used for recombinant protein production systems and with potential application in artificial cell control mechanisms [16]. The transporter's structure and activity are strongly influenced by bilayer chemical and mechanical properties [17, 18]; therefore, lipid mixtures are chosen that balance protein activity in vitro with DIB stability [19, 20]. The fluorescent compound 4-methylumbelliferyl-β-galactopyranoside (MUG) is a substrate of LacY [21] and can

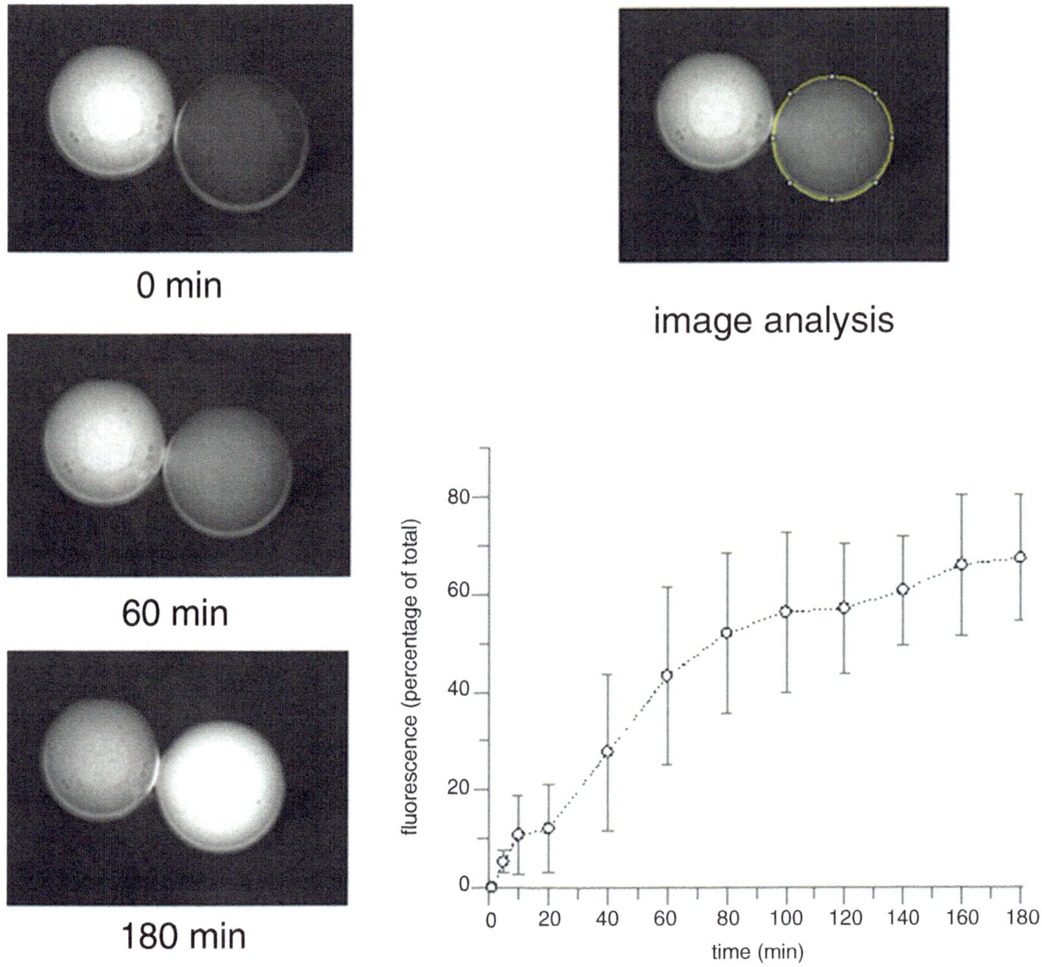

Fig. 2 Cell-free expression and active transport of LacY in a simple droplet interface bilayer system. Fluorescent substrate MUG is loaded in the left-hand droplet and the cell-free reaction in the right-hand droplet. Over time, the substrate is transported across the interface, past equilibrium, and up a concentration gradient. The fluorescence signal in the right-hand droplet is integrated and plotted as a fraction of total fluorescence. The errors are ±SD of $n \geq 3$. The reaction reaches an equal distribution of substrate between droplets after approximately 1 h and then continues to transport up the concentration gradient, driven by the pH gradient. (Adapted from Findlay et al. (2016) [5])

be used to directly monitor movement between droplet compartments by fluorescence microscopy. Analysis of a series of fluorescence images allows for the calculation of rates of transport and the comparison of different buffer, lipid, or expression conditions.

2 Materials

All chemicals are purchased from Sigma unless otherwise stated.

2.1 Preparation of Liposomes and Proteoliposomes

1. 1,2-Dioleoyl-sn-glycero-3-phosphocholine (DOPC), 1,2-dioleoyl-sn-glycero-3-phosphoethanolamine (DOPE), and 1,2-dioleoyl-sn-glycerol-3-[phosphor-rac-(1-glycerol)] (DOPG) as powder from Avanti Polar Lipids.
2. Cyclohexane (anhydrous).
3. 50 mM sodium phosphate buffer pH 7.5.
4. Mini-Extruder kit from Avanti Polar Lipids with 100 nm polycarbonate filters from Whatman.
5. Octyl-β-D-glucopyranoside (OG) and dodecyl-β-D-maltopyranoside (DDM) from Anatrace ≥97% purity.
6. Purified LacY protein in sodium phosphate buffer including 0.02% (w/v) DDM at a concentration of approximately 2 mg/ml (purification protocols for LacY protein can be found in Harris et al. (2014) [22].
7. Detergent Removal Spin Columns from Pierce.

2.2 Droplet Interface Bilayers

1. A small dish made of glass or UV transparent acrylic.
2. Hexadecane (anhydrous).
3. Inverted fluorescence microscope such as a Nikon Eclipse Ti with a 4× objective lens with a UV excitation and emission filter cube, preferably with a temperature-controlled stage.
4. 50 mM sodium phosphate buffer pH 6.5.
5. 4-Methylumbelliferyl-β-galactopyranoside (MUG).
6. PURExpress cell-free expression system from New England Biolabs.
7. Plasmid containing the *lacY* gene under control of a T7 promoter, in this example cloned into pET28 from Novagen.
8. 40 mM HEPES buffer pH 7.6.
9. PURE expression buffer pH 6.6 as a 2.5× stock solution. This concentrated stock solution contains 100 mM HEPES pH 6.6, 200 mM potassium glutamate, 26 mM magnesium acetate, 4 mM spermidine, 2 mM DTT, 6 mM amino acids (0.3 mM of each).

3 Methods

3.1 Preparation of Liposomes and Proteoliposomes

1. Dissolve each lipid in cyclohexane to create stock solutions of 50 mg/ml of each of DOPC, DOPE, and DOPG (*see* **Note 1**).
2. Combine the lipids at the correct molar ratio in small glass vials, freeze over liquid nitrogen, and lyophilize overnight under vacuum. A mix of 4:4:2 of DOPC:DOPE:DOPG is used in this example.
3. Rehydrate the lipid films in 50 mM sodium phosphate buffer pH 7.5 to a final concentration of 10 mg/ml lipid and mix at room temperature for 15 min with regular vortexing to achieve an even suspension of multi-lamellar liposomes.
4. Using an Avanti Mini-Extruder pass through 100 nm polycarbonate filters a minimum of 11 times to create unilamellar vesicles of the correct size (*see* **Note 2**).
5. For proteoliposomes, pre-saturate the bilayer with detergent by adding OG to the liposomes at a final concentration of 1% (w/v) and incubate at room temperature for 10 min with gentle mixing on a tube rotator or roller.
6. Add the purified LacY (in buffer containing DDM) to a concentration of 3000:1 lipid:protein molar ratio. Incubate at room temperature for a further 1 h.
7. Remove the excess detergent using Detergent Removal Spin Columns, pre-equilibrated in sodium phosphate buffer, according to the manufacturer's instructions (*see* **Note 3** for alternatives).
8. To remove any residual detergent from the bilayer, dilute the vesicles at least tenfold in sodium phosphate buffer, incubate at room temperature for 30 min with gentle mixing, and recover the proteoliposomes by centrifugation at 100,000 × g for 1 h (*see* **Note 4**).
9. Resuspend the pellet in sodium phosphate buffer at a concentration of 10 mg/ml lipid (*see* **Note 5**).

3.2 Assembly of Droplet Interface Bilayers

1. Fill the dish with hexadecane to a depth of at least 5 mm (*see* **Note 6**). All steps are undertaken at room temperature unless specifically indicated otherwise.
2. Prepare the solutions for the two aqueous droplets, one with proteoliposomes diluted to 1 mg/ml in sodium phosphate buffer pH 7.5 and the second with empty liposomes diluted to 1 mg/ml in sodium phosphate buffer pH 6.5 with 50 μM MUG (*see* **Note 7**). The difference in pH between the two droplets provides the proton gradient necessary to drive uphill transport of galactosidase substrate.

3. Separately, pipette 0.4 μl of each solution into the hexadecane and leave to equilibrate for 15 min to allow the formation of the lipid monolayer. The droplets should move freely in the solvent. Place the dish on the microscope stage. Both droplets should fit comfortably in the field of view with a 4× objective lens.
4. Use a needle or plastic cocktail stick to gently push the two droplets together until touching. A bilayer interface should form over the course of a few minutes. The "zipping up" of the interface can be observed in a phase contrast image of the droplet pair, as shown in Fig. 1.
5. Collect fluorescent images of the DIB at regular time intervals until the transport reaches equilibrium or the interface collapses and the droplets fuse (*see* **Notes 8–10**).
6. Quantify the fluorescence using analysis software such as ImageJ [23]. On the image of the first time point, draw a circle that covers a droplet size as shown in Fig. 2. Use the integrated density function to calculate the total fluorescence in that droplet. Move the circle to a blank, non-DIB, area and integrate again to give a background value. Repeat for the second droplet (*see* **Note 11**). Repeat the exercise for each frame of the time series.
7. At every time point, calculate the percentage of fluorescence in each droplet, compared to the total fluorescence across the entire DIB system. Plot this percentage against time (in Fig. 2, only the values from acceptor droplet is shown). The transport activity will be downhill and more rapid until the fluorescence in the acceptor droplet reaches 50% of system fluorescence. As the amount of MUG added at the start of the experiment is known, and the increase in fluorescence is linear to this point, the rate of transport in nmol/min is readily calculated by a simple linear fit. In the presence of a pH gradient, secondary transporters such as LacY can continue to transfer substrate up the concentration gradient, albeit at a slower rate that should be calculated separately.

3.3 Cell-Free Expression in Droplet Interface Bilayer

1. Prepare liposomes largely as detailed above, but in a lipid mixture of 1:2:1 DOPC:DOPE:DOPG, rehydrated in HEPES buffer 7.6 and extruded at a concentration of 20 mg/ml lipid (*see* **Note 12**).
2. Prepare a small-scale cell-free synthesis reaction using the PURExpress In Vitro Protein Synthesis Kit according to the manufacturer's instructions, supplemented with liposomes at a final concentration of 5 mg/ml lipid.

3. Make a second solution with liposomes at 5 mg/ml in PURE expression buffer pH 6.6 (diluted from the 2.5× stock) and supplemented with 50 μM MUG.
4. Add 100 ng of *lacY* plasmid to the PURExpress solution and incubate at 37 °C for 5 min to initiate transcription (*see* **Note 13**).
5. Pipette a 0.6 μl droplet from each solution into room temperature hexadecane and incubate for a further 5 min to allow for monolayer formation.
6. Push the two droplets together to form a bilayer interface and record the fluorescence as described above.

4 Notes

1. With a relatively high freezing point of 6 °C, cyclohexane is a good choice of solvent for this step as it is removed more efficiently from the lipid film during lyophilization compared to the more commonly used chloroform. The disadvantage is that many lipids have lower solubility in cyclohexane. Gently warming on a hotplate can help dissolve the lipid powder. If you do need to use chloroform to prepare your lipid stocks, thoroughly dry the lipid films under a stream of nitrogen gas before the vacuum step, as chloroform vapor that escapes into the pump will damage the equipment.
2. A homogenous sample of unilamellar liposomes will result in better efficiency during protein reconstitution, and more passes may be required to achieve this. Lipid mixtures containing high amounts of non-bilayer-forming lipids (e.g. DOPE) are especially likely to require an increased number. The average size and distribution of the sample can be easily measured by dynamic light scattering which should show a tight distribution around the pore size of the filter used, in this case 100 nm.
3. Proteoliposome samples up to a volume of 200 μl can be used with these spin columns, which is usually sufficient for DIB assays. For larger volumes, use Biobeads to remove most of the excess detergent. Prepare the Biobeads by washing in methanol and then water and equilibrating with sodium phosphate buffer. Add approximately 200 mg wet beads per milliliter vesicles and incubate on a tube rotator or roller for 1 h at room temperature. Spin down the beads in a benchtop microcentrifuge and retain the supernatant. Repeat this step twice.
4. It is important to remove as much of the excess detergent as possible from the reconstituted sample to avoid unstable monolayer and leakage across the bilayer in downstream experiments. Visibly, the sample should no longer contain bubbles.

The amount of residual detergent can be measured using a phenol and sulfuric acid concentration assay [24] and should be reduced to 0.01% (w/v) or less.

5. The reconstitution efficiency can be measured with a protein concentration assay that can be used in the presence of lipids or detergents. Both the Markwell-Lowry assay [25] and the amido black assay involve a precipitation step in trichloroacetic acid that will wash away these non-proteinaceous elements and are suitable. Compare the value with the amount of protein originally added to the sample.
6. For this assay, the fluorescent substrate excites and emits in the ultraviolet range so both the dish and the solvent must be UV transparent. Acrylic has reduced wettability and is a good choice. For other fluorescent compounds that emit in the visible spectrum, a greater range of solvents are possible, such as silicone oil and mineral oil, which have been used successfully to create stable DIBs.
7. 50 μM MUG is a concentration comfortably within the detection limits of the microscope and filter cube used in this example. Some optimization of substrate concentration may be necessary on alternative equipment. Too low a concentration and detection is difficult, and the reaction may go to completion too swiftly. Too high a concentration can also lead to difficulties, due to issues with substrate solubility or effects on droplet monolayer or interface stability.
8. Many factors affect the stability and longevity of the DIBs. In this example, droplets were generally stable for at least 2–3 h. The solvent, temperature, lipid composition and concentration, and aqueous buffer content including osmolarity and the presence of divalent cations all impact the stability of the individual droplet monolayer and the interface formed between them. Optimization is likely to be required. For example, it was necessary to increase the lipid concentration from 1 mg/ml to 5 mg/ml and increase the droplet size when using the cell-free system compared to the reconstituted system described here to obtain equally stable DIBs.
9. MUG fluorescence can be detected using the most standard DAPI filter cubes, though long integration times of at least 10 s are likely to be required. If available, a far UV filter cube with a lower excitation wavelength will result in a brighter signal and increased sensitivity of the assay and will require shorter integration times.
10. If transport activity continues for a long time, so needs extended monitoring, photobleaching of the fluorescent moiety can be an issue. Reducing the frequency of image collection whilst closing the shutters on the microscope will minimize the effect.

11. If the droplets form a larger interface, the circle tool may no longer fit your image. In this case, use the freeform tool to surround your droplet. Then instead of directly using the integrated density function, calculate your total fluorescence by measuring the mean gray value and the area and multiply them.
12. For the cell-free expression version of this assay, it is preferable that all aqueous buffers should be prepared in RNAse-free water to prevent degradation of RNA molecules that are present in the transcription/translation reaction.
13. The PURExpress system produces very little LacY protein at room temperature. This short pre-incubation step at 37 °C to initiate transcription results in a substantial increase in total protein expression even when the remaining translation occurs at room temperature. A microscope with a temperature-controlled stage could in principal be used to run the whole reaction at an elevated temperature, though this would have knock-on effects on droplet and interface stability. Other cell-free extract-based expression systems may also be used that are more efficient at room temperature, though they can have a significant problem with both liposome and monolayer behavior.

References

1. Kurihara K, Tamura M, Shohda K et al (2011) Self-reproduction of supramolecular giant vesicles combined with the amplification of encapsulated DNA. Nat Chem 3 (10):775–781. https://doi.org/10.1038/nchem.1127
2. Peters RJRW, Marguet M, Marais S et al (2014) Cascade reactions in multicompartmentalized polymersomes. Angew Chem Int Ed Engl 53(1):146–150. https://doi.org/10.1002/anie.201308141
3. Bayley H, Cronin B, Heron A et al (2008) Droplet interface bilayers. Mol BioSyst 4 (12):1191–1208. https://doi.org/10.1039/b808893d
4. Leptihn S, Castell OK, Cronin B et al (2013) Constructing droplet interface bilayers from the contact of aqueous droplets in oil. Nat Protoc 8(6):1048–1057. https://doi.org/10.1038/nprot.2013.061
5. Findlay HE, Harris NJ, Booth PJ (2016) *In vitro* synthesis of a major facilitator transporter for specific active transport across droplet interface bilayers. Sci Rep 6:39349. https://doi.org/10.1038/srep39349
6. Carreras P, Law RV, Brooks N et al (2014) Microfluidic generation of droplet interface bilayer networks incorporating real-time size sorting in linear and non-linear configurations. Biomicrofluidics 8(5):054113. https://doi.org/10.1063/1.4897495
7. Villar G, Graham AD, Bayley H (2013) A tissue-like printed material. Science 340 (6128):48–52. https://doi.org/10.1126/science.1229495
8. Haylock S, Friddin MS, Hindley JW et al (2020) Membrane protein mediated bilayer communication in networks of droplet interface bilayers. Commun Chem 3(1):77. https://doi.org/10.1038/s42004-020-0322-1
9. Holden MA, Needham D, Bayley H (2007) Functional bionetworks from nanoliter water droplets. J Am Chem Soc 129 (27):8650–8655. https://doi.org/10.1021/ja072292a
10. Elfaramawy MA, Fujii S, Uyeda A et al (2018) Quantitative analysis of cell-free synthesized membrane proteins at the stabilized droplet interface bilayer. Chem Commun 54 (86):12226–12229. https://doi.org/10.1039/c8cc06804f
11. Venkatesan GA, Lee J, Farimani AB et al (2015) Adsorption kinetics dictate mono layer self-assembly for both lipid-in and lipid-out approaches to droplet Interface bilayer formation. Langmuir 31(47):12883–12893.

https://doi.org/10.1021/acs.langmuir.5b02293
12. Hwang WL, Chen M, Cronin B et al (2008) Asymmetric droplet interface bilayers. J Am Chem Soc 130(18):5878–5879. https://doi.org/10.1021/ja802089s
13. Syeda R, Holden MA, Hwang WL et al (2008) Screening blockers against a potassium channel with a droplet interface bilayer array. J Am Chem Soc 130(46):15543–15548. https://doi.org/10.1021/ja804968g
14. Booth MJ, Schild VR, Graham AD et al (2016) Light-activated communication in synthetic tissues. Sci Adv 2(4):e1600056. https://doi.org/10.1126/sciadv.1600056
15. Kaback HR, Smirnova I, Kasho V et al (2011) The alternating access transport mechanism in LacY. J Membr Biol 239(1–2):85–93. https://doi.org/10.1007/s00232-010-9327-5
16. Miller D, Booth PJ, Seddon JM et al (2013) Protocell design through modular compartmentalization. J R Soc Interface 10 (87):20130496. https://doi.org/10.1098/rsif.2013.0496
17. Dowhan W, Bogdanov M (2011) Lipid-protein interactions as determinants of membrane protein structure and function. Biochem Soc Trans 39(3):767–774. https://doi.org/10.1042/BST0390767
18. Vitrac H, Bogdanov M, Dowhan W (2013) *In vitro* reconstitution of lipid-dependent dual topology and postassembly topological switching of a membrane protein. Proc Natl Acad Sci 110(23):9338–9343. https://doi.org/10.1073/pnas.1304375110
19. Allen-Benton M, Findlay HE, Booth PJ (2019) Probing membrane protein properties using droplet interface bilayers. Exp Biol Med (Maywood) 244(8):709–720. https://doi.org/10.1177/1535370219847939
20. Findlay HE, Booth PJ (2017) The folding, stability and function of lactose permease differ in their dependence on bilayer lipid composition. Sci Rep 7(1):13056. https://doi.org/10.1038/s41598-017-13290-7
21. Sandermann H Jr (1977) beta-D-Galactoside transport in Escherichia coli: substrate recognition. Eur J Biochem 80(2):507–515. https://doi.org/10.1111/j.1432-1033.1977.tb11906.x
22. Harris NJ, Findlay HE, Simms J et al (2014) Relative domain folding and stability of a membrane transport protein. J Mol Biol 426 (8):1812–1825. https://doi.org/10.1016/j.jmb.2014.01.012
23. Schneider CA, Rasband WS, Eliceiri KW (2012) NIH image to ImageJ: 25 years of image analysis. Nat Methods 9(7):671–675. https://doi.org/10.1038/nmeth.2089
24. Dubois M, Gilles KA, Hamilton JK et al (1956) Colorimetric method for determination of sugars and related substances. Anal Chem 28 (3):350–356. https://doi.org/10.1021/ac60111a017
25. Markwell MAK, Haas SM, Bieber LL et al (1978) Modification of Lowry procedure to simplify protein determination in membrane and lipoprotein samples. Anal Biochem 87 (1):206–210. https://doi.org/10.1016/0003-2697(78)90586-9

Chapter 3

Membrane Protein Engineering with Rosetta

Rebecca F. Alford and Jeffrey J. Gray

Abstract

Protein engineering can yield new molecular tools for nanotechnology and therapeutic applications through modulating physiochemical and biological properties. Engineering membrane proteins is especially attractive because they perform key cellular processes including transport, nutrient uptake, removal of toxins, respiration, motility, and signaling. In this chapter, we describe two protocols for membrane protein engineering with the Rosetta software: (1) $\Delta\Delta G$ calculations for single point mutations and (2) sequence optimization in different membrane lipid compositions. These modular protocols are easily adaptable for more complex problems and serve as a foundation for efficient membrane protein engineering calculations.

Key words Implicit membrane, Lipid composition, Monte Carlo, Rosetta, Protein design

1 Introduction

Membrane proteins perform key cellular processes including transport, nutrient uptake, removal of toxins, respiration, motility, and signaling. They constitute over 30% of proteins and are targets for over 60% of pharmaceutical drugs [1, 2]. Understanding their structures will provide insight into cellular function and the molecular basis of disease and enable design of proteins for therapeutic and nanotechnology applications. Protein engineering is a promising tool for this purpose [3]. However, experimental methods to characterize designs in a lipid bilayer are onerous, time consuming, and expensive [4]. Thus, computational tools provide a promising route to accelerate efforts in membrane protein engineering [5].

The past 20 years of computational membrane protein engineering has largely focused on reprogramming natural membrane protein functions [6–8]. Chen et al. [9] applied structural and bioinformatics calculations to generate G-protein coupled receptors that decrease agonist affinity via stabilization. To ease experimental difficulties, membrane proteins have been redesigned to assemble and function in the aqueous phase [10, 11]. Young

Irina S. Moreira et al. (eds.), *Computational Design of Membrane Proteins*, Methods in Molecular Biology, vol. 2315,
https://doi.org/10.1007/978-1-0716-1468-6_3,

et al. [12] engineered the dopamine D2 receptor to favor different chemical pathways. Further, both α-helical and β-barrel proteins have been repurposed as pores for industrial-scale separation of water and other solutes including glucose, sucrose, and salt [13, 14]. This handful of examples demonstrates the versatility of computational methods to yield new molecular tools.

Membrane protein engineers have also pursued the holy grail of de novo design: generating new proteins from physical principles with novel sequences [15]. De novo design can inform how proteins work [16] and create a wider array of tools outside of what biology has already provided such as drug delivery [17] and process control [18, 19]. For membrane proteins, there have been two significant de novo milestones. DeGrado and coworkers designed the PRIME and ROCKR alternating access transporters formed by two tightly interacting pairs of helices [20, 21]. Lu et al. also achieved de novo design of helical bundles that insert into the membrane [22].

The aforementioned successes leverage many approaches for scoring and sampling both sequences and structural conformations. A key point of differentiation is how these methods capture the variable affinity of side chains for the heterogeneous lipid membrane environment. Knowledge-based algorithms that derive depth-dependent side chain probabilities from known structures in the Protein Databank are realistic [23–25]; yet, they cannot capture different lipid compositions. Thermodynamic-based scales offer a first principles approach [26–28]; yet, until recently, these scales were derived from side chain analogs in organic solvents which do not resemble the behavior of biological membranes [29]. Further, both the approaches fell short of fine-tuning the lipid-facing regions [30] and were subject to over-designing hydrophobic residues [31]. In contrast, biological transmembrane domains can be marginally hydrophobic and depend on local sequence context for insertion [32].

Recently, we developed an approach for membrane protein design using a thermodynamics-centered approach [33]. This advance builds upon our prior work developing a framework for membrane protein modeling in Rosetta [34, 35]. There were three key developments: (1) models of different lipid compositions, (2) efficient computation of pore dimensions, and (3) the Moon and Fleming side chain hydrophobicity scale in phospholipid bilayers in the context of a real membrane protein [36]. This model is an addition to the existing useful set of mathematical rules used to discriminate near-native from non-native conformations: the energy function. First, we establish a central term based on the Lazaridis formalism for the water-to-bilayer transfer energy (Eq. 1).

$$\Delta G_{\text{memb}} = \left(1 - f_{\text{hyd}}\right)\left(\Delta G_{\text{lipid}} - \Delta G_{\text{water}}\right) \quad (1)$$

In this framework, ΔG_{lipid} is the free energy of a side chain in lipid and ΔG_{water} is the free energy of a side chain in the aqueous phase. The term f_{hyd} captures the shape of the bilayer. Here, the membrane is modeled as an implicit continuum [37]. The equation for f_{hyd} again follows the Lazaridis formalism of a mixture model between the shape in the *x*, *y*, and *z* dimensions (Eq. 2).

$$f_{\text{hyd}} = f_{\text{thk}} + f_{\text{pore}} - f_{\text{thk}} f_{\text{pore}} \tag{2}$$

The term f_{thk} models the thickness of the hydrocarbon core and the steepness of the transition between the aqueous and lipid phase for different phospholipid bilayer compositions. The second term f_{pore} represents cavities in the protein. Capturing cavities is key because a central function of membrane proteins is responding to extracellular signals or transporting materials by exposing the interior of a protein to water.

Next, we integrated the water-to-bilayer term with the current Rosetta all-atom energy function: ref2015 [38]. The Rosetta energy function is a hybrid model that aims to approximate the Gibbs free energy through a combination of physics-based, empirically derived, and knowledge-based terms. Currently, these terms include van der Waals, Coulomb electrostatics, Lazaridis–Karplus solvation, hydrogen bonding, and knowledge-based torsion terms. The weights on each term were determined through a Nelder–Mead optimization scheme [39]. We created the membrane version, called *franklin2019*, by adding the water-to-biolayer term and assigning a weight of 0.5. The weight was determined by iteratively running four scientific benchmark tests: peptide tilt angle, $\Delta\Delta G$ of mutation, sequence recovery, and native structure discrimination.

The energy function provides a foundation for choosing optimal sequences and conformations. The second key step of macromolecular modeling is to sample both the sequence and conformational space. To do so, we use Rosetta3's existing Monte Carlo-plus-minimization sampling framework [34]. Here, the system degrees of freedom are perturbed, energy minimized, and then changes are accepted or rejected according to the Metropolis criteria: given that $A = \min(1, \exp(\Delta E/k_{\text{B}}T)$ accept if $A \geq U(0, 1)$, where $U(0, 1)$ is a random number between zero and one (inclusive); otherwise, reject the changes. Importantly, Rosetta samples degrees of freedom as internal coordinates (ϕ, ψ, ω) rather than Cartesian coordinates (*x*, *y*, *z*). Bond lengths and angles can be sampled but are usually kept fixed. Finally, the calculation is stochastic, meaning the program is run many times and the ensemble of low energy models represents the final solution.

The basic recipe of Monte Carlo-plus-minimization forms a modular framework for creating different design protocols. In this chapter, we discuss two membrane protein engineering protocols with *franklin2019*: (1) a $\Delta\Delta$G calculation for single point mutations and (2) a fixed-backbone design protocol for optimizing sequences in bilayers with different lipid composition. These protocols can be adapted and expanded for wide-ranging membrane protein engineering calculations.

2 Materials

This chapter demonstrates membrane protein design calculations in the Rosetta macromolecular modeling package. The tutorial will require a Unix-based operating system (Linux or Mac OS). Windows users will need to install the Linux subsystem. We have included examples for both the UNIX command line and Python environments. Additionally, the tutorial requires Python Version 3.6 or 3.7. The protocols were tested with PyRosetta revision #249 and Rosetta revision #61215.

To run this tutorial, the user will need to download the following software:

- Rosetta (https://www.rosettacommons.org/software/license-and-download).
- PyRosetta (https://www.pyrosetta.org/dow).
- PyMOL (https://pymol.org/2/).

Additionally, this tutorial uses the Orientations of Proteins in Membranes database (https://opm.phar.umich.edu). We have also provided the tutorial as an interactive Python notebook and standalone Python scripts on GitHub. To run the interactive session, the user will need to download the following software:

- Anaconda (https://anaconda.org/anaconda).
- Jupyter Interactive Python notebooks (https://jupyter.org/install).

There are several additional resources available for building membrane protein engineering protocols within the Rosetta software suite. Several resources are listed below:

1. Rosetta Forums (Q&A): https://www.rosettacommons.org/forum.
2. Documentation: https://www.rosettacommons.org/docs/latest/Home.
3. PyRosetta Workshops: http://www.pyrosetta.org/.

3 Methods

In this chapter, we demonstrate two membrane protein engineering calculations: (1) predicting the $\Delta\Delta G$ of single point mutations and (2) sequence optimization in the context of bilayers with different lipid composition. Both the protocols can be executed on standard computers and do not require specialized hardware. The execution time varies from minutes to hours depending on the hardware and protein size.

3.1 Protocol #1: $\Delta\Delta G$ of Single Point Mutations

Accurately estimating the thermodynamic cost of a mutation is a fundamental step of protein engineering and design. This task is especially challenging for membrane proteins because the calculations must account for the heterogeneous membrane environment. Here, we will walk through the protocol for estimating the $\Delta\Delta G$ of point mutations at lipid facing positions using RosettaMP [35] and the *franklin2019* energy function [33]. The protocol is available as an interactive Jupyter notebook at https://github.com/RosettaCommons/PyRosetta.notebooks (see Notebook #14.02).

As an example, we will examine mutations in the integral membrane enzyme PagP. PagP is a β-barrel protein that transfers a palmitoyl group from the *sn*1 position of a glycerophospholipid to the endotoxin of lipopolysaccharide [40]. This enzyme provides bacterial resistance to pathogens such as antimicrobial peptides [41]. Recently, Marx and Fleming measured the energetic cost of point mutations at site 111 on PagP [42]. In this tutorial, we will perform the same set of mutations with Rosetta and compare the predictions with the experimentally measured values. Text following the > symbols represent command lines that are entered in a Python interactive session or Jupyter notebook. Text following the $ symbols represent command lines that are entered into a shell terminal.

3.1.1 Preparing the PyRosetta Working Environment

To initialize the working environment, we first download the coordinates for PagP, clean the PDB file, and initialize the PyRosetta working environment. If using the PyRosetta notebooks GitHub repository, the initial PDB file is available in the `inputs/` directory.

1. Download the coordinates for PagP (PDB: 3GP6) from the Orientations of Proteins in Membranes database (https://opm.phar.umich.edu/).
2. Clean the PDB file by removing all "non-ATOM" lines.

```
$ grep '^ATOM' inputs/3gp6.pdb > inputs/3gp6_clean.pdb
```

3. Open a Python3 interactive environment or Jupyter notebook and load PyRosetta.

```
> from pyrosetta import *
> init( extra_options="-mp:lipids:has_pore false")
```

4. Load the PDB coordinate file into a `Pose` object. The `Pose` is a data structure that stores the coordinates and chemical information for the system.

```
> pose = pose_from_pdb("inputs/3gp6_A.pdb")
```

5. Initialize the protein in the membrane using `AddMembraneMover`. The protein is already oriented in the bilayer so we can estimate the transmembrane spans from the protein coordinates. For this reason, we use the `from_structure` option to initialize the transmembrane span information. The default lipid composition is a 1,2-dilauroyl-*sn*-glycero-3-phosphocholine (DLPC) bilayer which matches the experimental context of a DLPC vesicle.

```
> from pyrosetta.rosetta.protocols.membrane import *
> add_memb = AddMembraneMover("from_structure")
> add_memb.apply(pose)
```

3.1.2 Computing the ΔΔG of Mutation

Next, we will compute the ΔΔG for several point mutations in PagP. Note, PyRosetta residue numbering may differ from PDB numbering because PyRosetta requires continuous numbering for calculations. We obtain the PyRosetta residue number through the `PDBInfo` object.

1. Create a `ScoreFunction` object with the `franklin2019` energy function.

```
> sfxn = create_score_function("franklin2019")
```

2. Load the `predict_ddG` package and use the `mutate_residue` function to create the native conformation. Note: in the Marx and Fleming experiment, position 111 is first mutated from valine to alanine. Then, the V111A pose is used as the reference state for the remaining ΔΔ*G* calculations.

```
> import predict_ddG as pd
> site = pose.pdb_info().pdb2pose(111)
> ref_pose = pd.mutate_residue(pose, site, "A", 8.0, sfxn)
```

3. Score the alanine reference pose.

```
> score_A111 = sfxn.score(ref_pose)
```

4. Generate a tryptophan mutant and score the new pose.

```
> pose_W111 = pd.mutate_residue(pose, site, "W", 8.0, sfxn)
> score_W111 = sfxn.score(pose_W111)
```

5. Compute the $\Delta\Delta G$ of mutation as the difference between the score of the mutated conformation and the native conformation.

```
> ddG = score_W11 - score_A11
> print(ddG)
```

The $\Delta\Delta G$ for mutating alanine to tryptophan at position 111 is −1.84 Rosetta Energy Units (REU). A Rosetta Energy Unit is an arbitrary unit for the Rosetta energy function.

6. Ultimately, we would like to compute the $\Delta\Delta G$ for mutating alanine to all 19 canonical amino acids. We will use a function that encapsulates this code. Then, we write a loop that computes the $\Delta\Delta G$ for all canonical amino acids and store the result in a Python dictionary. This step will take 10 min to compute depending on the available hardware and protein size.

```
> amino_acids = ['A', 'C', 'D', 'E', 'F', 'G', 'H', 'I', 'K', 'L',
'M', 'N', 'P', 'Q', 'R', 'S', 'T', 'V', 'W', 'Y']
> ddG_data = {}
> for aa in amino acids:
>   ddG = compute_ddG(reference_pose, "A", 104, aa, sfxn)
>    ddG_data[ aa ] = ddG
```

3.1.3 Comparison Between Predictions and Experimentally Measured Values

The next step is to compare the $\Delta\Delta$G predictions to the experimentally measured values from Marx and Fleming. The experimental data are located in a file called `PagP_Marx_Fleming_set.dat`. This file can be downloaded from the `inputs/` directory in the PyRosetta notebook GitHub repository.

1. Parse the file and import the values into a Python Dictionary.

```
> with open('inputs/PagP_Marx_Fleming_set.dat', 'rt') as f:
    data = f.readlines()
    data = [x.strip() for x in data]
    data = [x.split(' ') for x in data]
> exp_ddG_data = {}
> for I in range(1, len(data)):
    exp_ddG_data[data[i][2]] = float(data[i][3])
```

2. Convert the dictionary format to numpy arrays.

```
> import numpy as np
> mutations = np.asarray( ddG_data.keys() )
> pred_values = np.asarray( list(ddG_data.values()) )
> exp_values = np.asarray( list(exp_ddG_data.values()) )
```

3. Compute the correlation coefficient between the experimentally measured and predicted values.

```
> corr = np.corrcoeff(exp_values, pred_values)
> print(corr[0,1])
```

4. We find that the correlation coefficient is low (0.376). Importantly, the Pearson correlation coefficient is easily skewed by outliers. Thus, we write and execute a function to identify any outliers.

```
> def find_outliers(x):
>     outliers = []
>      upper = np.percentile(x, 75)
>      lower = np.percentile(x, 25)
>      IQR = (upper - lower)
>      quartile_set = (lower - IQR, upper + IQR)
>      for y in x.tolist():
>        if (y < quartile_set[0]) or (y > quartile_set[1]):
>           outliers.append(y)
>
> outliers = find_outliers(ddG_values)
```

5. Using this function, we find that proline is an outlier. We will investigate this more later. For now, we remove it from the set and recompute the correlation coefficient (note: Proline is the 13th amino acid of 20).

```
> outlier_idx = list(ddG_data.values()).index(outliers[0])
> exp_data_no_P = []
> pred_data_no_P = []
> for i in range(0, 20):
    if (i != outlier_idx):
    exp_data_no_P.append(list(exp_ddG_data.values())[i])
    pred_data_no_P.append(list(ddG_data.values())[i])

> exp_data_no_P = np.asarray( exp_data_no_P )
> pred_data_no_P = np.asarray( pred_data_no_P )
> corr = np.corrcoef( exp_data_no_P, pred_data_no_P )
> print(corr[0,1])
```

The new value of $R = 0.692$ is more encouraging! We also visualized the predicted vs. experimentally measured values with a scatterplot. The code for generating this plot is provided in the Jupyter notebook.

3.1.4 Examining Contributions to the $\Delta\Delta G$ of Mutation

Finally, we would like to use the models to learn why some mutations stabilized PagP, whereas other side chains did not.

1. We need a metric for identifying the most confident predictions, especially since the correlation coefficient is not perfect. Thus, we compute the residuals from the line of best fit and set an empirical cutoff of 1.5 REU. This value is aligned with the experimental uncertainty of 1.5 kcal/mol.

```
> import seaborn as sns
> resid = sns.residplot(exp_data_no_P, pred_data_no_P, color="b")
> resid.set_ylabel("Residual")
> resid.set_xlabel("Exp (kcal/mol)")
```

   We find five values that are predicted outside of the cutoff: G, L, V, W, and T.

2. Next, we will hypothesize a mechanism for an example correct prediction (lysine) and rationalize incorrect predictions for proline and leucine. First, we quantify which energy components make the largest contribution to the overall $\Delta\Delta G$ of mutation. The function `get_energy_components()` is defined in the Jupyter Notebook.

```
>mutant_tyr = pd.mutate_residue(pose, site, "Y", 8.0, sfxn)
>mutant_lys = pd.mutate_residue(pose, site, "K", 8.0, sfxn)
>mutant_leu = pd.mutate_residue(pose, site, "L", 8.0, sfxn)

> labels, tyr_ddGs = get_energy_components(reference_
pose, mutant_tyr, sfxn )
> labels, lys_ddGs = get_energy_components(reference_
pose, mutant_lys, sfxn )
> labels, leu_ddGs = get_energy_components(reference_
pose, mutant_leu, sfxn )
```

3. Finally, we make bar graphs to visualize the individual contributions. These are shown in Fig. 1a.
4. Further, we dump the models to PDB files for visualization in PyMOL (Fig. 1b–d).

```
> mutant_tyr.dump_pdb( "PagP_A111Y.pdb" )
> mutant_lys.dump_pdb( "PagP_A111K.pdb" )
> mutant_leu.dump_pdb( "PagP_A111L.pdb" )
```

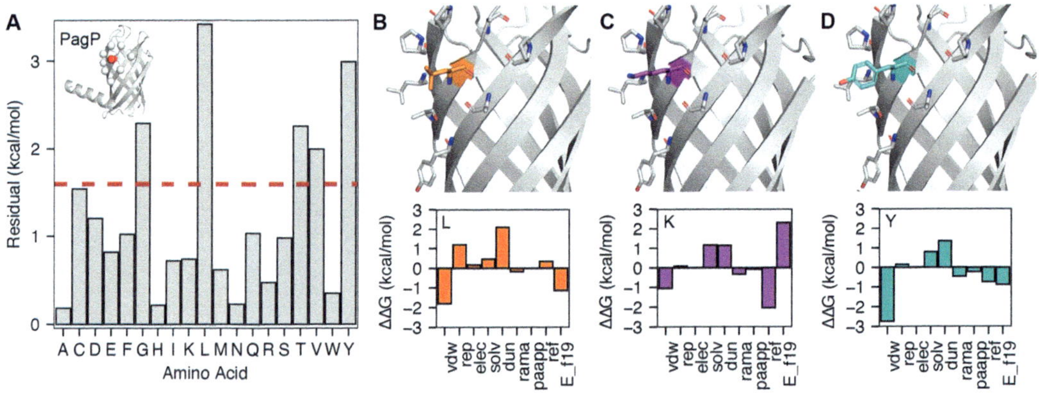

Fig. 1 Models for evaluating the mechanism underlying single point mutations in PagP. (**a**) Residuals between the predicted and experimentally measured $\Delta\Delta G_{mut}$ for mutations from alanine to all canonical amino acids at position 111 (104). The dotted red line represents the accuracy cutoff of 1.5 kcal/mol. The complete structure of PagP (PDB 3gp6) is shown in the top left corner of the plot, and position 111 is highlighted in red. Models of the A111L, A11K, and A11Y are shown in panels (**b**), (**c**), and (**d**), respectively. Within each panel, the top sub-panel shows the structural model of the mutated PagP with focus on the mutated site. The bottom sub-panel shows the contribution of individual energy terms to the overall $\Delta\Delta G$ for terms that contribute greater than 0.01 kcal/mol

3.2 Protocol #2: Sequence Optimization in Different Lipid Compositions

The ultimate goal of membrane protein engineering is to search for a sequence that achieves a new protein stability, structure, or function. This task may involve changing a part of the sequence or allowing full protein redesign and optimization. Here, we will demonstrate a design calculation with a goal of optimizing the sequence of a membrane protein in different lipid compositions.

As an example, we will redesign the structure of the eukaryotic calcium/proton exchanger VDXC1 (PDB 4K1C) [43]. VDXC1 is part of the CAX family of proteins whose members maintain cytosolic calcium homeostasis during steep rises in intracellular calcium, or following signal transduction caused by hyperosmotic shock or hormone responses [44, 45]. As a eukaryotic protein, we anticipate that the designed sequence will be more optimal in a thicker phospholipid bilayer, such as 1-palmitoyl-2-oleoyl-glycero-3-phosphocholine (POPC). To test this hypothesis, we will perform fixed-backbone redesign in both lipid compositions and compare the resulting sequences.

3.2.1 Preparing the Rosetta Working Environment

To initialize the working environment, we first download the coordinates of the crystal structure of VDXC1, clean the PDB file, generate a spanning topology file, and configure the working directory. This tutorial assumes that Rosetta is already installed in a directory that we will refer to as `ROSETTA` (see Materials for installation instructions). The "$" lines denote commands in a bash or other shell terminal environment.

1. Download the coordinates for VDXC1 (PDB 4K1C) from the Orientations of Proteins in Membranes database (https://opm.phar.umich.edu/).
2. Clean the PDB file by removing all "non-ATOM" lines.

```
$ grep '^ATOM' 4k1c.pdb > 4k1c_clean.pdb
```

3. Generate a file listing the transmembrane segments in the protein using the `mp_span_from_structure` application. The resulting spanning topology file will be called `out.span`.

```
$ROSETTA/main/source/bin/mp_span_from_pdb.linuxgccrelease
    -in:file:s 4k1c_clean.pdb
     -out:no_output true
```

4. Rename the spanning topology file to match the PDB name:

```
$ mv out.span 4k1c.span
```

3.2.2 Optimizing the Sequence of VDXC1 in Different Lipid Compositions

To perform sequence optimization, we use a Monte Carlo fixed-backbone design protocol that samples sequence space using a full protein rotamer-and-sequence optimization and multi-cool annealer-simulated annealing protocol [46]. Each protein is initialized in the orientation from the Orientation of Proteins in Membranes Database [47], and the orientation remains fixed during sequence optimization. Then, we compute properties of the sequence using additional Rosetta applications.

1. Run the fixed-backbone design protocol on the crystal structure of VDXC1 with a lipid composition of DLPC (a short chain lipid). This step will require 1–2 h of CPU time on a standard computer. We also repeat this step using a lipid composition of POPC.

```
$ ROSETTA/main/source/bin/fixbb.linuxgccrelease
    -in:file:s 4k1c_clean.pdb
     -mp:setup:spanfiles 4k1c_clean.span
     -score:weights franklin2019
     -in:membrane
     -mp:lipids:composition DLPC
     -mp:lipids:temperature 37
     -out:prefix 4k1c_DLPC_design_
     -out:file:scorefile 4k1c_DLPC_design.sc

$ ROSETTA/main/source/bin/fixbb.linuxgccrelease
    -in:file:s 4k1c_clean.pdb
     -mp:setup:spanfiles 4k1c_clean.span
     -score:weights franklin2019
```

```
-in:membrane
-mp:lipids:composition POPC
-mp:lipids:temperature 37
-out:prefix 4k1c_POPC_design_
-out:file:scorefile 4k1c_POPC_design.sc
```

2. Generate a list of native and designed files to work with the format of the sequence recovery application.

```
$ ls 4k1c_DLPC_design_4k1c_clean_0001.pdb > 4k1c_DLPC.list
$ ls 4k1c_POPC_design_4k1c_clean_0001.pdb > 4k1c_POPC.list
$ ls 4k1c_clean.pdb > 4k1c_native.list
```

3. Compute the number of amino acids recovered overall and within each environment (e.g., buried vs. surface exposed, and lipid-facing vs. aqueous).

```
$ ROSETTA/main/source/bin/mp_seqrecov.linuxgccrelease
    -mp:setup:spanfiles 4k1c.span
    -native_pdb_list 4k1c_native.list
    -redesign_pdb_list 4k1c_DLPC.list
    -seq_recov_filename 4k1c_DLPC_fixbb.txt

$ ROSETTA/main/source/bin/mp_seqrecov.linuxgccrelease
    -mp:setup:spanfiles 4k1c.span
    -native_pdb_list 4k1c_native.list
    -redesign_pdb_list 4k1c_POPC.list
    -seq_recov_filename 4k1c_POPC_fixbb.txt
```

4. Compute key metrics including sequence recovery and Kullback–Leibler divergence for all residues, subsets of amino acids, and for classes of amino acids using the `process_protein_design_results.py` script. The script is located in the PyRosetta notebooks GitHub repository in the directory called `additional_scripts`. An example analysis of design results is also shown in Fig. 2.

```
$ python process_protein_design_results.py
    --energy_fxn f19
    --seqrecov_file 4k1c_DLPC_fixbb.txt
    --prefix 4k1c_DLPC_
    --basedir $(pwd)

$ python process_protein_design_results.py
    --energy_fxn f19
    --seqrecov_file 4k1c_POPC_fixbb.txt
    --prefix 4k1c_POPC_
    --basedir $(pwd)
```

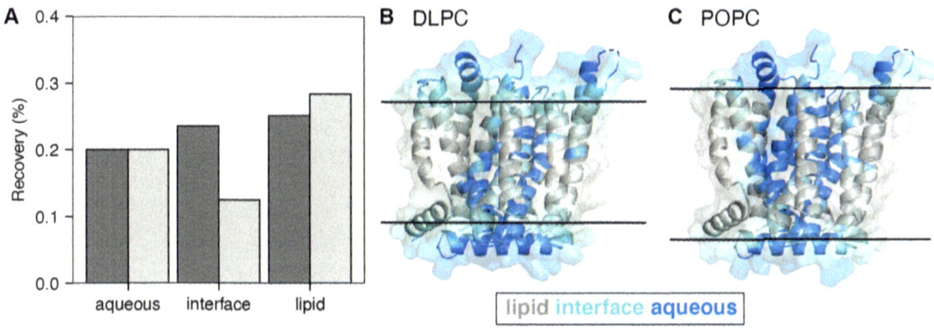

Fig. 2 Optimizing the sequence of a eukaryotic calcium/proton exchanger in short- and long-chain phospholipid compositions. (**a**) Sequence recovery after fixed-backbone redesign of the VDXC1 calcium/proton exchanger (PDB 4k1c) subdivided by exposure of side chains to the aqueous, interface, and lipid phases. Buried positions are excluded from the calculation. Dark gray bars represent recovery in DLPC, and light gray bars represent recovery in POPC. (**b**) and (**c**) show the redesigned models of VDXC1 in DLPC and POPC, respectively. The color of each position indicates the exposure phase, with blue representing the aqueous phase, teal representing the interface, and gray representing the lipid phase

Acknowledgments

R.F.A. is funded by a Hertz Foundation Fellowship and a National Science Foundation Graduate Research Fellowship. R.F.A. and J.J.G. are also funded by NIH Grant GM-078221. We also thank Priyamvada Prathima and Kathy Le for testing the protocols.

References

1. Tan S, Tan HT, Chung MCM (2008) Membrane proteins and membrane proteomics. Proteomics 8:3924–3932. https://doi.org/10.1002/pmic.20080059
2. Overington JP, Al-Lazikani B, Hopkins AL (2006) How many drug targets are there? Nat Rev Drug Discov 5:993–996. https://doi.org/10.1038/nrd2199
3. Samish I, MacDermaid CM, Perez-Aguilar JM, Saven JG (2011) Theoretical and computational protein design. Annu Rev Phys Chem 62:129–149. https://doi.org/10.1146/annurev-physchem-032210-103509
4. Bill RM, Henderson PJF, Iwata S et al (2011) Overcoming barriers to membrane protein structure determination. Nat Biotechnol 29:335–340. https://doi.org/10.1038/nbt.1833
5. Koehler Leman J, Ulmschneider MB, Gray JJ (2015) Computational modeling of membrane proteins. Proteins 83:1–24. https://doi.org/10.1002/prot.24703
6. Perez-Aguilar JM, Saven JG (2012) Computational design of membrane proteins. Structure 20:5–14. https://doi.org/10.1016/j.str.2011.12.003
7. Barth P, Senes A (2016) Toward high-resolution computational design of the structure and function of helical membrane proteins. Nat Struct Mol Biol 23:475–480. https://doi.org/10.1038/nsmb.3231
8. Slusky JS (2017) Outer membrane protein design. Curr Opin Struct Biol 45:45–52. https://doi.org/10.1016/j.sbi.2016.11.003
9. Chen K-YM, Zhou F, Fryszczyn BG, Barth P (2012) Naturally evolved G protein-coupled receptors adopt metastable conformations. Proc Natl Acad Sci U S A 109:13284–13289. https://doi.org/10.1073/pnas.1205512109
10. Slovic AM, Kono H, Lear JD et al (2004) Computational design of water-soluble analogues of the potassium channel KcsA. Proc Natl Acad Sci U S A 101:1828–1833. https://doi.org/10.1073/pnas.0306417101

11. Perez-Aguilar JM, Xi J, Matsunaga F et al (2013) A computationally designed water-soluble variant of a G-protein-coupled receptor: the human mu opioid receptor. PLoS One 8:e66009. https://doi.org/10.1371/journal.pone.0066009
12. Young M, Dahoun T, Sokrat B et al (2018) Computational design of orthogonal membrane receptor-effector switches for rewiring signaling pathways. Proc Natl Acad Sci U S A 115:7051–7056. https://doi.org/10.1073/pnas.1718489115
13. Kumar M, Grzelakowski M, Zilles J et al (2007) Highly permeable polymeric membranes based on the incorporation of the functional water channel protein Aquaporin Z. Proc Natl Acad Sci U S A 104:20719–20724. https://doi.org/10.1073/pnas.0708762104
14. Chowdhury R, Ren T, Shankla M et al (2018) PoreDesigner for tuning solute selectivity in a robust and highly permeable outer membrane pore. Nat Commun 9:1–10. https://doi.org/10.1038/s41467-018-06097-1
15. Huang PS, Boyken SE, Baker D (2016) The coming of age of de novo protein design. Nature 537:320–327. https://doi.org/10.1038/nature19946
16. Baker D (2019) What has de novo protein design taught us about protein folding and biophysics? Protein Sci 28:678–683. https://doi.org/10.1002/pro.3588
17. King NP, Sheffler W, Sawaya MR et al (2012) Computational design of self-assembling protein nanomaterials with atomic level accuracy. Science 336:1171–1174. https://doi.org/10.1126/science.1219364
18. Langan RA, Boyken SE, Ng AH et al (2019) De novo design of bioactive protein switches. Nature 572:205–210. https://doi.org/10.1038/s41586-019-1432-8
19. Humphris EL, Kortemme T (2007) Design of multi-specificity in protein interfaces. PLoS Comput Biol 3:1591–1604. https://doi.org/10.1371/journal.pcbi.0030164
20. Joh NH, Wang T, Bhate MP et al (2014) De novo design of a transmembrane zn2+-transporting four-helix bundle. Science 346:1520–1524. https://doi.org/10.1126/science.1261172
21. Korendovych IV, Senes A, Kim YH et al (2010) De novo design and molecular assembly of a transmembrane diporphyrin-binding protein complex. J Am Chem Soc 132:15516–15518. https://doi.org/10.1021/ja107487b
22. Lu P, Min D, DiMaio F et al (2018) Accurate computational design of multipass transmembrane proteins. Science 359:1042–1046. https://doi.org/10.1126/science.aaq1739
23. Koehler Leman J, Bonneau R, Ulmschneider MB (2018) Statistically derived asymmetric membrane potentials from α-helical and β-barrel membrane proteins. Sci Rep 8:4446. https://doi.org/10.1038/s41598-018-22476-6
24. Senes A, Chadi DC, Law PB et al (2007) Ez, a depth-dependent potential for assessing the energies of insertion of amino acid side-chains into membranes: derivation and applications to determining the orientation of transmembrane and interfacial helices. J Mol Biol 366:436–448. https://doi.org/10.1016/j.jmb.2006.09.020
25. Yarov-Yarovoy V, Schonbrun J, Baker D (2005) Multipass membrane protein structure prediction using Rosetta. Proteins 62:1010–1025. https://doi.org/10.1002/prot.20817
26. Lazaridis T (2003) Effective energy function for proteins in lipid membranes. Proteins Struct Funct Genet 52:176–192. https://doi.org/10.1002/prot.10410
27. Lazaridis T, Karplus M (1999) Effective energy function for proteins in solution. Proteins 35:133–152. https://doi.org/10.1002/(sici)1097-0134(19990501)35:2<133::aid-prot1>3.0.co;2-n
28. Barth P, Schonbrun J, Baker D (2007) Toward high-resolution prediction and design of transmembrane helical protein structures. Proc Natl Acad Sci 104:15682–15687. https://doi.org/10.1073/pnas.0702515104
29. MacCallum JL, Bennett WFD, Tieleman DP (2008) Distribution of amino acids in a lipid bilayer from computer simulations. Biophys J 94:3393–3404. https://doi.org/10.1529/biophysj.107.112805
30. Kroncke BM, Duran AM, Mendenhall JL et al (2016) Documentation of an imperative to improve methods for predicting membrane protein stability. Biochemistry 55:5002–5009. https://doi.org/10.1021/acs.biochem.6b00537
31. Duran AM, Meiler J (2018) Computational design of membrane proteins using Rosetta-Membrane. Protein Sci 27:341–355. https://doi.org/10.1002/pro.3335
32. De Marothy MT, Elofsson A (2015) Marginally hydrophobic transmembrane α-helices shaping membrane protein folding. Protein Sci 24:1057–1074. https://doi.org/10.1002/pro.2698

33. Alford RF, Fleming PJ, Fleming KG, Gray JJ (2020) Protein structure prediction and design in a biologically realistic implicit membrane. Biophys J 118(8):2042–2055. https://doi.org/10.1016/j.bpj.2020.03.006
34. Leaver-Fay A, Tyka M, Lewis SM et al (2011) Rosetta3: an object-oriented software suite for the simulation and design of macromolecules. Methods Enzymol 487:545–574
35. Alford RF, Koehler Leman J, Weitzner BD et al (2015) An integrated framework advancing membrane protein modeling and design. PLoS Comput Biol 11:e1004398. https://doi.org/10.1371/journal.pcbi.1004398
36. Moon CP, Fleming KG (2011) Side-chain hydrophobicity scale derived from transmembrane protein folding into lipid bilayers. Proc Natl Acad Sci U S A 108:10174–10177. https://doi.org/10.1073/pnas.1103979108
37. Feig M (2008) Implicit membrane models for membrane protein simulation. Methods Mol Biol 443:181–196. https://doi.org/10.1007/978-1-59745-177-2_10
38. Alford RF, Leaver-Fay A, Jeliazkov JR et al (2017) The Rosetta all-atom energy function for macromolecular modeling and design. J Chem Theory Comput 13:3031–3048. https://doi.org/10.1021/acs.jctc.7b00125
39. Park H, Bradley P, Greisen P et al (2016) Simultaneous optimization of biomolecular energy functions on features from small molecules and macromolecules. J Chem Theory Comput 12:6201–6212. https://doi.org/10.1021/acs.jctc.6b00819
40. Cuesta-Seijo JA, Neale C, Khan MA et al (2010) PagP crystallized from SDS/Cosolvent reveals the route for phospholipid access to the hydrocarbon ruler. Structure 18:1210–1219. https://doi.org/10.1016/j.str.2010.06.014
41. Guo L, Lim KB, Poduje CM et al (1998) Lipid A acylation and bacterial resistance against vertebrate antimicrobial peptides. Cell 95:189–198. https://doi.org/10.1016/S0092-8674(00)81750-X
42. Marx DC, Fleming KG (2017) Influence of protein scaffold on side-chain transfer free energies. Biophys J 113:597–604. https://doi.org/10.1016/j.bpj.2017.06.032
43. Waight AB, Pedersen BP, Schlessinger A et al (2013) Structural basis for alternating access of a eukaryotic calcium/proton exchanger. Nature 499:107–110. https://doi.org/10.1038/nature12233
44. Shigaki T, Rees I, Nakhleh L, Hirschi KD (2006) Identification of three distinct phylogenetic groups of CAX cation/proton antiporters. J Mol Evol 63:815–825. https://doi.org/10.1007/s00239-006-0048-4
45. Hirschi KD, Zhen RG, Cunningham KW et al (1996) CAX1, an H+/Ca2+ antiporter from Arabidopsis. Proc Natl Acad Sci U S A 93:8782–8786. https://doi.org/10.1073/pnas.93.16.8782
46. Leaver-Fay A, O'Meara MJ, Tyka M et al (2013) Scientific benchmarks for guiding macromolecular energy function improvement. Methods Enzymol 523:109–143
47. Lomize MA, Pogozheva ID, Joo H et al (2012) OPM database and PPM web server: resources for positioning of proteins in membranes. Nucleic Acids Res 40(Database issue): D370–D376. https://doi.org/10.1093/nar/gkr703

Chapter 4

Engineering of Biological Pathways: Complex Formation and Signal Transduction

Philipp Junk and Christina Kiel

Abstract

The rational in silico design of interface mutations within protein complexes is a synthetic biology tool that enables—when introduced into biological systems—the artificial rewiring of biological pathways. Here we describe the three-dimensional structure-based design of "rewiring" mutations using the FoldX force field. Specifically, we provide the protocol for the design and selection of interface mutations in three Ras-effector complex structures (PDB entries 3KUD, 4K81, and 6AMB). Ras mutations that impair binding to some but not all interacting partners are selected.

Key words Protein engineering, FoldX, Binding energies, Interface mutations, Networks, Position energy matrices

1 Introduction

Protein design is a valuable molecular tool in synthetic biology, which can be either used to study how natural signaling networks function or to build synthetic networks with specific functionalities [1]. Mutations impacting complex formations can be used to alter the affinities and specificities of an interaction, thereby inducing network "rewiring." Provided that high-resolution three-dimensional (3D) structures of protein complexes are available, the impact of mutations can be quickly and inexpensively assessed using in silico design algorithms such as FoldX [2, 3]. Protein interface modelling using FoldX has been used previously to predict the impact of interface mutations (Ala scan) with good correlations between in silico prediction and in vitro experiments [4]. Likewise, interface mutations with changed electrostatic properties and altered kinetics of complex formations were designed using FoldX

Supplementary Information The online version of this chapter (https://doi.org/10.1007/978-1-0716-1468-6_4) contains supplementary material, which is available to authorized users.

Irina S. Moreira et al. (eds.), *Computational Design of Membrane Proteins*, Methods in Molecular Biology, vol. 2315, https://doi.org/10.1007/978-1-0716-1468-6_4,

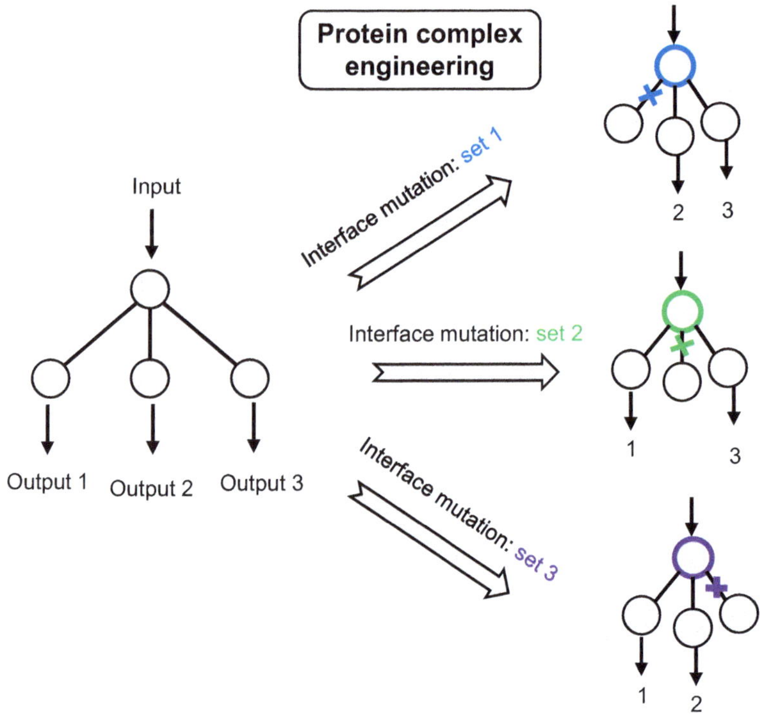

Fig. 1 Illustration of rewiring mutations introduced in Ras binding to three different effectors. In each round, the goal is to disrupt binding of one of the effectors to Ras but keep binding of the other two effectors unchanged

and subsequently engineered and tested in human cell lines [5] and yeast cells [6].

Here, we provide the protocol for designing rewiring mutations using FoldX for three Ras-effector complexes, with the aim to disrupt binding of one of the effectors to Ras but keep binding of the other two effectors unchanged for all three effectors (Fig. 1). The workflow contains five steps: (1) preparation of 3D complex structures, (2) identification of interface residues, (3) evaluation of binding energy changes on mutation, (4) validation of protein stability, and (5) selection of rewiring mutations of interest.

2 Materials

2.1 Software

In this protocol, FoldX is used to calculate energies based on protein structures [2, 3]. It can calculate both Gibbs free energy of a protein (ΔG), evaluating its stability (*see* **Note 1**), and Gibbs free energy of binding (ΔG_{bind}), evaluating the interaction of two proteins in a complex (*see* **Note 2**). FoldX has two main functionalities; the first one is the evaluation of the energy of a protein using the FoldX force field. A force field is a set of theoretical and/or empirical energy terms based on which a total energy for a protein structure can be calculated (*see* **Note 1**). The second functionality is the adaptation of protein structures by exploring different structural configurations of the amino acid side chains, called rotamers. This can improve the energy of an experimentally determined structure. The inbuild rotamer library is also used for the actual step of in silico mutagenesis: FoldX can mutate all 20 natural amino acids. It has long been a difficult task to calculate absolute energies, and this task is still not adequately accomplished so far. Instead, relative energies are usually considered. This means that for every mutant structure of interest, a reference structure must be created, calculated, and compared to. This provides changes in Gibbs free energies ($\Delta\Delta G = \Delta G_{\mathrm{Mut}} - \Delta G_{\mathrm{WT}}$) and changes in Gibbs free energies of binding ($\Delta\Delta G_{\mathrm{bind}} = \Delta G_{\mathrm{bind-Mut}} - \Delta G_{\mathrm{bind-WT}}$) (*see* **Note 3**). The FoldX software can be obtained on the official web page of the project: http://foldxsuite.crg.eu/. To install, please follow the instructions of the official manual.

2.2 Databases

FoldX requires protein structures as input. There are multiple ways of experimentally determining the structure of a protein (X-ray crystallography, NMR, cryo-EM), and experimentally determined protein structures have been invaluable in explaining the functions and mechanisms of action of proteins (*see* **Note 4**). All protein structures published are deposited in a common database, the Protein Data Base (PDB). There, the parameters of the experimental procedure are available for download alongside the structures itself, all under a unique identifier, the PDB ID. The PDB can be accessed at http://rcsb.org/. For this example protocol, three protein structures from the PDB will be used (Table 1).

2.3 Scripts

All code examples given here are bash code. While the FoldX commands will be transferable without adaptation to the Windows command line or the MacOS console (*see* **Note 5**), the bash commands used (such as `grep`, `tr` or `mkdir`) are bash specific and their respective counterparts should be used. The full bash scripts describing the data acquisition for the three structures are available

Table 1
Overview over the structures used in this example. Their PDB IDs as well as which structure is stored under which chain identifier are listed in this table

PDB ID	Chain A	Chain B	Reference
4K81	Grb14	Ras	[7]
3KUD	Ras	Raf (A85K)	[8]
6AMB	Ras	AF6	[9]

in the supplementary materials (Supplementary Files 1, 2, and 3). All data analysis and visualization were performed with R, ggplot2, and tidyverse [7–9]. A data analysis script for this protocol is available in the supplementary materials (Supplementary File 4).

3 Methods

3.1 Structure Preparation

Before working with the protein structures, there are some minor modifications to be made.

1. Download the complex structures of interest from the PDB. The PDB identifiers are 3KUD, 4K81, and 6AMB [10–12] (Table 1). The structures will be downloaded in PDB file format. Place the structures in the current working directory and name them [PDB ID].pdb, i.e., 4K81.pdb.
2. Prepare the structure for working with FoldX. Delete all crystal water and all unnecessary protein chains from the PDB file (*see* **Note 6**). Replace the character ' with * for the correct parametrization of the GTP in the structure. This can be done manually, with a structure editor, or with text processing tool. As an example, this is the command for the structure 4K81, using simple bash tools:

```
# deletion of all HOH entries from the file
grep -v 'HETATM.*HOH' 4K81.pdb | \
# deletion of all ATOM, HETATM and TER entries for the chains C-H
     grep -v 'ATOM.................[C,D,E,F,G,H]' | \
     grep -v 'HETATM...............[C,D,E,F,G,H]' | \
     grep -v 'TER..................[C,D,E,F,G,H]' | \
# replacement of ' with *
     tr "'" "*" > 4K81_clean.pdb
```

3. Repair of the structure with FoldX. During the repair of the structure, the amino acid side chains of the structure will be reoriented in order to minimize the energy of the structure. Energy calculations made with FoldX are more accurate after repairing a structure. The repair will create a report and a repaired structure 4K81_clean_Repair.pdb. The command is:

```
foldx --command=RepairPDB --pdb=4K81_clean.pdb
```

4. Next, Ras-only structures will be obtained from the complex structures. This will be necessary for assessing the influence of Ras mutations on protein stability (Sect. 3.4). To do this, the effector chain will be deleted from the PDB file. For example, for the 4K81 structure:

```
# deletion of all ATOM, HETATM and TER entries for the chain A
grep -v 'ATOM.................[A]' 4K81_clean.pdb | \
     grep -v 'HETATM...............[A]' | \
     grep -v 'TER..................[A]' > ./4K81_clean_Ras.pdb
```

5. Repair of the Ras-only structures with FoldX analogously to the repair of the complex structures.

3.2 Identification of Interface Residues

In order to reduce the computational cost, the mutations will be introduced only in residues of Ras that are participating in the interaction between Ras and its effectors, the so-called interface residues.

1. Analysis of the complex structure with FoldX. The command is:

```
foldx --command=AnalyseComplex --pdb=4K81_clean_Repair.pdb
```

2. Extraction of the interface residues from the output file **Interface_Residues_4 K81_clean_Repair_AC.fxout**. The residues are named in the following format: [Amino Acid 1 Letter Code][Chain ID][Residue Number], i.e., DB33 describes the aspartic acid at residue 33 of chain B.
3. Repeat for the other structures (3KUD, 6AMB) and compare the interface residues. For the following analysis, all interface residues from the three structures are pooled (*see* **Note 7**). Figure 2 shows the occurrence of interface residues in all three structures.

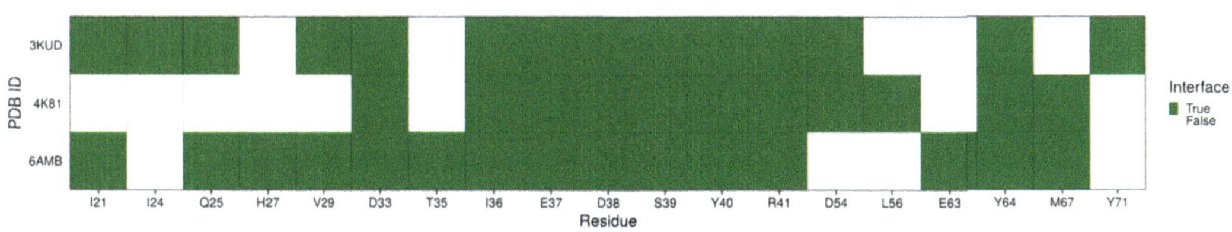

Fig. 2 Interface residues determined by FoldX in all complex structures

3.3 Evaluation of Binding Energy Changes on Mutation

In order to identify mutations in the interface of Ras and its effectors that selectively inhibit one interaction, a so-called position energy matrix for binding energy $\Delta\Delta G_{\text{bind}}$ will be created for each of the complex structures. A position energy matrix is a matrix that contains all changes to the energy upon all mutations for all residues of interest.

1. Create an output directory, i.e., on a Linux operating system:

```
mkdir -p output_4K81
```

2. Run the PSSM command in FoldX to evaluate mutations in all amino acids of interest with regard to their influence on the binding energy between Ras and its effector. Amino acids of interest are all the interface residues identified for all three structures previously. FoldX expects amino acid inputs in the format [Amino Acid 1 Letter Code][Chain ID][Residue Number][Mutation(s)], i.e., IB21a means that isoleucine at position 21 of chain B will be mutated to all canonical amino acids. Depending on the computational infrastructure this is run on, mutating 18 amino acids to all 20 canonical amino acids might take multiple hours. The command for 4K81 is:

```
foldx --command=Pssm \
      --pdb=4K81_clean_Repair.pdb \
      --output-dir=output_4K81 \
      --analyseComplexChains=A,B \
      --positions=IB21a,IB24a,QB25a,VB29a,DB33a,TB35a,IB36a, \
EB37a,DB38a,SB39a,YB40a,RB41a,DB54a,LB56a,EB63a,YB64a,MB67a, \
YB71a
```

While the PSSM command produces a lot of output files, the main information of interest can be found in **PSSM_4K81_clean_Repair.txt**. This file contains all changes of interaction energy of all mutations for all amino acids of interest in matrix format.

3. Repeat **steps 1** and **2** for 3KUD and 6AMB. A visualization of the results from all three structures can be found in Fig. 3.

3.4 Validation of Ras Protein Stability

After determining the effect of mutations on binding energy, a necessary validation step is to check whether a mutation significantly influences the stability of our protein (Ras). To do this, we will create another position energy matrix investigating the changes of Gibbs free energy $\Delta\Delta G$ upon mutation of our residues of interest.

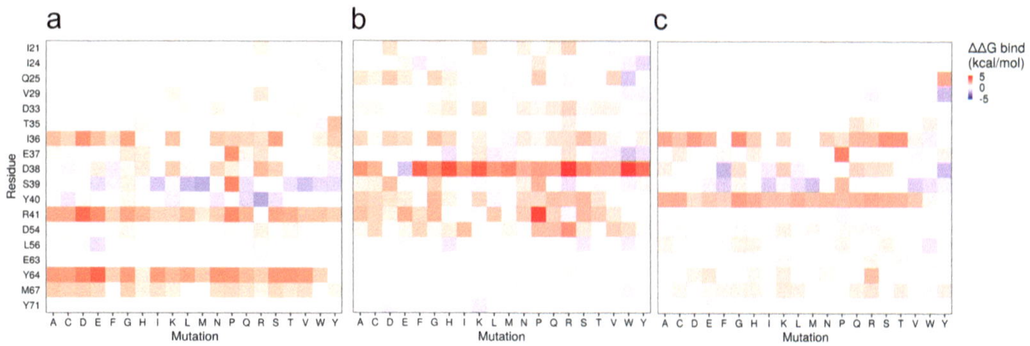

Fig. 3 Heatmap of binding energy $\Delta\Delta G_{\text{bind}}$ for all mutations for all residues. (**a**) Position energy matrix for 4K81. (**b**) Position energy matrix for 3KUD. (**c**) Position energy matrix for 6AMB

1. Create an output directory, i.e., on a Linux operating system:

```
mkdir -p output_4K81_Ras
```

2. Run the PositionScan command in FoldX. Similar to the PSSM command, it evaluates a list of mutations against the WT structure and generates a summarized output. In contrast to PSSM, which evaluates Gibbs free energy of binding (interaction), PositionScan evaluates Gibbs free energy (protein stability).

```
cd output_4K81_Ras
foldx --command=PositionScan \
    --pdb=4K81_clean_Ras_Repair.pdb \
    --pdb_dir=./.. \
    --positions=IB21a,IB24a,QB25a,VB29a,DB33a,TB35a,IB36a, \
EB37a,DB38a,SB39a,YB40a,RB41a,DB54a,LB56a,EB63a,YB64a,MB67a, \
YB71a
```

   Again, many output files will be created. An overview over the Gibbs free energy changes can be found in **PS_4K81_clean_Ras_Repair_scanning_output.txt**.

3. Repeat **step 2** for 3KUD and 6AMB. A visualization of the results from all three structures can be found in Fig. 4.

3.5 Identification of Mutations of Interest

After obtaining both the changes to the binding energy $\Delta\Delta G_{\text{bind}}$ and the changes to the Gibbs free energy (stability) $\Delta\Delta G$ for all mutations of interest, the next step will be to select one or multiple mutations of interest. As stated earlier, the objective for this exercise is to generate mutants that specifically interrupt one Ras–Effector interaction, while simultaneously not affecting the other

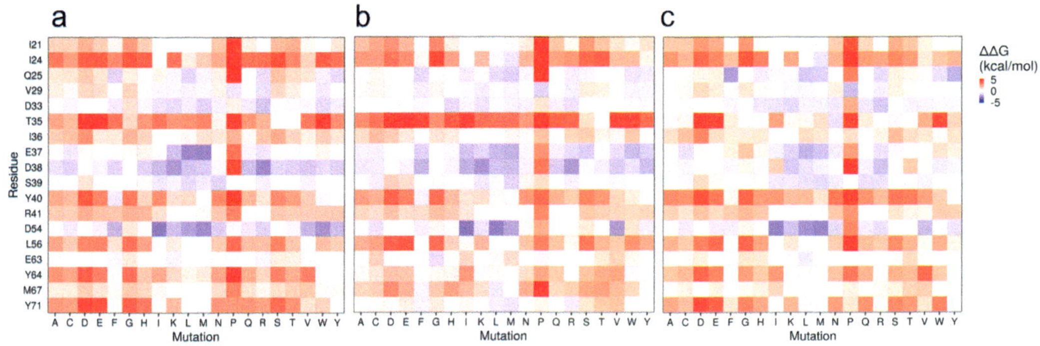

Fig. 4 Heatmap of stability $\Delta\Delta G$ for all mutations for all residues. (**a**) Position energy matrix for 4 K81. (**b**) Position energy matrix for 3KUD. (**c**) Position energy matrix for 6AMB

interactions. Therefore, the selection of mutations will be based on the following three criteria that have to be evaluated and balanced for each mutation:

1. Change of binding energy in the structure of interest $\Delta\Delta G_{\text{bind}-i}$: as high as possible.
2. Change of binding energy in the other structures $\Delta\Delta G_{\text{bind}-j}$: as close to zero as possible.
3. Change of Gibbs free energy of Ras structures $\Delta\Delta G$: as close to zero as possible.

While it would be possible to identify mutations of interest by simply studying the heatmaps, this is a complicated task. However, the selection process can be simplified and made objective by evaluating a scoring function based on the three criteria.

$$\text{score} = \Delta\Delta G_{\text{bind-interest}} - \text{mean}(|\Delta\Delta G_{\text{bind-other}}|) - \text{mean}(|\Delta\Delta G|)$$

The scoring function is implemented like this:

$$\begin{aligned}\arg\max_{m\in M} \text{score}(m) = {} & a(\Delta\Delta G_{\text{bind}-i}(m)) \\ & - b\left(\frac{1}{n-1}\sum_{j\in X;j\neq i}|\Delta\Delta G_{\text{bind}-j}(m)|\right) \\ & - c\left(\frac{1}{n}\sum_{k\in X}|\Delta\Delta G_k(m)|\right)\end{aligned}$$

with $\Delta\Delta G_{\text{bind}}$ being the changes to the binding energy upon mutation, $\Delta\Delta G$ being the changes to the Gibbs free energy upon mutation, $M = \{\text{I21A}, \text{I21C}, \ldots, \text{Y71Y}\}$ is the set of analyzed mutations, $X = \{\text{3KUD}, \text{4K81}, \text{6AMB}\}$ is the set of analyzed structures, n is the cardinality of the set X, and a, b, c being manually chosen weights for each of the three criteria.

Table 2
Overview over the top scoring mutations for the structure 4 K81. The weights used in for the scoring procedure were a = b = c = 1. All energy values are given in kcal/mol

Mutation	Score	$\Delta\Delta G_{bind}$ 4K81	$\Delta\Delta G_{bind}$ 3KUD	$\Delta\Delta G_{bind}$ 6AMB	$\Delta\Delta G$ 3KUD	$\Delta\Delta G$ 4K81	$\Delta\Delta G$ 6AMB
Y64L	1.63	2.279	−0.013	−0.271	0.474	0.665	0.382
R41D	1.443	3.54	0.653	0.262	1.554	1.872	1.491
Y64F	1.228	1.432	−0.013	−0.106	−0.057	−0.332	−0.046
R41Q	1.145	1.949	1.259	0.042	0.185	0.185	0.089
Y64A	0.887	2.526	0.101	−0.047	0.506	2.586	1.601

Table 3
Overview over the top scoring mutations for the structure 3KUD. The weights used in for the scoring procedure were a = b = c = 1. All energy values are given in kcal/mol

Mutation	Score	$\Delta\Delta G_{bind}$ 3KUD	$\Delta\Delta G_{bind}$ 4K81	$\Delta\Delta G_{bind}$ 6AMB	$\Delta\Delta G$ 3KUD	$\Delta\Delta G$ 4K81	$\Delta\Delta G$ 6AMB
D38W	3.474	4.759	0.46	−0.235	−0.934	−1.083	−0.794
D38H	2.829	4.072	0.783	0.284	−0.526	−1.174	−0.429
D38R	2.114	5.607	1.816	0.998	−2.212	−2.5	−1.546
D38G	1.805	2.836	0.472	1.087	−0.041	−0.176	0.539
D38M	1.771	3.002	0.003	0.106	−1.58	−1.325	−0.622

Table 4
Overview over the top scoring mutations for the structure 6AMB. The weights used in for the scoring procedure were a = b = c = 1. All energy values are given in kcal/mol

Mutation	Score	$\Delta\Delta G_{bind}$ 6AMB	$\Delta\Delta G_{bind}$ 3KUD	$\Delta\Delta G_{bind}$ 4K81	$\Delta\Delta G$ 3KUD	$\Delta\Delta G$ 4K81	$\Delta\Delta G$ 6AMB
Y40M	1.112	1.685	−0.235	−0.037	0.359	−0.049	0.904
I36T	0.943	2.564	1.135	−0.319	0.54	1.128	1.015
Q25Y	0.821	2.143	0.163	0.089	−0.74	−1.024	−1.822
I36Q	0.553	1.835	0.95	1.229	0.115	0.396	−0.069
Y40F	0.544	1.34	−0.285	−0.053	−0.628	−0.793	0.461

Tables 2, 3, and 4 show the five top scoring mutations for each of the investigated structures according to our scoring function using the simple weights of a, b, $c = 1$. Considering the candidates in the light of our objective, there are multiple mutations that our approach identified as possible candidates. For the Ras-Grb14 structure (4K81), the mutations Y64L and Y64F, ranked first and third, respectively, seem promising candidates (Table 2). Both offer relatively high disruption of binding energy $\Delta\Delta G_{bind}$ (*see* **Note 8**), while simultaneously not impacting the stability $\Delta\Delta G$ (*see* **Note 9**).

The Ras-Raf structure (3KUD) is more challenging to evaluate. All mutations suggested affect the residue D38. This residue has been identified as a strong contributor to the interaction of Ras and Raf, and in this particular structure, due to an interface mutation in Raf (A85K), the importance of D38 to this interaction has only been enhanced [11, 13]. This results in very high disruptions of binding energy $\Delta\Delta G_{bind}$ for mutations in D38, that dominate the scoring function (Table 3). While the mutations D38W and D38G, scoring first and fourth, respectively, are interesting candidates, another approach would be to adjust the weights in the scoring function such that $a < b = c$. This case highlights that the scoring function can and should be tailored to the problem at hand.

For 6AMB, the calculated scores are lower than for the other two structures (Table 4). Nonetheless, the mutations Y40M, Q25Y, and Y40F, scoring first, third, and fifth, respectively, seem promising.

4 Notes

1. The FoldX force field calculates the free energy of unfolding (ΔG, free energy difference between the folded and unfolded proteins) of a protein using the following equation [2, 3]:

$$\Delta G = a \times \Delta G_{vdw} + b \times \Delta G_{solvH} + c \times \Delta G_{solvP} + d \times \Delta G_{hbond} + e \times \Delta G_{wb} + f \times \Delta G_{el} + g \times \Delta G_{kon} + h \times \Delta S_{mc} + i \times \Delta S_{sc} + j \times \Delta G_{clash}$$

where ΔG_{vdw} is sum of Van der Waals contributions of all atoms, ΔG_{solvH} is the difference in solvation energy for apolar groups, ΔG_{solvP} is the difference in solvation energy for polar groups, ΔG_{hbond} is the free energy difference between the formation of an intra-molecular hydrogen-bond compared to inter-molecular hydrogen-bond formation with solvent, ΔG_{wb} is the extra stabilizing free energy provided by water molecules making more than one hydrogen bond to the protein, ΔG_{el} is the electrostatic contribution of charged groups, ΔG_{kon} is the effect of electrostatic interactions on the association rate constant (k_{on}), ΔS_{mc} is the entropy cost for fixing the backbone in the folded state, ΔS_{sc} is the entropic cost of fixing a side chain in a specific conformation, and ΔG_{clash} is a measure of unfavorable steric overlaps between atoms in the structure. Parameters ($a \ldots j$) are relative weights of the different energy terms.

2. For protein complexes, FoldX calculates the interaction energy (ΔG_{bind}) between two proteins, A and B, using the following equation [2, 3]:

$$\Delta G_{\text{bind}} = \Delta G_{\text{AB}} - (\Delta G_{\text{A}} + \Delta G_{\text{B}})$$

3. Validation of energy changes calculated by FoldX with energies measured experimentally found that FoldX energy changes have a standard deviation of $\sigma = \pm 0.8$ kcal/mol [2]. When evaluating FoldX energies, it is important to keep in mind that these values are in silico calculations that come with a certain error.
4. FoldX works most reliable when high-resolution 3D structures are used, which are typically obtained from X-ray crystallography.
5. The FoldX software has been designed as a command line tool and has been released for all major operating system. As mentioned, the commands are transferable. Recently, a graphical interface to FoldX using the YASARA protein visualization program has been released [14, 15]. The protocol described here can also be replicated using the graphical interface; however, this protocol focusses on giving instructions for using the command line interface to FoldX. The FoldX plugin for YASARA along with instructions for the installation can be found on the official homepage of the FoldX project: http://foldxsuite.crg.eu/.
6. Only one biological assembly should be kept in the asymmetric unit, i.e., for the 4K81 structure, the chains C-H are removed. Sometimes, the biological assemblies are slightly different, so it might pay off to look through all of them.
7. Interface residues that do not contribute much to binding energy could nevertheless be good candidates for mutations as they might disrupt binding energy by introducing steric clashes in the interface.
8. A reasonable energy threshold for considering whether a mutation does disrupt a protein–protein interaction would be $\Delta\Delta G_{\text{bind}} > 1 - 2$ kcal/mol. This would reduce the binding affinity of an interaction by a factor of 5–10, which would be enough to dramatically reduce the number of complexes formed. This is especially true in a highly competitive system such as the Ras–Effector system.
9. Disruption of stability should be evaluated in a different way from disruptions of binding energy. In general, proteins are more resistant to mutations affecting stability. A reasonable energy threshold would be a disruption of $\Delta\Delta G < 1.6$ kcal/mol, at which proteins would be affected. This threshold of twice the standard error of FoldX is commonly applied (*see* **Note 3**).

Acknowledgment

This work is part of the research program "Quantitative and systems analysis of (patho)physiological signalling networks" with project number 16/FRL/3886, which is financed by Science Foundation Ireland (SFI) (to C. Kiel).

References

1. Kiel C, Yus E, Serrano L (2010) Engineering signal transduction pathways. Cell 140:33–47. https://doi.org/10.1016/j.cell.2009.12.028
2. Guerois R, Nielsen JE, Serrano L (2002) Predicting changes in the stability of proteins and protein complexes: a study of more than 1000 mutations. J Mol Biol 320:369–387. https://doi.org/10.1016/S0022-2836(02)00442-4
3. Schymkowitz J, Borg J, Stricher F et al (2005) The FoldX web server: an online force field. Nucleic Acids Res 33(Web Server issue): W382–W388. https://doi.org/10.1093/nar/gki387
4. Kiel C, Serrano L, Herrmann C (2004) A detailed thermodynamic analysis of ras/effector complex interfaces. J Mol Biol 340:1039–1058. https://doi.org/10.1016/j.jmb.2004.05.050
5. Kiel C, Serrano L (2009) Cell type-specific importance of ras-c-raf complex association rate constants for MAPK signaling. Sci Signal 2:ra38. https://doi.org/10.1126/scisignal.2000397
6. Stojanovski K, Ferrar T, Benisty H et al (2017) Interaction dynamics determine signaling and output pathway responses. Cell Rep 19:136–149. https://doi.org/10.1016/j.celrep.2017.03.029
7. R Core Team (2019) R: A Language and Environment for Statistical Computing. R Foundation for Statistical Computing, Vienna, Austria. https://www.R-project.org/
8. Wickham H (2016) ggplot2: elegant graphics for data analysis. Springer-Verlag, New York
9. Wickham H, Averick M, Bryan J et al (2019) Welcome to the tidyverse. J Open Source Softw 4:1686. https://doi.org/10.21105/joss.01686
10. Qamra R, Hubbard SR (2013) Structural basis for the interaction of the adaptor protein Grb14 with activated Ras. PLoS One 8: e72473. https://doi.org/10.1371/journal.pone.0072473
11. Filchtinski D, Sharabi O, Rüppel A et al (2010) What makes Ras an efficient molecular switch: a computational, biophysical, and structural study of Ras-GDP interactions with mutants of Raf. J Mol Biol 399:422–435. https://doi.org/10.1016/j.jmb.2010.03.046
12. Smith MJ, Ottoni E, Ishiyama N et al (2017) Evolution of AF6-RAS association and its implications in mixed-lineage leukemia. Nat Commun 8:1099. https://doi.org/10.1038/s41467-017-01326-5
13. Nassar N, Horn G, Herrmann C et al (1996) Ras/Rap effector specificity determined by charge reversal. Nat Struct Biol 3:723–729. https://doi.org/10.1038/nsb0896-723
14. Van Durme J, Delgado J, Stricher F et al (2011) A graphical interface for the FoldX forcefield. Bioinformatics 27:1711–1712. https://doi.org/10.1093/bioinformatics/btr254
15. Krieger E, Vriend G (2014) YASARA view—molecular graphics for all devices—from smartphones to workstations. Bioinformatics 30:2981–2982. https://doi.org/10.1093/bioinformatics/btu426

Part III

Structure Prediction and Oligomerization

Chapter 5

Homology Modeling of Class A G-Protein-Coupled Receptors in the Age of the Structure Boom

Asma Tiss, Rym Ben Boubaker, Daniel Henrion, Hajer Guissouma, and Marie Chabbert

Abstract

With 700 members, G protein-coupled receptors (GPCRs) of the rhodopsin family (class A) form the largest membrane receptor family in humans and are the target of about 30% of presently available pharmaceutical drugs. The recent boom in GPCR structures led to the structural resolution of 57 unique receptors in different states (39 receptors in inactive state only, 2 receptors in active state only and 16 receptors in different activation states). In spite of these tremendous advances, most computational studies on GPCRs, including molecular dynamics simulations, virtual screening and drug design, rely on GPCR models obtained by homology modeling. In this protocol, we detail the different steps of homology modeling with the MODELLER software, from template selection to model evaluation. The present structure boom provides closely related templates for most receptors. If, in these templates, some of the loops are not resolved, in most cases, the numerous available structures enable to find loop templates with similar length for equivalent loops. However, simultaneously, the large number of putative templates leads to model ambiguities that may require additional information based on multiple sequence alignments or molecular dynamics simulations to be resolved. Using the modeling of the human bradykinin receptor B1 as a case study, we show how several templates are managed by MODELLER, and how the choice of template(s) and of template fragments can improve the quality of the models. We also give examples of how additional information and tools help the user to resolve ambiguities in GPCR modeling.

Key words Homology modeling, MODELLER, GPCR, Membrane receptor, Loop modeling, Computational biology

1 Introduction

Class A (rhodopsin-like) G protein-coupled receptors (GPCRs) form the largest family of transmembrane receptors in the human genome [1, 2]. They include about 300 non-olfactory receptors classified into a dozen of subfamilies (Fig. 1) and 400 olfactory receptors. These receptors allow the transfer of information from an

Asma Tiss and Rym Ben Boubaker contributed equally to this work.

Irina S. Moreira et al. (eds.), *Computational Design of Membrane Proteins*, Methods in Molecular Biology, vol. 2315, https://doi.org/10.1007/978-1-0716-1468-6_5,

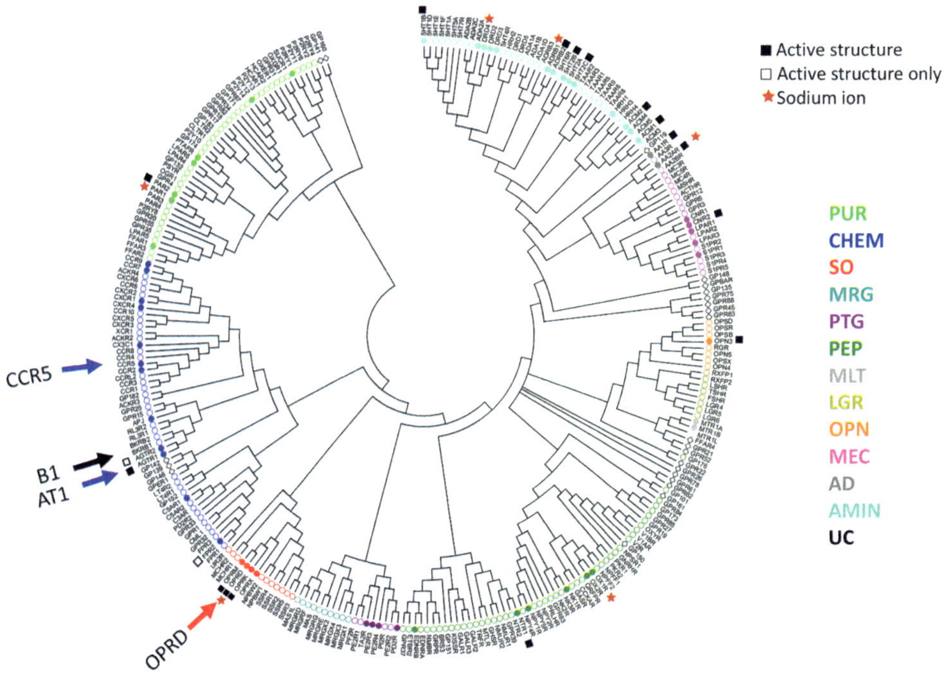

Fig. 1 Evolutionary tree of human class A GPCRs, indicating structurally resolved receptors by subfamily. The color code of the circles indicates the GPCR subfamily (PUR: light green, CHEM: blue, SO: red, MRG: teal, PTG: violet, PEP: dark green, MLT: light gray, LGR: khaki, OPN: orange, MEC: magenta; AD: dark gray; AMIN: cyan; UC: black). The circle is open when the receptor has no PDB structure and closed when the receptor has at least one PDB structure. Receptors with at least one active state structure are indicated by squares (black squares for both inactive and active state structures, open squares for active state structures only). Receptors with a sodium-bound structure are indicated by red stars. For the resolved structures, orthologous receptors from any organism are considered. The receptor names in the tree correspond to the UniProt identification name without the "HUMAN" extension. The arrows indicate the B1 target and the CCR5, AT1, and OPRD templates that are discussed in the text

extracellular signal to the cell cytoplasm. The extracellular signal is usually a ligand that, after binding to the receptor, induces a conformational change from an inactive to an active conformation, which in turn binds to and activates effector proteins such as G proteins and arrestins. GPCRs participate in numerous physiopathological processes and are the target of about 30% of presently available drugs [3].

The pharmacological importance of GPCRs explains the considerable effort spent to resolve their molecular structure. These receptors share a common fold comprising seven transmembrane helices (TM), with highly conserved anchor positions in each helix [4]. The first structure of a GPCR, rhodopsin, was resolved in 2000 [5]. The resolution of a second receptor, the β2 adrenergic receptor, required 7 years [6]. Since then, several technical locks were broken, and new GPCR structures from different families followed almost non-stop. Now in 2020, the structures of 57 unique class A

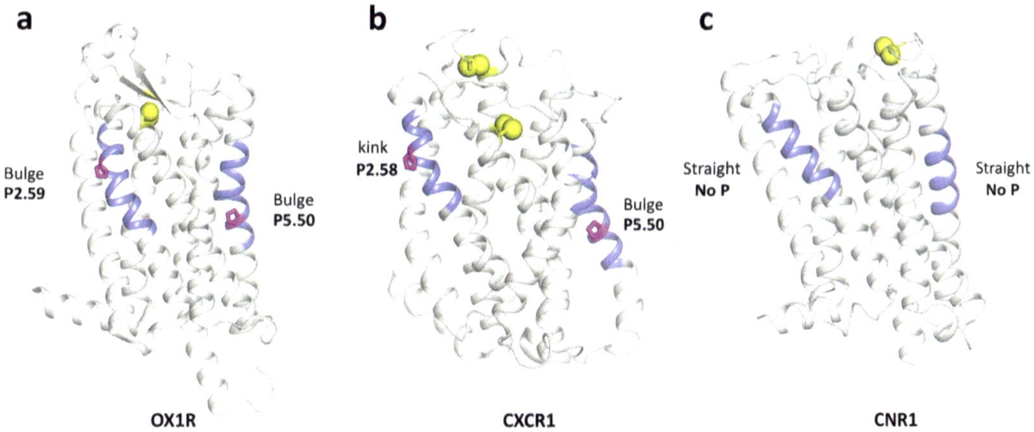

Fig. 2 Structural diversity of class A GPCRS. (**a**) The orexin receptor 1, OX1R, has a bulge in TM2 (P2.59 pattern), a bulge in TM5 (P5.50 pattern), and a disulfide bond linking the extracellular terminus of TM3 to ECL2; (**b**) The receptor CXCR1 has a kink in TM2 (P2.58 pattern), a bulge in TM5 (P5.50 pattern), the TM3-ECL2 disulfide bond, and an addition disulfide bond linking the extracellular terminus of TM7 to the N-terminus of the receptor; (**c**) The cannabinoid receptor 1, CNR1, has two straight TM2 and TM5 helices (no proline residue in these helices) and an unusual disulfide bond in ECL2. In the three cases, the receptors are in an inactive state. The TM2 and TM5 proline residues are magenta. The PDB codes are 6TOD (OX1R), 2NLN (CXCR1) and 5TGZ (CNR1)

receptors (all of them being non-olfactory receptors), totalizing more than 300 structures in different complexes or activation states, have been resolved by X-ray crystallography, serial femtosecond crystallography, cryo-electron microscopy, or solid state NMR, providing a deeper understanding of the mechanism of action of GPCRs [7–10]. Nevertheless, in spite of this avalanche of structures, about 80% of the non-olfactory GPCRs still do not have resolved structures.

Among the 57 class A receptors with at least one structure, 39 receptors have been resolved in inactive state only, two receptors in active state only, and 16 receptors in different activation states. These structures reveal the structural diversity of the transmembrane fold (Fig. 2) and have evidenced the large conformational change occurring upon activation (e.g., the type 1 angiotensin II receptor, AT1, in Fig. 3a), with a pivotal motion of TM6 that opens an intracellular cavity allowing effector binding [11]. They have also revealed an allosteric binding site for the sodium ion (e.g., the δ opioid receptor, OPRD, in Fig. 3b) that acts as an allosteric modulator [12] and should be taken into account in molecular docking to GPCRs [13]. Inactive states of GPCRs are adapted for drug design or virtual screening of antagonists/inverse agonists, whereas active states should be adapted for drug design of agonists [14, 15]. Design of biased agonists (agonists specific of a signaling pathway) raises an additional level of complexity [16]. Deorphanisation of orphan receptors often relies on virtual screening

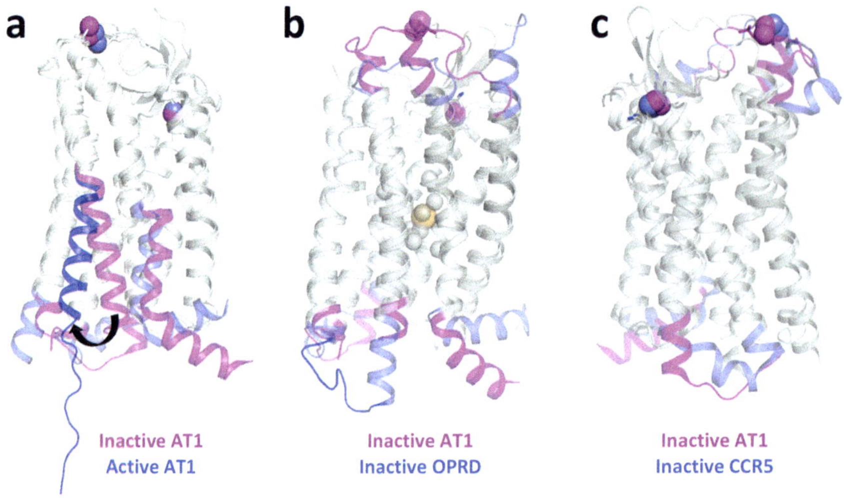

Fig. 3 Comparison of templates for B1 modeling. The inactive structure of AT1 (PDB 4YAY) is superposed on the structure of (**a**) active AT1 (PDB 6OS0), (**b**) inactive OPRD (PDB 4N6H), and (**c**) inactive CCR5 (PDB 4MBS). The structures are shown as white ribbons with differences highlighted in magenta for inactive AT1 and slate for the other structures. The sulfur atoms of the disulfide bonds are shown as magenta and slate spheres for inactive AT1 and the other receptors, respectively. In (**a**), the arrow indicates the pivotal motion of TM6 upon activation. The tilted orientation of H8 in the inactive structure is not observed in the active structure. In (**b**), the sodium ion and the neighbor water molecules present in the structure of OPRD are shown as yellow and gray spheres, respectively. The orientation highlights the tilted orientation of H8 in AT1 and the structure of ICL3 in OPRD. In (**c**), the orientation highlights the difference in the structures of ICL2 and the tilted orientation of the TM6 extracellular terminus in CCR5

[17, 18]. In addition, understanding the mechanisms of action of GPCRs requires molecular dynamics (MD) simulations of receptors in different activation states. These studies still rely heavily on molecular modeling. Thus, even in the age of the GPCR structure boom, molecular modeling by homology is still necessary.

This chapter discusses basic and advanced features of molecular modeling of class A GPCRs with the homology-based MODELLER software [19, 20]. The concept of homology modeling is based on evolution. Proteins are homologous when they share a common ancestor, which results in structure and, to a lesser extent, to sequence and function similarity [21]. The unknown structure of the target protein is modeled from the known structure of (at least) one homologous protein (the template) and the sequence alignment of the target versus the template(s). With keeping in mind that homology modeling programs always give a model, the main questions concern the quality of the model(s) and the way(s) to improve them. The inputs have to be carefully prepared and the outputs carefully evaluated, with critical assessment of available information and assumptions.

Difficulties of homology modeling depend on the sequence similarities between the target and the template. It is generally assumed that, above a threshold of 30% of identity, homology modeling may be quite straightforward, even if caveats can occur [22]. Below this limit, modeling usually becomes increasingly difficult because of structural variations. In spite of their common fold, each GPCR structure is unique [7–10], with structural variations to adapt to the variety of ligands (Fig. 2). It is thus mandatory to carefully select template(s) to correctly translate sequence into structural similarities. Moreover, as active structures are far less frequent than inactive ones (Fig. 1), modeling of active states requires special attention.

Here, we detail the procedure of molecular modeling of rhodopsin-like GPCRs and the customization of the modelling process that is possible with standalone MODELLER. We emphasize the importance of evolutionary information [23–26] for template selection and model customization. Using the human bradykinin receptor B1 with bound sodium ion in the inactive state as a case study, we show how MODELLER manages multiple templates and how the quality of the resulting model(s) can be improved by the careful choice of multiple template(s) and of template fragment(s).

2 Materials

2.1 Hardware

A computer running Linux/Unix, Apple Mac OS X (10.6 or later), or Microsoft Windows (XP or later).

2.2 Software

The MODELLER 9.23 program [19, 27] can be downloaded and installed from http://salilab.org/modeller. It is written in Fortran 90 and uses Python for its control language. Thus, all input scripts to MODELLER are Python scripts. In addition to MODELLER, several tools are required:

1. A text editor capable of outputting plain text files, such as the free and open-source software `gedit`, available for Linux, Windows and Mac OS X.
2. A molecular viewing tool, such as the PyMOL Molecular Graphics System, Schrodinger, LLC, or UCSF Chimera [28]. The structural alignment tools they include are usually sufficient for comparison of templates and models.
3. A multiple sequence alignment (MSA) program, such as ClustalW [29], T-COFFEE [30], or MUSCLE [31].
4. A visual software for MSA editing. We recommend GeneDoc (Multiple Sequence Alignment Editor, Analyzer, and Shading Utility) developed at Pittsburgh Supercomputing Center's National Resource for Biomedical Supercomputing [32].

2.3 Input Files

MODELLER needs three types of input files:

1. The .py script file written in Python.
2. At least one .pdb file containing the structure of one template.
3. The .ali file indicating the alignment between the target and the template(s). This alignment file has a specific format for MODELLER.

2.4 Additional Tools and Websites

1. For template structures: The Protein Data Bank (PDB) (https://www.rcsb.org/).
2. For the search of homologous templates: direct mining of the PDB with the Blastp utility in UniProt (https://www.uniprot.org/), Expasy (https://www.expasy.org/), or NCBI (https://blast.ncbi.nlm.nih.gov/Blast.cgi).
3. For general information on receptors: UniProt, including Swiss-Prot with reviewed entries and UniProtKB with automatic entries. Swiss-Prot centralizes functional information on proteins with detailed annotations that are curated by experts.
4. For secondary structure prediction: JPred4 [33], accessible at http://www.compbio.dundee.ac.uk/jpred4/index.html.
5. For quality check of the model: PROCHECK [34], accessible at https://www.ebi.ac.uk/thornton-srv/software/PROCHECK.
6. For building non-redundant homologous sequence sets: the nrdb90.pl perl script [35] that can be found at ftp://biodisk.org/Program/Perl/Bioperl/nrdb90.pl.

2.5 GPCR-Specific Websites

Two websites gather invaluable, updated information on GPCRs:

1. The GPCRDB database [10, 36], accessible at https://gpcrdb.org, which gathers sequences, structures, genetic variations, mutational data, and structure-based alignments, with a classification based on ligands.
2. The Zhang Lab website at the university of Michigan (https://zhanglab.ccmb.med.umich.edu) with online services for GPCRs (GPCR-EXP: Experimental structures, GPCR-RD: Experimental restraints, GLASS: GPCR-ligand association database).

3 Methods

The general principles of homology modeling by satisfaction of spatial restraints as implemented in MODELLER are detailed in **Note 1**. Here we detail the steps necessary for homology modeling of class A GPCRs in different activation states with MODELLER. For clarity purpose, as a case study, we will model the human bradykinin receptor B1 (UniProt access code: BKRB1_HUMAN, thereafter B1), in an inactive state with a bound sodium ion at the allosteric binding site.

3.1 Gaining Information on the Target Receptor

Before initiating the modeling project, gather available information on the target. In addition to literature, check UniProt. In its reviewed part, each entry includes not only core data (amino acid sequence, name, citations, etc.) but also classifications, cross-references, mutational data, natural variants, amino acid modifications (modified residues, glycosylation, lipidation, disulfide bonds, etc.), binary interactions, subunit structure, and, possibly, 3D structures or models. In addition to UniProt, the GPCRDB database [36] and the Zhang Lab website [37] provide updated, GPCR-specific information (see Materials).

3.2 Gaining Information on the Receptor Subfamily

Class A receptors are classified into a dozen of evolutionary-based subfamilies that make consensus in the literature [2, 25]. Sequence identity is usually around 25–30% within subfamilies, but may be as low as 15% between subfamilies. In addition, around 20 receptors cannot be related to any subfamily and are unclassified (UC). Evolutionary classification gives information on several sequence and structural patterns (*see* **Note 2**). A first pattern concerns two transmembrane helices, TM2 and TM5, which have variable proline motifs and, thus, structural variability [24, 25, 38], as exemplified in Fig. 2. A second pattern concerns the disulfide bonds stabilizing the receptor extracellular domain.

In addition, careful analysis of the multiple sequence alignment of the target sequence with receptors from its subfamily may reveal unusual patterns that should be taken into account in the modeling procedure and/or the subsequent computational studies, in particular (1) indels in transmembrane helices [24, 38], (2) mutations in the highly conserved sequence motifs of class A GPCRs (DRY in TM3, CWXP in TM6, NPXXY in TM7) [4], and (3) mutations in the sodium binding site [12, 26].

For example, in our case study, the receptor B1 belongs to the subfamily of chemotaxic (CHEM) receptors (Fig. 1). It is closely related to a set of receptors for vasoactive peptide receptors (the bradykinin receptor B2, the angiotensin II receptors AT1 and AT2, the apelin receptor) and to the chemokine receptors. As these receptors, it possesses the P2.58/P5.50 proline pattern characterized by a proline kink in TM2 and a proline bulge in TM5. In addition, as most chemotaxic receptors, it possesses two disulfide bonds between TM3 and extracellular loop 2 (ECL2) and between the N-terminus and ECL3. Please note that, in UniProt, it is classified as class A GPCR (IPR000276 GPCR_Rhodpsn), bradykinin receptor B1 (IPR001186 Brdyknn_1_rcpt), and bradykinin receptor (IPR000496 Brdyknn_rcpt which includes B1 and B2 receptors). The intermediary subfamily level is not provided by UniProt (*see* **Note 3**) but may be found in the literature [2, 24].

3.3 Searching for Suitable Template(s)

The primary requirement for homology modeling is the identification of at least one known structure with similarity to the target sequence to be used as a template. Now, numerous structures of GPCRs are available and the choice of the "best" template(s) has to be done carefully, keeping in mind that the selection of the closest homolog based on sequence identity does not guaranty model accuracy [39, 40]. Nevertheless, as each subfamily presents unique structural features that have to be considered for modeling, it is recommended to select at least one template from the same subfamily, with similar proline and cysteine patterns. This should be possible in most cases except for the LGR (leucine-rich repeat GPCRs) and MRG (Mas-related GPCRs) subfamilies for which no structure has been resolved (Fig. 1).

Search of close homologs with resolved structures in the Protein Data Bank can be carried out straightforwardly by blasting the target sequence from UniProtKB (hits with 3D structures or models), Expasy or NCBI (hits with 3D structures only). Homology modeling may be based on several templates, rather than a single one. Indeed, the use of several templates approximately equidistant from the target sequence, with a weighting based on sequence similarity, generally increases the model accuracy [41].

For our test case B1, the closest hits (in decreasing order of similarity) are AT1 (6 structures), AT2 (5), CCR5 (6), CCR9 (1), APJ (1), and then, the opioid receptors (6, 4, and 4 structures for delta, kappa, and mu, respectively). All these receptors have the same TM2 and TM5 proline patterns as B1 and all, except the opioid receptors, have the double disulfide bonds in the extracellular domain. All, except AT2, have inactive state structures. AT1, AT2, and the opioid receptors have active state structures. Among these structures, only the δ opioid receptor (OPRD) has an inactive structure with a sodium ion bound at the allosteric binding site [42]. In these structures, N-and C-termini and loops may be missing, because of truncation, intrinsic disorder, or replacement by a fusion protein for crystallization purposes.

3.4 Selecting Suitable Template(s)

Template selection depends on (1) the target state to be modeled (see the large outward motion of TM6 that differentiates the inactive from the active state of AT1 in Fig. 3a), (2) the resolution of the N- and C-termini and of the loops, and (3) the inclusion of bound ligands (receptor agonists or antagonists, ions, lipids, and water molecules). The experimental resolution of the crystal structure is usually not the most crucial factor. By contrast, the modifications done for crystallization (mutations, nanobodies, and insertions) and quaternary/crystal contacts may have an impact on the receptor structure and have to be carefully evaluated. The specific sequence/structure patterns of GPCRs also need careful evaluation. It is necessary to check what is "normal" or not in the structures under scrutiny, such as the curvature of helices, additional bends or kinks, unusual distortions (π or 3_{10} helices), unusual structural motifs, and helix orientations, in particular for the C-terminal helix 8 (H8).

Figure 3 displays four putative templates for the modeling of B1: AT1 (4YAY) [43], CCR5 (4MBS) [44], and OPRD (4N6H) [42] in inactive states, and AT1 in an active state (6OS0) [45]. Among these templates, only OPRD has a resolved ICL3 loop and a bound sodium ion with coordination water, but this receptor does not possess the second disulfide bond between the N-terminus and ECL3. Inactive AT1 has an unstructured ICL2 and an unconventional, tilted positioning of the C-terminus (also observed in CCR5, but not in active AT1). In CCR5, active AT1 and OPRD, ICL2 is structured as a helix. In addition, CCR5 has an unusual outward orientation of the C-terminus of TM6. Thus, different templates will be necessary to model the entire sequence of B1, but their selection will affect the resulting models.

3.5 Mining and Analyzing Receptor Homologs

In many cases, ambiguities on "best" templates are observed. For example, in the case of B1, should we model ICL2 with an α-helix, as observed in the OPRD, CCR5, and active AT1 templates, or as a coil as observed in the inactive AT1 template? Other examples of ambiguities can be found when the target receptor does not possess one of the anchor prolines in the TM helices [38] or when there are insertions/deletions in loops of the target versus the template(s). In ambiguous cases, analysis of orthologous or paralogous sequence sets may be informative [24, 38]. To build these sets, use the InterPro identifiers (*see* **Note 3**) and mine UniprotKB. The subsequent analysis may be facilitated by building a nonredundant set with sequence identities lower than 90%. This can be carried out easily with the nrdb90.pl perl script [35]. The multiple sequence alignment of the resulting set should be carefully checked and, if necessary, manually corrected with Genedoc [32]. This procedure considerably reduces the number of orthologous sequences and, subsequently, allows better visualization of key evolutionary events. In some cases, secondary structure prediction using JPred4 [33] and customized MSA may be informative (see example below).

3.6 Including Ligand or Nonprotein Residues

If the template contains a ligand, water molecules, ions, or other nonprotein residues (anything marked as HETATM in the PDB file), MODELLER can include them into the generated model. By default, all HETATM records are ignored. They are read when the `env.io.hetatm` and `env.io.water` Booleans are set to `TRUE`. Ions and water molecules are indicated by "i" or "w" in the alignment and participate in the refinement step. The unrecognized residues or ligands are indicated with the BLK (".") residue type (both in the template and the target sequences) to copy them as rigid bodies into the model. The atom coordinates are transferred, but the BLK residues are static and do not participate in the refinement steps.

The HETATM records are read from the templates, in the order they are written in the PDB file. Thus, they must be indicated with the appropriate symbol (".", "w", or "") in the *same order* as in the alignment. If the template includes extra HETATM ligands that must not be modeled in the target, manually delete them in the PDB file or align them with a gap ("-") in the target sequence. If a chain break ("/") is added immediately before the "." residues in the alignment, this will force the ligands to have a different chain identifier (ID). If you model a ligand peptide, you need to add a chain break ("/") between the receptor and the peptide ligand, resulting in different chain IDs.

To model the sodium binding site in an inactive receptor, use as a template the sodium binding site of another receptor with similar binding mode [12]. Depending on the similarities with the other templates, you may prefer using the entire receptor or the sodium binding fragment including residues from TM2, TM3, and TM7. Most importantly, do not forget to include water molecules in the sodium binding site (Fig. 3b). They will not be present in the model if you do not indicate them. Be cautious when you select water molecules to be included. A strong overlap between the preliminary target model and a water molecule may lead to the crash of the MODELLER job. In that case, carefully check the water molecules to be included.

3.7 Building Receptor Chimera

In several cases, it may be interesting to use different templates for different parts of the target. In addition to the sodium binding site detailed above, two other cases are worth mentioning:

1. *Modeling an active target*: Comparison of active/inactive structures reveals that the largest structural changes during activation occur in TM5, TM6, and TM7 [6]. As active templates are less frequent than inactive ones (Fig. 1), an active state target may be better modeled as a chimer of (parts of) closely related inactive templates and farther related active templates. This may be done by using either (a) both an active template for the overall fold (TM1 to TM7) and an inactive template for the most stable part (TM1 to TM4) or (b) inactive templates for TM1 to TM4 and active templates for TM5 to TM7. In this latter case, however, it will be necessary to include the active template TM3 to correctly adjust the orientation of the active and inactive templates. The precise determination of the active and inactive template regions to be used requires careful visual inspection.
2. *Modeling loops*: This is a difficult step in a modeling procedure because loops may be highly variable in sequences and structures, and they are frequently missing in the template structure (s). MODELLER proposes two functions (`loopmodel` and `dope_loopmodel`) to automatically generate/refine loops

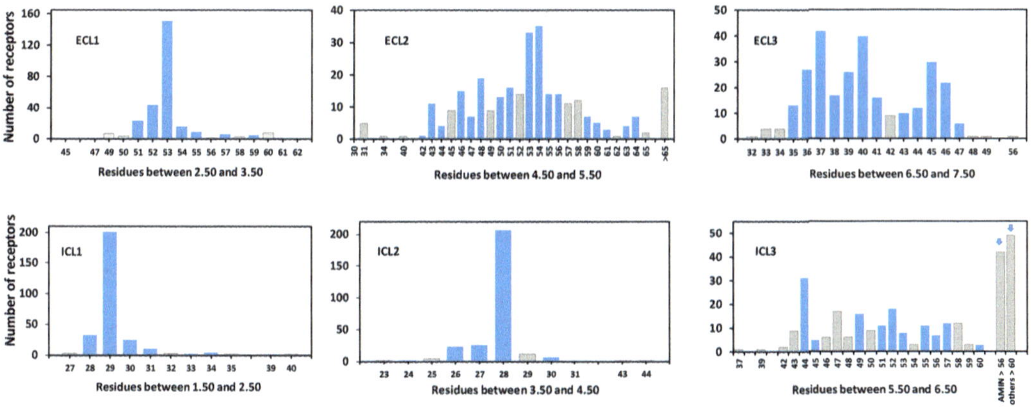

Fig. 4 Statistical analysis of the loop lengths in human GPCRs. The length is measured as the number of residues between the anchor residues *n*.50 present in each helix *n*. Blue bars indicate that at least one resolved loop of the indicated length is present in the available GPCR structures. Otherwise, bars are gray. The arrows for the long ICL3 loops indicate the resolution of two loops with 87 residues for the PEP receptor OX1R (6TP3) and 78 residues for the AMIN receptor ADRB2 (6MXT)

[46]. However, the loops obtained with these methods may markedly differ from the loops observed in GPCR structures. Now that a large set of GPCR structures are available, it is highly preferable to search for resolved equivalent loops with the same length and use them as templates. Figure 4 summarizes, for each loop, the lengths that have a structural template. In case of ambiguities, SS predictions based on judicious MSA may help. In addition, we note than insertions in loops have frequently an α-helical structure and may result in a protruding TM helix. A first example is given by the orexin receptor 1 (6TP3) in Fig. 2a, in which the long ECL3 (87 residues between the residues 5.50 and 6.50) is structured as a protruding N-terminal part of TM6. A second example is given by CXCR1, AT1, and CCR5 (Figs. 2 and 3) for which the long ECL3 is structured as a protruding N-terminal part of TM7.

3.8 Preparing Template File(s)

After selecting appropriate templates, prepare your PDB template file(s). MODELLER allows the user to select the first and last position of a *single contiguous* segment of the template to be used in the modeling procedure, but this selection is not possible in case of *discontinuous* segments. In this latter case, due for example to insertion of a fusion protein in the template, there is no alternative to manual editing of the PDB file to avoid long and unmanageable insertions in the alignment file. Breaks in the structure have to be indicated with a "/" symbol in the alignment file. In any case, even when this is not strictly necessary, manual editing of the PDB file (s) to excise the regions of the template that will not be used in the modeling procedure is recommended.

3.9 Aligning Model Sequence with the Template(s) and Preparing .ali File

This is a strategic part of the modeling procedure that has to be done very carefully. Several points have to be kept in mind:

1. Identification of the best templates generally involves the alignment of the target sequence with a set of available template sequences. However, the "best" alignments obtained by automatic alignment programs depend on the parameters used and may not be "optimal." They do not consider user's additional information that may improve the alignment and the resulting model(s). Thus, manual corrections of the alignment may be necessary.
2. In MODELLER, the alignment file creates *spatial restraints* (*see* **Note 1**). Thus, it may be useful to "de-align" residues to remove spatial restraints, increase flexibility and let MODELLER deal with stereochemical restraints only. The easiest way to remove structural restraints is adding gaps in the .ali file.
3. The template sequences in the .ali and .pdb files must *exactly* match. Thus, unresolved parts missing in the .pdb file must be removed from the sequence in the .ali file.
4. The .ali file has a special format. For each sequence, the first line gives the sequence name in the pir format (`>P1; name`). The third line gives the sequence and ends by the terminating character "*". Each chain break must be indicated by a single "/".
5. The second line gives information on the nature of the sequence and the region to be used. There must be 10 fields, separated by 9 colons (":"). The first field indicates if the sequence is a template (structure, structureX, structureN..) or the target (sequence), the second field indicates the sequence name, the fields 7–10 are optional, but the colon characters ":" are mandatory. The fields 3–6 indicate first position and chain, last position and chain. If the template is a single contiguous segment, simply specify here the beginning and ending residues that will be used for modeling. Examples can be found in the MODELLER tutorial. However, when templates contain non-contiguous segments, the easiest way to avoid problems is to edit the templates in order to remove any atom or heteroatom that will not be used in the modeling procedure and then to use the "." character between the colons as in "`structure: name:.:.:.:.: : : :`." This indicates that all the residues of the template .pdb file have to be read.

3.10 Adding or Suppressing Restraints

Two commands in MODELLER allow either adding restraints (`restraints.add()`) or removing restraints (`restraints.unpick()`). Different types of restraints can be added in a MODELLER script. However, for GPCR modeling, most useful ones are:

1. Secondary structure restraints. They allow extension of a TM helix, modeling of helix 8 which is often not resolved in templates, structuration of a missing loop as an α-helix or forcing the β-strand structure of ECL2.
2. Dihedral constraints to reorient side chains or favor interactions.
3. Distance restraints to maintain or create interactions.

Disulfide bonds not present in the template(s) may be added by the `special_patches()` command.

Suppressing restraints is also possible, in particular for structural restraints built from the alignment. This can be done with the `unpick(*atom_ids)` command. Alternatively, it can be obtained by de-aligning the alignment with the introduction of gaps. For example, before adding dihedral restraints on a sidechain, the residue can be de-aligned to avoid conflicts between restraints. It is worth to note that these commands work on the atoms or residues of the *target*, so that their correct identifiers in the target (and not in the template) have to be provided.

3.11 Model Building

This is the easiest step in the procedure. Select the number of models and the refinement procedure of models, and, optionally of loops (not recommended in GPCR modeling except for short loops or loop regions). Among the five procedures proposed ("none," "very.fast," "fast," "slow," "very.slow"), the "fast" and "slow" options give optimal results. From 20 to 50 models usually ensure a good representability of the available conformational space. A good compromise is to initiate the modeling procedure with 20 models/fast refinement option and to terminate with 50 models/slow refinement option, after optimization of restraints.

3.12 Evaluating Models

Once a set of models has been generated, there are different ways to assess them. The DOPE [47] score provided by MODELLER allows a comparison of model and template profiles and the visualization of putative problematic regions in the alignment. However, for GPCRs, when several models are generated from the same alignment, the DOPE scores are very similar, and the MODELLER molpdf scores [19] that indicate the violations of the restraints are more discriminative. Several steps have to be carried out to evaluate models:

1. Check the log file from the modeling run for runtime errors and restraint violations. The global molpdf scores are indicated at the end of the log file, which allows determining the "best" models that will be used for further analysis.
2. Visually inspect the models to ensure that no coarse mistake has been done in the alignment file or in the script. Superpose all the models to visualize the conformational space available, especially for loops.

3. Now focus on the "best" models (typically 5 out of 20 or 50 generated models). The sums of the violations for each residue are indicated at the B factor position in the .pdb files of the models (*B999*.pdb). Visualize them on the model structures. For regions with higher violations, inspect the *V999* files that give the type of restraints that are violated. If necessary, modify the alignment or the restraints accordingly.
4. Additional methods can be used to assess the quality of the model, for example, PROCHECK [34] that verifies the stereochemistry of the model.
5. In the selection of the "optimal" model used for further computational studies, be very careful to the orientation of key residues or motifs such as W4.50, W6.48, or DRY in TMH3. Be also very careful to the positions of N- and C-termini and to the structures of the loops.

3.13 Return to Our Test Case

Here we analyze the molecular models of the human B1 receptor, with bound sodium ion, obtained from different templates. The aim is highlighting MODELLER practice to help future users in template selection. In Fig. 3, we have shown four representative structures of homologs: AT1 in the inactive state (closest homolog), OPRD (resolved ICL3 and closest homolog with bound sodium ion), CCR5 (homolog with resolved ECL2), and AT1 in an active state (resolved ECL2). Among these putative templates, we removed CCR5 because of the unusual outward positioning of TM6 on the extracellular side.

First, we used inactive AT1 (4YAY) and OPRD (4N6H) as templates. Since, in OPRD, there is no disulfide bond linking the N-terminus and ECL3, we aligned the sequences of the templates with the target, except for the N-terminus and ECL3. Thus, both templates are used for ICL1, ICL2, ECL1, ECL2, and H8. Nevertheless, in the resulting models of B1 (Fig. 5a), the ICL1, ICL2 and ECL1 loops of the target match those of the AT1 template only (closest template). The ICL3 and ECL3 loops match their unique templates, respectively, OPRD and AT1. The models differ strongly in the modeling of the ECL2 hairpin, since there is no template for this segment. Interestingly, concerning H8, MODELLER does not privilege one template over the other one, and the resulting models have either the tilted orientation observed in AT1 (4YAY) or the horizontal orientation observed in OPRD (4N6H), but not an average orientation.

In the second procedure, we aimed at improving the modeling of the ECL2 hairpin (Fig. 5b). As the length between TMH4 and the ECL2 cysteine is identical in B1 and AT1, we selected a fragment of active AT1 (6OS0) in which ECL2 is resolved. The fragment includes TM3 and TM4 for proper positioning, ICL2 (in helical conformation) and ECL2 up to the Cys residue. With

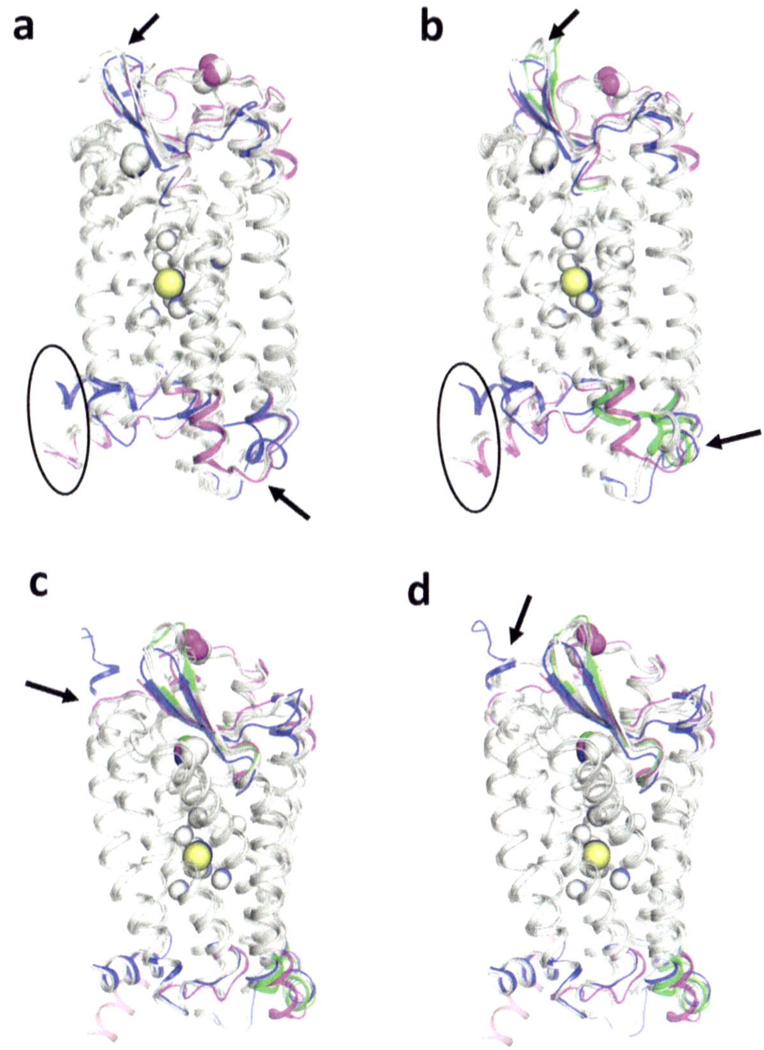

Fig. 5 Influence of templates on the resulting B1 models. In each panel, the top five models (out of 20) obtained with MODELER are superposed with the templates used in modeling procedure. For clarity purpose, all the structures are shown as white ribbon, except the regions of interest in the templates. In (**a**), B1 is modeled from OPRD (blue) and inactive AT1 (magenta). In (**b**), B1 is modeled from OPRD (blue), inactive AT1 (magenta), and a fragment from active AT1 encompassing TM3, ICL2, TM4, and ECL2 up to the Cys residue (green). In (**c**), B1 is modeled as in (**b**) except that the C-terminus of AT1, in light pink, has been "de-aligned." In (**d**), B1 has been modeled as in (**c**), except for the N-terminus. In (**a–c**), the N-terminus has been modeled from the AT1 template only. In (**d**), the OPRD template has been included, resulting in an additional helical turn at the N-terminus of TM1

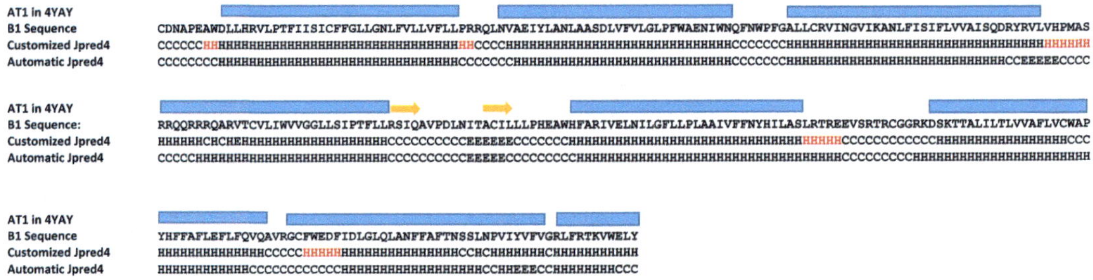

Fig. 6 Secondary structure predictions for B1 using JPred4. The automatic Jpred4 prediction is based on automatic BLAST search starting from the B1 sequence. The customized prediction is based on the MSA of 52 human receptors with the two disulfide bonds in the extracellular domain as observed in B1. The SS drawing corresponds to the experimental SS of AT1 in the 4YAY structure

the addition of this fragment, there are three templates for ICL2. Now, MODELLER privileges the helical structure of ICL2 and the tilted orientation of H8 in the B1 models.

These different models rise the issues of the orientation of H8 and of the structure of ICL2. Concerning H8, the influence of H8 modeling on subsequent MD simulations is detailed in **Note 4**, with the example of AT1, and we recommend to privilege the horizontal orientation of H8. Concerning ICL2, we note that, in both active AT1 and inactive CCR5, ICL2 has a helical structure. To help resolve the uncertainty between the helical and coil conformations of ICL2, we carried out secondary structure prediction using JPred4 [33]. Figure 6 displays SS predictions for B1 using either automatic BLAST search or the MSA of the 52 human sequences that share two properties. First, they belong to the evolutionary related CHEM and PUR subfamilies [24] and, second, they possess the double disulfide bonds in the extracellular domain. The former approach is not informative. By contrast, the latter approach with a customized MSA predicts a helical structure for ICL2, supporting the helical conformation.

In the third procedure, the C-terminus of inactive AT1 was "de-aligned" to exclude it from structural restraints in MODELLER. In this case, the resulting models have a helical ICL2 and a horizontal H8 (Fig. 5c). Finally, in the fourth procedure, we were concerned with the modeling of the N-terminus from the cysteine (first residue in our models) to TMH1. This segment in B1 is one residue longer than AT1 or CCR5. In the first to third procedures, it was modeled with the insertion of one residue compared to the AT1 template. In this fourth procedure, since SS predictions suggest an N-terminal extension of TM1, we checked whether the length of this segment in B1 is compatible with a helical extension. Thus, we also aligned the N-terminus of OPRD in the alignment file. In this latter case, the B1 models do have an additional helical turn at the extracellular side of TM1 without violations of stereochemical restraints (Fig. 5d).

3.14 Concluding Remarks

In this chapter, we have shown how the choice of the template(s) is determinant for the resulting target models and how MODELLER deals with multiple templates. Rather than an average conformation, MODELLER estimates the probability of each conformation and may propose several clearly different models, as exemplified by the H8 orientation in B1 modeling. We emphasize the importance of generating and comparing several models to estimate the conformational space compatible with the restraints provided by the template(s) and the alignment.

We have also shown that carefully selected receptor fragments can greatly improve modeling. Template-based combination of fragments is a powerful approach for molecular modeling of GPCRs that has been implemented in several websites for automatic GPCR modeling, such as GPCR-I-TASSER [48] and GPCR-SFFE [49]. The automatic approaches are very efficient for the modeling of the transmembrane domain but do not provide to the user the possibility to use target-specific information to guide modeling, especially for loops, now that a large variety of templates is available (Fig. 4). Understanding the details of the homology modeling procedure by MODELLER will help the user to make rational choices that will allow improving the quality of customized GPCR models.

4 Notes

1. Modeling by satisfaction of spatial restraints.

   The modeling program MODELLER works by satisfying restraints on the structure of the target sequence [19, 20]. The program is designed to take into account:

   (a) *Spatial homology-derived restraints:* Distance and dihedral angle restraints on the target sequence are derived from the template structure and the sequence alignment between target and template. Distance restraints are obtained by assuming that the corresponding distances between *aligned residues in the alignment file* are similar. Dihedral angle restraints are also derived from *aligned residues* in the alignment file. The forms of the restraints are based on empirical knowledge of structural evolution in homologous protein families.

   (b) *Stereochemical restraints:* Additional restraints based on the CHARMM22 force field are added to enforce proper stereochemistry [50].

   (c) *User added restraints:* Additional structural or functional information on the target that cannot be derived from the templates, such as a suspected disulfide, cross-linking restraints, site directed-mutagenesis results, and predicted secondary structure, can be included as additional restraints. Most commonly added restraints for GPCRs are distance restraints, dihedral angle restraints, and secondary structure restraints (α-helix restraints for helix termini and β-strand/sheet restraints for extracellular loop 2). Information can also be obtained from careful examination of sequence alignment of the target orthologs which may provide general knowledge about receptor specificities. Adding restraints can be easily done with the special_restraints() and special_patches() functions of MODELLER and may markedly improve the quality of the target structure.

(d) *User removed restraints:* MODELLER can unselect all the restraints on specified atoms with the restraints.unpick() command. It is also very easy to remove alignment-based spatial restraints by modifying the alignment with the addition of gaps.

Finally, all the restraints are combined in an objective function that is optimized in Cartesian space by using the variable target function method [51] to obtain the model. Then the model is refined by using conjugate gradients and simulated annealing [19]. For a receptor with about 350 residues and 2500 atoms, MODELLER builds and optimizes about 50,000 restraints. Several slightly different models can be calculated by varying the initial structure with a random shift in the position of each atom. This allows gaining information on regions with putative structural flexibility or variability.

2. GPCR classification.

GPCRs are present in fungi, amoeba, animals, and possibly plants [23, 52]. They underwent a stupendous evolutionary success in animals where they highly diversified. Species-specific subfamilies make difficult a general classification among the animal reign but, in vertebrates, the GRAFS classification into five families has been widely adopted [2, 52]. Among these five families, the rhodospsin-like or class A is the largest one with 700 out of the 800 total human receptors. These 700 receptors include 400 olfactory and 300 nonolfactory receptors. Finally, the 300 human nonolfactory receptors can be further classified into a dozen of evolutionary based subfamilies, to which can be added the UC class of "unclassified" receptors. The nomenclature for these subfamilies is as follows: PUR (purinergic receptors), CHEM (chemotaxic receptors, including vasoactive peptide receptors), SO (somatostatin and opioid receptors), PEP (peptide receptors), MLT (melatonin receptors), OPN (opsins), PTG (prostaglandin receptors), MEC (melanocortin,

EDG, and cannabinoid receptors), AD (adenosine receptors), AMIN (amine receptors), MRG (MAS-related receptors), and LGR (leucine-rich repeat receptors). However, the evolutionary relationships between these subfamilies are not obvious. The classification into four α, β, γ, and δ groups [2] has been widely adopted in the literature but is not consistent with the order of subfamily apparition during evolution [24, 52]. Based upon a multidimensional scaling analysis, we proposed a model of radial evolution of GPCRs from ancestral peptide receptors with three main evolutionary pathways [25].

The model of radial evolution provides a framework to rationalize the variable proline patterns in TM2 and TM5 observed in class A GPCRs and is consistent with the order of subfamily apparition. Peptide (PEP) receptors are among the most ancient GPCRs. Most PEP receptors are characterized by proline residues at position 2.59 and 5.50 and bulges in TM2 and TM5, as observed for the orexin receptor 1 [53] shown in Fig. 2a. The first evolutionary pathway is related to the deletion of one residue in TM2 and led the SO, CHEM, and PUR receptors. This deletion occurred in an ancestor of the SO receptors that led to the CHEM and PUR receptors by divergence [24]. The SO, CHEM, and PUR subfamilies are characterized by the P2.58 and P5.50 proline patterns, corresponding to a kink in TM2 and a bulge in TM5, respectively, as observed for CXCR1 [54] (Fig. 2b). The second pathway is related to the divergence of amine (AMIN) and adenosine (AD) receptors, characterized by the same sequence and structural patterns as PEP receptors. The third pathway corresponds to independent evolution of several subfamilies with mutations of the proline residues in TM2 and TM5, which are often correlated. These mutations can lead to straight TM2 and TM5, as observed in the cannabinoid receptor 1 [55] from the MEC subfamily (Fig. 2c). Prostaglandin receptors (PTG) present an example of sequence variability for TM2 (P2.59 or no proline) with a conserved bulge observed in PE2R3 (P2.59), PD2R (P2.59) and TA2R (noP), whereas the absence of proline in TM5 led to a bulged TM5 in PD2R and to straight helices in PE2R3 and TA2R [56–58]. The MRG and LGR subfamilies have no proline residues in TM2 and TM5 and no structurally resolved member. Thus, no a priori hypothesis can be done on the structure of these helices. Combination of cellular biology experiments, extensive sequence analysis, and molecular dynamics simulations were necessary to propose a bulged TM2 and a straight TM5 in TSHR, a LGR receptor [38].

In addition to the proline pattern in TM2 and TM5, a second sequence pattern presents interesting features in class A GPCRs. It concerns the disulfide bond(s) stabilizing the receptor extracellular domain. A first disulfide bond links the N-terminal part of TM3 to the extracellular loop 2 (ECL2) and is present in most subfamilies, except MEC and MRG. A second disulfide bond links the N-terminus to an extracellular extension of TM7 in some purinergic and chemotaxic receptors. Alternative disulfide bonds have been observed in MEC receptors, whereas an additional disulfide bond is frequently found in ECL3 of amine receptors. Thus, each subfamily presents unique structural features which have to be considered for modeling.

It is worth noting that GPCRDB [59] uses a classification based on the chemical nature of the ligands because it is focused on molecular docking and drug design. However, this may be confusing for homology modeling because similar ligands may bind receptors from different subfamilies. For example, the lysophosphatidic acid receptors LPAR1-3 and LPAR4-6 are members of the MEC and of the PUR subfamilies, respectively, with no proline in TM2 and TM5 for LPAR1-3 and two proline residues at P2.58 and P5.50 for LPAR4-6.

3. Mining GPCRs in UniProt.

In Uniprot, class A GPCRs can be easily mined by searching entries with the correct family identifiers. They are identified as PF00001 (7tm_1) in Pfam, IPR000276 (GPCR_Rhodpsn) or IPR017452 (GPCR_Rhodpsn_7TM) in InterPro, PS00237 (G_PROTEIN_RECEP_F1_1) or PS50262 (G_PROTEIN_RECEP_F1_2) in PROSITE. In InterPro, the IPR017452 identifier is broader than the IPR000276 identifier and includes a number of taste and vomeronasal receptors. In PROSITE, the PS50262 identifier, based on sequence profiles, is of higher quality than the PS00237 identifier, based on motifs. The three identifiers PF00001, IPR000276, and PS50262 are equivalent. Olfactory receptors can be identified as IPR000725 (Olfact_rcpt).

When the user wishes to go deeper in the classification tree, InterPro provides several useful levels of classification. For example, the angiotensin II receptors that include type 1 and type 2 have the reference IPR000248 (ATII_rcpt), type 2 has the reference IPR000147 (ATII_AT2_rcpt), allowing hierarchical selection. However, the intermediary level of classification corresponding to the 12 subfamilies is usually considered neither in UniProt nor in GPCRDB, based on the nature of the ligand (*see* **Note 2**). Lists of receptors belonging to the 12 subfamilies that make consensus can be found in the literature [2, 24].

4. Positioning of helix 8.

   The user should never avoid critical assessment of the template(s). For example, in the structure of inactive AT1 shown in Fig. 3 (4YAY), the orientation of H8 is tilted. Such tilted orientation has been observed in other GPCR structures, e.g., for CCR5 (4MBS). However, it has not been observed in other structures of AT1, such as the recently resolved active structures of AT1 in complex with angiotensin II (6OS0) or angiotensin derivatives (6OD1, 6OS1, 6OS2) [45]. This suggests that the tilted orientation observed in the inactive structure of AT1 might result from artifacts, due to truncation of the C-terminus and experimental conditions.

   To answer this question, we carried out molecular dynamics simulations of AT1 (Fig. 7). Starting from the 4YAY structure of AT1, we built two models of the receptor, the first one with H8 positioned as in the 4YAY structure and the second one with H8 positioned as observed in the OPRD 4N6H structure. The models were inserted into a POPC bilayer using the charmm-gui interface (www.charmm-gui.org) and then underwent a short equilibration step (1 ns procedure with progressive release of structural constraints) followed by 280 ns of production run with NAMD [60]. When H8 was modeled in an orientation parallel to the membrane bilayer, it remained stable in this orientation during the 280 ns of the simulations. When the starting structure had a tilted H8 orientation, this one was not stable. H8 underwent a large seesaw motion that induced strong perturbations not only of H8 but also of the intracellular sides of TM6, TM7, and TM1 (Fig. 7c). These results highlight the importance of the orientation of H8 on the stability of MD simulations and strongly suggests that the canonical horizontal orientation of H8 should be privileged in molecular modeling.

Acknowledgments

This study was supported by institutional grants from INSERM, CNRS and University of Angers. This work was granted access to HPC resources of IDRIS (GENCI grant 100567 to MC). MC is supported by CNRS. AT is supported by a fellowship from the University of Carthage (Tunisia). RB is supported by a fellowship from the University of Angers (France).

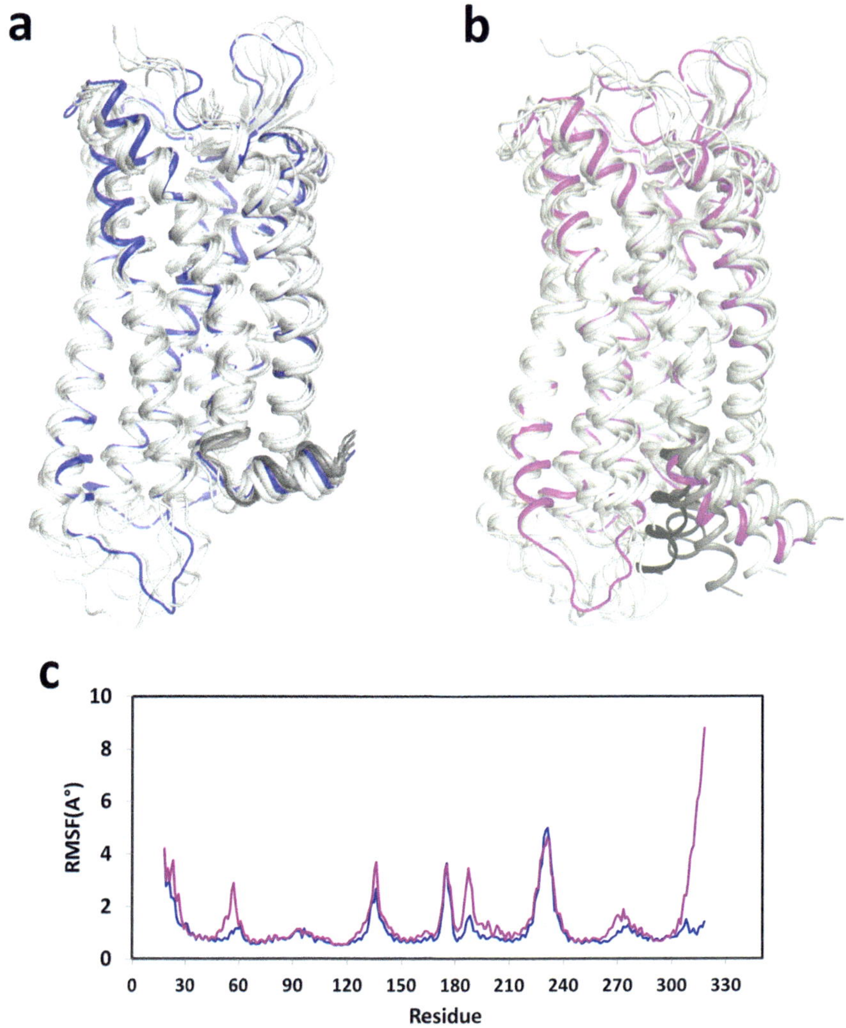

Fig. 7 Consequences of the orientation of H8 on MD simulations of the AT1 receptor. MD simulations were run for 280 ns with NAMD, starting from an AT1 model with horizontal H8 (**a**) and tilted H8 (**b**). The starting conformations are blue in (**a**) and magenta in (**b**). Representative snapshots from the beginning to the end of the production run indicate the conformational space sampled by H8 during the simulations. The snapshots are shown as white ribbons, with increasing graying of the C-terminus from white to dark gray with the simulation time. In (**c**), comparison of RMSF of AT1 when the trajectory was started with the horizontal (blue) and tilted orientation (magenta) of H8

References

1. Bockaert J, Pin JP (1999) Molecular tinkering of G protein-coupled receptors: an evolutionary success. EMBO J 18(7):1723–1729. https://doi.org/10.1093/emboj/18.7.1723
2. Fredriksson R, Lagerstrom MC, Lundin LG, Schioth HB (2003) The G-protein-coupled receptors in the human genome form five main families. Phylogenetic analysis, paralogon groups, and fingerprints. Mol Pharmacol 63(6):1256–1272. https://doi.org/10.1124/mol.63.6.1256
3. Overington JP, Al-Lazikani B, Hopkins AL (2006) How many drug targets are there? Nat Rev Drug Discov 5(12):993–996. https://doi.org/10.1038/nrd2199

4. Sealfon SC, Chi L, Ebersole BJ et al (1995) Related contribution of specific helix 2 and 7 residues to conformational activation of the serotonin 5-HT2A receptor. J Biol Chem 270 (28):16683–16688. https://doi.org/10.1074/jbc.270.28.16683
5. Palczewski K, Kumasaka T, Hori T et al (2000) Crystal structure of rhodopsin: a G protein-coupled receptor. Science 289 (5480):739–745. https://doi.org/10.1126/science.289.5480.739
6. Cherezov V, Rosenbaum DM, Hanson MA et al (2007) High-resolution crystal structure of an engineered human beta2-adrenergic G protein-coupled receptor. Science 318 (5854):1258–1265. https://doi.org/10.1126/science.1150577
7. Xiang J, Chun E, Liu C et al (2016) Successful strategies to determine high-resolution structures of GPCRs. Trends Pharmacol Sci 37 (12):1055–1069. https://doi.org/10.1016/j.tips.2016.09.009
8. Garcia-Nafria J, Tate CG (2020) Cryo-electron microscopy: moving beyond x-ray crystal structures for drug receptors and drug development. Annu Rev Pharmacol Toxicol 60:51–71. https://doi.org/10.1146/annurev-pharmtox-010919-023545
9. Katritch V, Cherezov V, Stevens RC (2012) Diversity and modularity of G protein-coupled receptor structures. Trends Pharmacol Sci 33 (1):17–27. https://doi.org/10.1016/j.tips.2011.09.003
10. Munk C, Mutt E, Isberg V et al (2019) An online resource for GPCR structure determination and analysis. Nat Methods 16 (2):151–162. https://doi.org/10.1038/s41592-018-0302-x
11. Rasmussen SG, DeVree BT, Zou Y et al (2011) Crystal structure of the beta2 adrenergic receptor-Gs protein complex. Nature 477 (7366):549–555. https://doi.org/10.1038/nature10361
12. Katritch V, Fenalti G, Abola EE et al (2014) Allosteric sodium in class A GPCR signaling. Trends Biochem Sci 39(5):233–244. https://doi.org/10.1016/j.tibs.2014.03.002
13. Margiotta E, Deganutti G, Moro S (2018) Could the presence of sodium ion influence the accuracy and precision of the ligand-posing in the human A2A adenosine receptor orthosteric binding site using a molecular docking approach? Insights from Dockbench. J Comput Aided Mol Des 32 (12):1337–1346. https://doi.org/10.1007/s10822-018-0174-2
14. Nygaard R, Frimurer TM, Holst B et al (2009) Ligand binding and micro-switches in 7TM receptor structures. Trends Pharmacol Sci 30 (5):249–259. https://doi.org/10.1016/j.tips.2009.02.006
15. Congreve M, Dias JM, Marshall FH (2014) Structure-based drug design for G protein-coupled receptors. Prog Med Chem 53:1–63. https://doi.org/10.1016/B978-0-444-63380-4.00001-9
16. Shonberg J, Lopez L, Scammells PJ et al (2014) Biased agonism at G protein-coupled receptors: the promise and the challenges—a medicinal chemistry perspective. Med Res Rev 34(6):1286–1330. https://doi.org/10.1002/med.21318
17. Diaz C, Angelloz-Nicoud P, Pihan E (2018) Modeling and deorphanization of orphan GPCRs. Methods Mol Biol 1705:413–429. https://doi.org/10.1007/978-1-4939-7465-8_21
18. Stockert JA, Devi LA (2015) Advancements in therapeutically targeting orphan GPCRs. Front Pharmacol 6:100. https://doi.org/10.3389/fphar.2015.00100
19. Sali A, Blundell TL (1993) Comparative protein modelling by satisfaction of spatial restraints. J Mol Biol 234(3):779–815. https://doi.org/10.1006/jmbi.1993.1626
20. Webb B, Sali A (2016) Comparative protein structure modeling using MODELLER. Curr Protoc Protein Sci 86:2 9 1–2 9 37. https://doi.org/10.1002/cpps.20
21. Devos D, Valencia A (2000) Practical limits of function prediction. Proteins 41(1):98–107. https://doi.org/10.1002/1097-0134(20001001)41:1<98::AID-PROT120>3.0.CO;2-S
22. Sanchez R, Sali A (1997) Advances in comparative protein-structure modelling. Curr Opin Struct Biol 7(2):206–214. https://doi.org/10.1016/s0959-440x(97)80027-9
23. Chabbert M, Castel H, Pele J et al (2012) Evolution of class A G-protein-coupled receptors: implications for molecular modeling. Curr Med Chem 19(8):1110–1118. https://doi.org/10.2174/092986712799320600
24. Deville J, Rey J, Chabbert M (2009) An indel in transmembrane helix 2 helps to trace the molecular evolution of class A G-protein-coupled receptors. J Mol Evol 68(5):475–489. https://doi.org/10.1007/s00239-009-9214-9
25. Pele J, Abdi H, Moreau M et al (2011) Multidimensional scaling reveals the main evolutionary pathways of class A G-protein-coupled receptors. PLoS One 6(4):e19094. https://doi.org/10.1371/journal.pone.0019094

26. Taddese B, Deniaud M, Garnier A et al (2018) Evolution of chemokine receptors is driven by mutations in the sodium binding site. PLoS Comput Biol 14(6):e1006209. https://doi.org/10.1371/journal.pcbi.1006209
27. Webb B, Sali A (2016) Comparative protein structure modeling using MODELLER. Curr Protoc Bioinformatics 54:5 6 1–5 6 37. https://doi.org/10.1002/cpbi.3
28. Pettersen EF, Goddard TD, Huang CC et al (2004) UCSF chimera—a visualization system for exploratory research and analysis. J Comput Chem 25(13):1605–1612. https://doi.org/10.1002/jcc.20084
29. Larkin MA, Blackshields G, Brown NP et al (2007) Clustal W and Clustal X version 2.0. Bioinformatics 23(21):2947–2948. https://doi.org/10.1093/bioinformatics/btm404
30. Notredame C, Higgins DG, Heringa J (2000) T-Coffee: a novel method for fast and accurate multiple sequence alignment. J Mol Biol 302 (1):205–217. https://doi.org/10.1006/jmbi.2000.4042
31. Edgar RC (2004) MUSCLE: multiple sequence alignment with high accuracy and high throughput. Nucleic Acids Res 32 (5):1792–1797. https://doi.org/10.1093/nar/gkh340
32. Nicholas KB, Nicholas HB Jr, Deerfield DWI (1999) GeneDoc: analysis and visualization of genetic variation. EMBNEWNEWS 4:14
33. Drozdetskiy A, Cole C, Procter J, Barton GJ (2015) JPred4: a protein secondary structure prediction server. Nucleic Acids Res 43(W1): W389–W394. https://doi.org/10.1093/nar/gkv332
34. Laskowski RA, Rullmannn JA, MacArthur MW et al (1996) AQUA and PROCHECK-NMR: programs for checking the quality of protein structures solved by NMR. J Biomol NMR 8 (4):477–486. https://doi.org/10.1007/bf00228148
35. Holm L, Sander C (1998) Removing near-neighbour redundancy from large protein sequence collections. Bioinformatics 14 (5):423–429. https://doi.org/10.1093/bioinformatics/14.5.423
36. Isberg V, Mordalski S, Munk C et al (2016) GPCRdb: an information system for G protein-coupled receptors. Nucleic Acids Res 44(D1): D356–D364. https://doi.org/10.1093/nar/gkv1178
37. Chan WK, Zhang H, Yang J et al (2015) GLASS: a comprehensive database for experimentally validated GPCR-ligand associations. Bioinformatics 31(18):3035–3042. https://doi.org/10.1093/bioinformatics/btv302
38. Chantreau V, Taddese B, Munier M et al (2015) Molecular insights into the transmembrane domain of the thyrotropin receptor. PLoS One 10(11):e0142250. https://doi.org/10.1371/journal.pone.0142250
39. Castleman PN, Sears CK, Cole JA et al (2019) GPCR homology model template selection benchmarking: global versus local similarity measures. J Mol Graph Model 86:235–246. https://doi.org/10.1016/j.jmgm.2018.10.016
40. Costanzi S, Skorski M, Deplano A et al (2016) Homology modeling of a class A GPCR in the inactive conformation: a quantitative analysis of the correlation between model/template sequence identity and model accuracy. J Mol Graph Model 70:140–152. https://doi.org/10.1016/j.jmgm.2016.10.004
41. Srinivasan N, Blundell TL (1993) An evaluation of the performance of an automated procedure for comparative modelling of protein tertiary structure. Protein Eng 6(5):501–512. https://doi.org/10.1093/protein/6.5.501
42. Fenalti G, Giguere PM, Katritch V et al (2014) Molecular control of delta-opioid receptor signalling. Nature 506(7487):191–196. https://doi.org/10.1038/nature12944
43. Zhang H, Unal H, Gati C et al (2015) Structure of the angiotensin receptor revealed by serial femtosecond crystallography. Cell 161 (4):833–844. https://doi.org/10.1016/j.cell.2015.04.011
44. Tan Q, Zhu Y, Li J et al (2013) Structure of the CCR5 chemokine receptor-HIV entry inhibitor maraviroc complex. Science 341 (6152):1387–1390. https://doi.org/10.1126/science.1241475
45. Wingler LM, Skiba MA, McMahon C et al (2020) Angiotensin and biased analogs induce structurally distinct active conformations within a GPCR. Science 367(6480):888–892. https://doi.org/10.1126/science.aay9813
46. Fiser A, Sali A (2003) Modeller: generation and refinement of homology-based protein structure models. Methods Enzymol 374:461–491. https://doi.org/10.1016/S0076-6879(03)74020-8
47. Shen MY, Sali A (2006) Statistical potential for assessment and prediction of protein structures. Protein Sci 15(11):2507–2524. https://doi.org/10.1110/ps.062416606
48. Zhang J, Yang J, Jang R, Zhang Y (2015) GPCR-I-TASSER: a hybrid approach to G protein-coupled receptor structure modeling and the application to the human genome. Structure 23(8):1538–1549. https://doi.org/10.1016/j.str.2015.06.007

49. Worth CL, Kreuchwig F, Tiemann JKS et al (2017) GPCR-SSFE 2.0-a fragment-based molecular modeling web tool for class A G-protein coupled receptors. Nucleic Acids Res 45(W1):W408–W415. https://doi.org/10.1093/nar/gkx399
50. MacKerell AD, Bashford D, Bellott M et al (1998) All-atom empirical potential for molecular modeling and dynamics studies of proteins. J Phys Chem B 102(18):3586–3616. https://doi.org/10.1021/jp973084f
51. Braun W, Go N (1985) Calculation of protein conformations by proton-proton distance constraints. A new efficient algorithm. J Mol Biol 186(3):611–626. https://doi.org/10.1016/0022-2836(85)90134-2
52. Fredriksson R, Schioth HB (2005) The repertoire of G-protein-coupled receptors in fully sequenced genomes. Mol Pharmacol 67 (5):1414–1425. https://doi.org/10.1124/mol.104.009001
53. Rappas M, Ali AAE, Bennett KA et al (2020) Comparison of orexin 1 and orexin 2 ligand binding modes using x-ray crystallography and computational analysis. J Med Chem 63 (4):1528–1543. https://doi.org/10.1021/acs.jmedchem.9b01787
54. Park SH, Das BB, Casagrande F et al (2012) Structure of the chemokine receptor CXCR1 in phospholipid bilayers. Nature 491 (7426):779–783. https://doi.org/10.1038/nature11580
55. Hua T, Vemuri K, Pu M et al (2016) Crystal structure of the human cannabinoid receptor CB1. Cell 167(3):750–762. e714. https://doi.org/10.1016/j.cell.2016.10.004
56. Fan H, Chen S, Yuan X et al (2019) Structural basis for ligand recognition of the human thromboxane A2 receptor. Nat Chem Biol 15 (1):27–33. https://doi.org/10.1038/s41589-018-0170-9
57. Wang L, Yao D, Deepak R et al (2018) Structures of the human PGD2 receptor CRTH2 reveal novel mechanisms for ligand recognition. Mol Cell 72(1):48–59.e4. https://doi.org/10.1016/j.molcel.2018.08.009
58. Morimoto K, Suno R, Hotta Y et al (2019) Crystal structure of the endogenous agonist-bound prostanoid receptor EP3. Nat Chem Biol 15(1):8–10. https://doi.org/10.1038/s41589-018-0171-8
59. Isberg V, Mordalski S, Munk C et al (2017) GPCRdb: an information system for G protein-coupled receptors. Nucleic Acids Res 45 (5):2936. https://doi.org/10.1093/nar/gkw1218
60. Phillips JC, Braun R, Wang W et al (2005) Scalable molecular dynamics with NAMD. J Comput Chem 26(16):1781–1802. https://doi.org/10.1002/jcc.20289

Chapter 6

Interface Prediction for GPCR Oligomerization Between Transmembrane Helices

Wataru Nemoto and Akira Saito

Abstract

Oligomers of G protein-coupled receptors (GPCRs) are closely related to their biochemical and biological functions and have been conserved during the course of molecular evolution. The mechanisms of GPCR interactions and the reason why GPCRs interact between themselves have remained elusive. Accurate interface prediction is useful to generate guidelines for mutation and inhibition experiments and would accelerate investigations of the molecular mechanisms of GPCR oligomerization and signaling. We have developed a method to predict the interfaces for GPCR oligomerization. Our method detects clusters of conserved residues along the surfaces of transmembrane helices, using a multiple sequence alignment and a target GPCR or closely related structure. This chapter outlines our method and introduces some problems that occur with it, along with our future direction to extend the method for interface predictions of general membrane proteins.

Key words Membrane proteins, Transmembrane helices, Conserved residues

1 Introduction

Oligomerization of G protein-coupled receptors (GPCRs) is strongly associated with their molecular and cellular functions [1–4]. Extensive evidence for GPCR oligomerization has been reported over the past few decades [5]. It is widely accepted that GPCRs form homodimers, heterodimers, or higher-order molecular complexes (oligomers). Heterodimerization often decreases the endogenous signals via GPCRs by reducing their affinities for agonists. Oligomerization is reportedly coupled to the regulation of GPCR internalization and recycling. Recent studies have revealed that GPCR oligomerization is involved in various diseases. Thus, GPCR oligomerization is now regarded as playing essential roles in cellular functions, and many researchers are investigating this

Irina S. Moreira et al. (eds.), *Computational Design of Membrane Proteins*, Methods in Molecular Biology, vol. 2315, https://doi.org/10.1007/978-1-0716-1468-6_6,

subject. Further details of the relationships between GPCR oligomerizations and functions are summarized in several reviews [1, 3, 4, 6].

Intermolecular interactions between transmembrane (TM) helices are important for oligomerization [7]. Class A GPCRs form oligomers primarily through interactions between the TM helices [8]. Class B and C GPCRs form oligomers stabilized by their large intracellular or extracellular domains [9]. However, the interactions between the TM helices are essential for the oligomerization of Class B and C GPCRs [10, 11]. We previously proposed the existence of a functional constraint for the intermolecular interactions along the surfaces of the TM helices, for the oligomerization of all GPCR classes [7].

Several methods have been applied to predict GPCR oligomerization interfaces by using the residue conservation over all of the sequences in the multiple sequence alignment (MSA) [12, 13]. However, these methods are problematic for the interface prediction of GPCR oligomers because the experimentally determined interfaces for GPCR oligomerization are not always conserved, even among closely related GPCR subtypes within the same subfamily [7]. For example, the D2 dopamine receptor (D2R) has interfaces on TMs 2 and 4, while they are on TM 1 in the D5 dopamine receptor. In such cases, the residue conservation over all of the sequences in the MSA is not an adequate indicator for interface prediction [7].

We previously developed a method to predict the GPCR oligomerization interfaces that are not always conserved even within the same subfamily but conserved within a subtype [7]. Our method, GRIP, successfully predicted all of the interfaces available at that time.

GRIP can predict interfaces for GPCR oligomerization by finding clusters of conserved residues along the surfaces of TM helices [7], as well as subtype-specific interfaces (Fig. 1). The GRIP web service is available at http://grip.b.dendai.ac.jp/~grip/, for the prediction of interfaces for GPCR oligomerization according to our method [14, 15]. A small example on how to use the GRIP web service is also described (*see* Subheading 3.4).

2 Materials

The GRIP method requires two types of data as input [7]. One is the target GPCR sequence or an MSA of the GPCR subtype of interest, which is used to calculate the conservation score at each alignment site. The other is the tertiary structure of the GPCR, on which the surface residues are used to map the conservation scores to facilitate searches for conserved residue clusters as predicted interfaces.

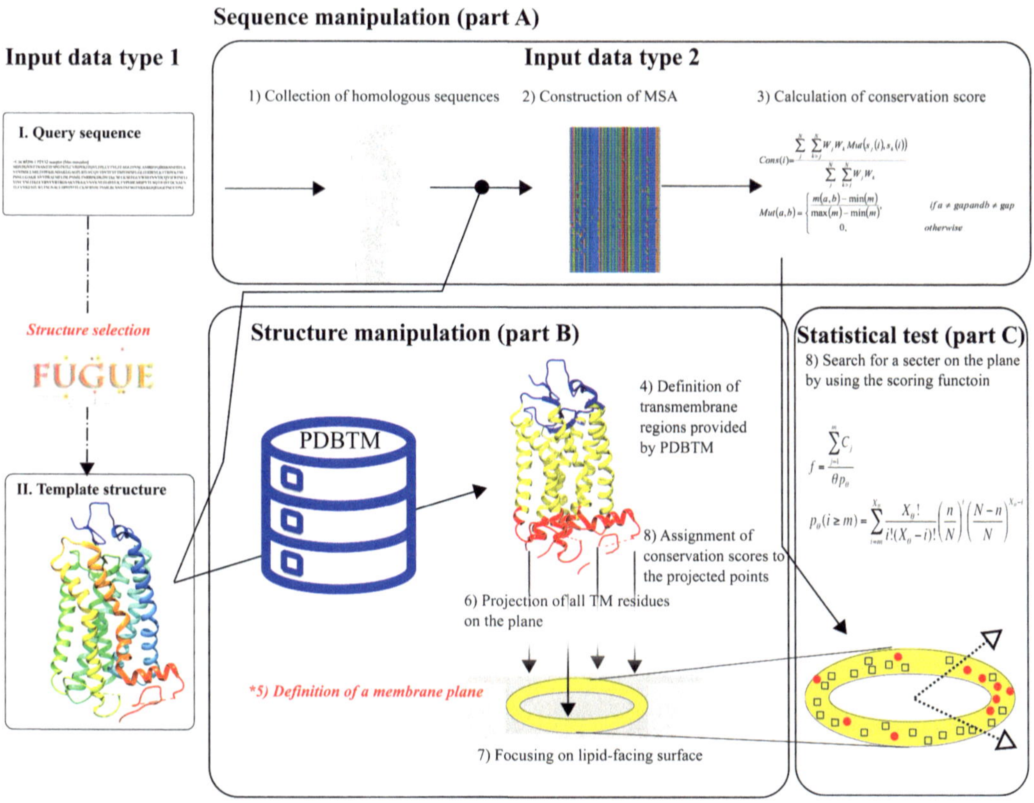

Fig. 1 Flowchart of data processing in GRIP (reproduced from ref. [14]) with permission from Taylor & Francis)

3 Methods

The flowchart of data processing in the GRIP server is shown in Fig. 1. The procedure consists of three parts: A, B, and C. The two types of input data, sequence and structure data, are processed by the following steps. Parts A and B manipulate the sequence and structure data, respectively. The information processed by the two parts is integrated into Part C, in which the search for interfaces of the given query sequence is performed. Details of each part are described below.

3.1 Sequence Data Manipulation (Part A)

1. Collect homologous sequences of a target GPCR sequence by using BLAST [16] or PSI-BLAST [17] with the default parameter settings, except for the Max target sequences and Database (*see* **Note 1**).
2. Remove disproportionately long or short sequences, as compared to the target GPCR sequence (*see* **Note 2**).
3. Construct an MSA of the collected homologous sequences (*see* **Note 3**).

4. Draw a phylogenetic tree of the collected sequences by using the MSA (*see* **Note 4**).
5. Select orthologous sequences of the target among the sequences corresponding to the operational taxonomic units (OTUs) in the phylogenetic tree (*see* **Note 5**).
6. Repeat steps 3 through 5 to generate a set of orthologous sequences.
7. Construct an MSA of the orthologous sequences of the target sequence (*see* **Note 3**).
8. Calculate conservation scores at all alignment sites (*see* **Note 6**).

3.2 Structure Data Manipulation (Part B)

1. Select a closely related structure of the target GPCR sequence in the PDB, with transmembrane region information available in the PDBTM [18] (*see* **Notes 7–9**).
2. Download a file containing structure data from the PDB [19] (*see* **Note 10**).
3. Select a chain in the structure data and remove all of the lines of other chains.
4. Remove all of the coordinates of heteroatoms in the coordinate file.
5. Remove all of the coordinates of the atoms corresponding to other proteins, such as T4 lysozyme, that are used to artificially replace the disordered loop regions of the structures and facilitate crystallization in X-ray crystallography.
6. Define planes by the least-square fitting of the 3D coordinates of the atoms constituting the extracellular and intracellular terminal residues of the seven TM helices. The coordinates of the terminus of a TM helix are obtained from the description in the XML file provided by PDBTM [18].
7. Define a plane that is assumed to be parallel to the membrane plane by averaging the two planes generated at **step 6**.
8. Project the geometric centers of the side chains of all residues on the defined two-dimensional plane. The projected point of the geometric center of the protein is used as the origin of the two vectors that define a sector. Consequently, the radius of the sector is sufficient to cover the distribution of the projected residues within the angle, θ, of the sector.
9. Remove the two-dimensional coordinates of the projected residues with relative accessible surface areas (rASAs) greater than or equal to 25% on the structure [7]. The remaining projected residues form a ring-like distribution on the plane, and the interface residues cluster in a sector region of the ring-like distribution.

3.3 Interface Prediction (Part C)

1. Generate many sectors by rotating the two vectors while changing the central angle, θ, of the sector (*see* **Note 11**). Calculate p_θ with the following formula to evaluate the statistical significance of the clustering of the conserved residues in a sector.

$$p_\theta = \sum_{i=m}^{X_\theta} \frac{X_\theta!}{i!(X_\theta - i)!} \left(\frac{n}{N}\right)^i \left(\frac{N-n}{N}\right)^{X_\theta - i}$$

   where N indicates the total number of surface residues of the selected structure. n is the number of conserved residues with conservation scores $\geq C_t$ found in the N residues. X_θ is the number of residues found in the sector (*see* **Note 12**).

2. Score each sector with the following *ad hoc* function to rank the sectors when $p_\theta < 0.05$.

$$f = \frac{\sum_{j=1}^{m} C_j}{p_\theta \theta}$$

   where m is the number of conserved residues found in the sector. A residue is regarded as being conserved when the alignment site corresponding to the residue has a conservation score greater than or equal to the automatically determined threshold value (C_t). C_j is the conservation score assigned to residue j of the selected structure (*see* **Note 13**).

3. Find a sector with the maximum f-value (*see* **Note 14**).
4. Map all residues corresponding to the projected points in the sector with the maximum f-value on the template structure (*see* **Note 15**).
5. Remove the projected points in the sector to find another interface after the identification of the first interface, because the GPCR of interest may have more than one interface for oligomerization.
6. Apply the same evaluation procedure to the remaining regions of the projected residues. Plural interfaces can be obtained by repeating the procedure (*see* **Note 16**).

   Visualize the predicted interfaces by mapping them on the adopted structure (*see* **Notes 17–19**).

3.4 Interface Prediction on GRIP Web Service

1. Input a target GPCR sequence in FASTA format to the text box (Fig. 2A).
2. Select a PDB ID of the structure, on which interfaces are searched for from the pull-down menu (Fig. 2B) (*see* **Note 9**).
3. Write a chain name of the adopted structure in the small text box if the default setting, A, needs to be changed (Fig. 2C).
4. Write E-value and percent identity in small text boxes if the default setting needs to be changed (Fig. 2D).

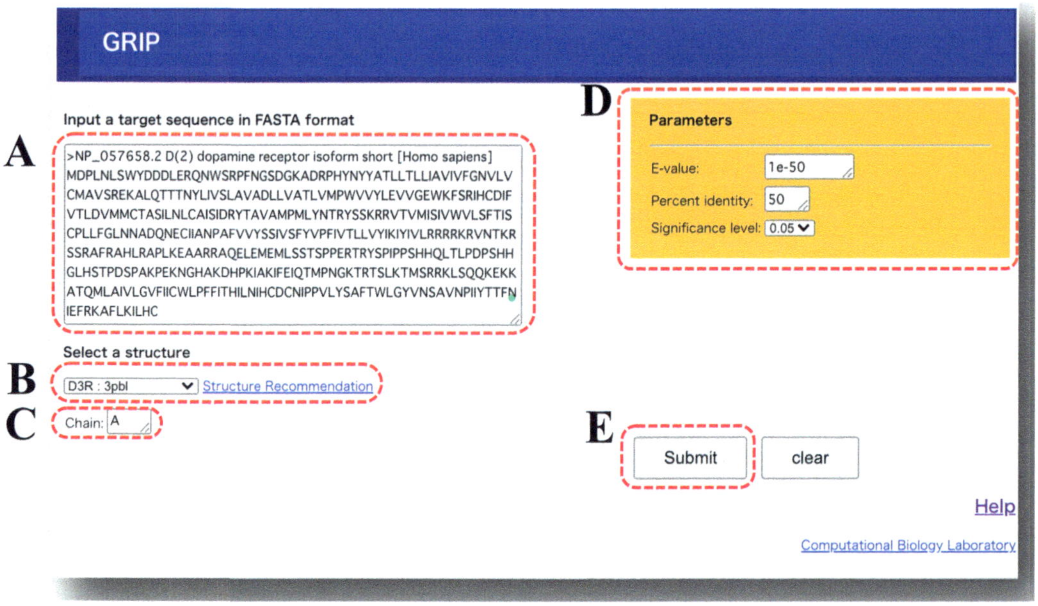

Fig. 2 Input page of the GRIP web service

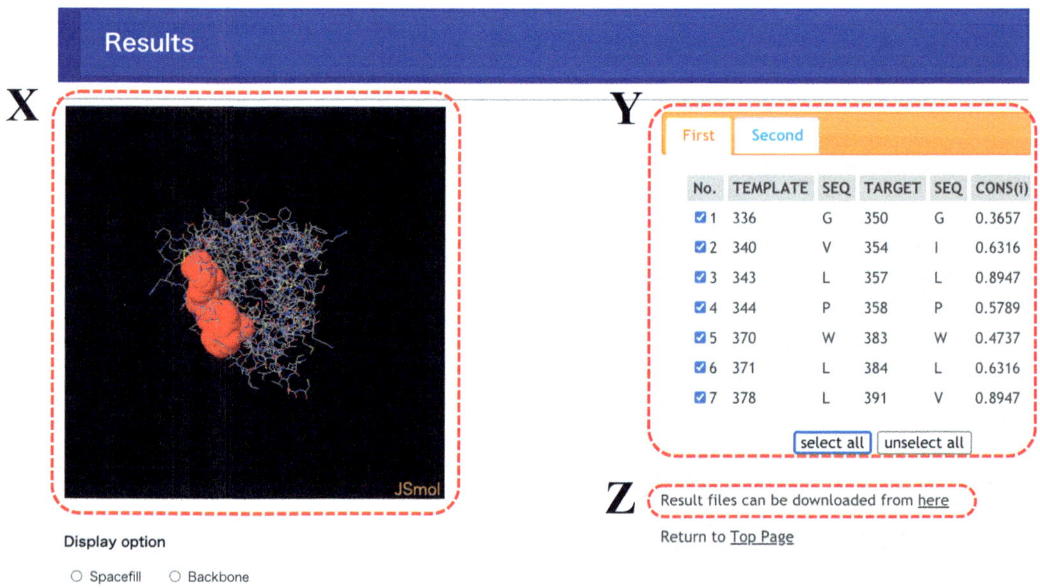

Fig. 3 Result page of the GRIP web service

5. Select either 0.05 or 0.01 as the significance level for statistical hypothesis testing (Fig. 2D).
6. Execute the prediction by pressing the Submit button.
7. The adopted structure is shown in the JSmol window (Fig. 3X).

8. Two tabs, "First" and "Second" corresponding to the two predicted interfaces in the Result page, appear after the prediction.
9. In order to show the predicted interface residues on the structure in the JSmol window, press "select all" button (Fig. 3Y). Then, the first interface residues are colored in red while the second interface residues are colored in yellow.
10. A compressed file containing all of the result files can be downloaded by pressing a link "here" on the right bottom part of the result page (Fig. 3Z). The file contains a chimera script to show the predicted interfaces on the adopted structure (*see* **Notes 15, 16**, and **18**).

4 Notes

1. Collect as many sequences as possible by setting the parameter, Max target sequence, to the largest possible value. Use the RefSeq database [20] or UniProt database [21] to retrieve only curated sequences.
2. We remove sequences shorter than 80% and longer than 120% of the target sequence.
3. Check the quality of the MSA by visual inspection. If there is a problem, such as amino acids that should be aligned with each other are not appropriately aligned, use software such as BioEdit to fine-tune the MSA.
4. Calculate and check the bootstrap values for assessing the significance of the branches and remove the sequences corresponding to the OTUs that lower the values. Note that the purpose of this step is not to create an accurate phylogenetic tree, but to identify sequences to be removed.
5. It is essential but difficult to construct a set of homologous sequences appropriate for functional region prediction [22]. An appropriate sequence set contains the sequences that share functions of interest. In general, the sizes of sequence spaces where functions are conserved strongly depend on the type of protein function. The active sites of enzymes tend to be more conserved than the interface regions for protein–protein interactions. It is essential to collect only the sequences that share identical functions of interest for functional region prediction.

   The same is true for interface prediction of GPCRs. For GPCRs, the appropriate set only contains the sequences with conserved oligomerization interfaces. As described previously, the interfaces for GPCR oligomerization are conserved among the sequences in a subtype but may differ among subtypes

[7]. Therefore, it is essential to collect only the sequences of the subtype to be analyzed.

It is challenging to know in advance how remotely the interface is conserved, relative to the distantly related sequences. At present, the only way to do this is to collect as many subtype sequences as possible. When we assumed that the interfaces are different for each subtype and compared the predicted interfaces among several subtypes, we found that the conserved interface regions are different for each subtype. Further details are described in the previous reports [7, 14].

6. There are many scores to evaluate evolutionary conservation at a site. We use Valdar and Thornton's method [23] with the linear method for matrix transformation, to distinguish invariant sites of rare and popular amino acids.
7. When we are unsure about which structure is appropriate for the prediction, fold recognition programs, such as FUGUE [24], are recommended to select an appropriate structure for the prediction. On our website, a simple and rapid procedure based on FUGUE can generate a list of candidate structures, on which the conservation scores are mapped to the residues. This procedure roughly lists the candidate structures for a given query more rapidly than the original version of FUGUE.
8. The structure data that are registered in the PDB [25] but not in the PDBTM [18] cannot be used for the prediction. If such a structure is needed for the prediction, then we implement the TMDET algorithm to prepare the XML file in the PDBTM.
9. The number of GPCR structures registered in the PDB has been increasing rapidly. As a result, more structures are available for use as prediction templates. Many agonist-bound conformations (active form) have been reported. We previously used only inactive forms, but now we can use active forms as well.

   The increase in the number of template structures raises the question of whether the structure of the most closely related sequence but in a different conformation or that of a distantly related sequence but in an identical conformation should be selected. At present, there is no clear answer as to whether active forms interact with each other, active forms interact with inactive forms, or inactive forms interact with each other. If agonist binding causes an interaction, then the active form would be better for the prediction. On the other hand, if the agonist binding separates the GPCR oligomer into monomers, then the inactive form would be better.
10. PDBx/mmCIF is the official working format of the wwPDB for coordinate files.
11. The procedure is repeated, while changing θ from 20° to 180°.

12. If p_θ is small enough, then the clustering of the conserved residues in the sector is regarded as being significant. We used 5% as the significance level of p_θ. To avoid the arbitrariness of the threshold setting, C_t was also changed from 0.5 to 1.0 with an interval of 0.1 in each iteration procedure.
13. The *f*-value was designed to satisfy four conditions. First, the sector is as narrow as possible. Second, the degree of conservation of the conserved residues within the sector is as high as possible. Third, p_θ is as small as possible. Last, the sector with an angle, θ, was rotated around the origin with intervals of 10°, from 20° to 180°.
14. The sector with the maximum *f*-value is considered to correspond to an interface of the subtype under consideration.
15. Not only the conserved residues but also the nonconserved residues included in the sector are regarded as the residues constituting the interface and are referred to as interface residues.
16. The method cannot detect more than three interfaces for many GPCRs, mainly because the number of projected points is too small to be statistically significant.
17. We performed interface predictions for 13 GPCRs with interface locations proposed by X-ray crystallography or biochemical experiments. We showed the predicted interfaces by our algorithm and experimentally suggested interfaces, as supplementary materials in our previous work [14]. In summary, TM helix IV seems to be conserved among class A GPCRs. In contrast, TM III is conserved in a class-specific manner, which might enable each GPCR to interact with its distinct partners. The interfaces observed in several crystal structures are at a set of TM helices IV and V, a set of TM helices I and II, or both. The interfaces suggested by in vitro experiments are at a set of TM helices I, IV, and V. As mentioned previously, the interfaces for GPCR oligomerization are thus conserved in a subtype-specific manner.

    It should be noted that these experimentally determined interfaces are considered to be for homo-oligomerization, because they have been observed in crystal structures of homo-oligomers or were suggested based on inhibition experiments using TM helix-derived peptides corresponding to the homo-oligomerization interfaces.
18. As described above, GPCRs form hetero-oligomers as well as homo-oligomers. In addition to GPCRs, various other molecules are assumed to be interaction partners for GPCRs. Some GPCRs form hetero-oligomers with non-GPCR membrane-associated proteins. The transmembrane helices of GPCRs reportedly interact with cholesterol and other types of lipids. The surface of the transmembrane region interacting with

various molecules would be functionally constrained. As a result, highly conserved amino acids may accumulate on the surface of the transmembrane region.

GRIP predicts one or two interfaces for GPCR oligomerization, based on the assumption that the clusters of highly conserved residues along the surfaces of transmembrane helices correspond to the interface. Nevertheless, GRIP does not provide information about the types of molecules that interact with the predicted interfaces. GRIP does not provide information about whether a GPCR forms a homo-oligomer or hetero-oligomer, or which protein is the partner if the GPCR forms a hetero-oligomer. Therefore, the predicted interfaces can participate in interactions with other molecules, including cholesterol or membrane lipids.

We can obtain experimentally determined or predicted interacting pairs for GPCR oligomerization through the Internet. To our knowledge, GPCR-HetNet is the only database in operation, and it provides 536 experimentally suggested GPCR–GPCR interaction pairs [26]. GPCR-OKB [27] and GRIPDB [28] are not presently available. GGIP is a bioinformatics method to predict interacting pairs for GPCR oligomerization and requires a pair of GPCR amino acids in the FASTA format as input [29].

19. The same strategy as that for GPCR oligomers can be applied to predict interfaces and interacting pairs for general membrane proteins. As described above, GPCRs interact with other membrane proteins. Other membrane proteins also interact with each other [30]. Accurate predictions of interfaces and interacting pairs would be useful to provide guidelines for mutation and inhibition experiments. They would accelerate investigations of the molecular mechanisms of oligomerization and signaling via general membrane proteins. Conserved residues along transmembrane regions are good indicators for interface predictions, while sequence and structural features are useful indicators for interaction pair predictions.

Acknowledgments

This work was supported by Grant-in-Aid for Scientific Research from the Ministry of Education Culture, Sports, Science and Technology of Japan (25870764, 18K06199).

References

1. Borroto-Escuela DO, Fuxe K (2019) Oligomeric receptor complexes and their allosteric receptor-receptor interactions in the plasma membrane represent a new biological principle for integration of signals in the CNS. Front Mol Neurosci 12:230. https://doi.org/10.3389/fnmol.2019.00230
2. Möller J, Isbilir A, Sungkaworn T et al (2020) Single-molecule analysis reveals agonist-specific dimer formation of μ-opioid receptors. Nat Chem Biol 16(9):946–954. https://doi.org/10.1038/s41589-020-0566-1
3. Milligan G, Canals M, Pediani JD et al (2006) The role of GPCR dimerisation/oligomerisation in receptor signalling. Ernst Schering Found Symp Proc (2):145–161. https://doi.org/10.1007/2789_2006_007
4. Nemoto W, Toh H (2006) Membrane interactive alpha-helices in GPCRs as a novel drug target. Curr Protein Pept Sci 7:561–575. https://doi.org/10.2174/138920306779025657
5. Ferré S, Casadó V, Devi LA et al (2014) G protein-coupled receptor oligomerization revisited: functional and pharmacological perspectives. Pharmacol Rev 66:413–434. https://doi.org/10.1124/pr.113.008052
6. Terrillon S, Bouvier M (2004) Roles of G-protein-coupled receptor dimerization. EMBO Rep 5:30–34. https://doi.org/10.1038/sj.embor.7400052
7. Nemoto W, Toh H (2005) Prediction of interfaces for oligomerizations of G-protein coupled receptors. Proteins 58:644–660. https://doi.org/10.1002/prot.20332
8. Gurevich VV, Gurevich EV (2008) GPCR monomers and oligomers: it takes all kinds. Trends Neurosci 31:74–81. https://doi.org/10.1016/j.tins.2007.11.007
9. Kniazeff J, Prézeau L, Rondard P et al (2011) Dimers and beyond: the functional puzzles of class C GPCRs. Pharmacol Ther 130:9–25. https://doi.org/10.1016/j.pharmthera.2011.01.006
10. Romano C, Miller JK, Hyrc K et al (2001) Covalent and noncovalent interactions mediate metabotropic glutamate receptor mGlu5 dimerization. Mol Pharmacol 59:46–53. https://doi.org/10.1124/mol.59.1.46
11. Pagano A, Rovelli G, Mosbacher J et al (2001) C-terminal interaction is essential for surface trafficking but not for heteromeric assembly of GABA(b) receptors. J Neurosci 21:1189–1202. https://doi.org/10.1523/JNEUROSCI.21-04-01189.2001
12. Lichtarge O, Bourne HR, Cohen FE (1996) An evolutionary trace method defines binding surfaces common to protein families. J Mol Biol 257:342–358. https://doi.org/10.1006/jmbi.1996.0167
13. Soyer OS, Dimmic MW, Neubig RR, Goldstein RA (2003) Dimerization in aminergic G-protein-coupled receptors: application of a hidden- site class model of evolution. Biochemistry 42:14522–14531. https://doi.org/10.1021/bi035097r
14. Saito A, Tsuchiya D, Sato S et al (2020) Update of the GRIP web service. J Recept Signal Transduct Res 40:348–356. https://doi.org/10.1080/10799893.2020.1734821
15. Nemoto W, Fukui K, Toh H (2009) GRIP: a server for predicting interfaces for GPCR oligomerization. J Recept Signal Transduct Res 29:312–317. https://doi.org/10.3109/10799890903295143
16. Altschul SF, Gish W, Miller W et al (1990) Basic local alignment search tool. J Mol Biol 215:403–410. https://doi.org/10.1016/S0022-2836(05)80360-2
17. Schäffer AA, Aravind L, Madden TL et al (2001) Improving the accuracy of PSI-BLAST protein database searches with composition-based statistics and other refinements. Nucleic Acids Res 29:2994–3005. https://doi.org/10.1093/nar/29.14.2994
18. Kozma D, Simon I, Tusnády GE (2013) PDBTM: Protein Data Bank of transmembrane proteins after 8 years. Nucleic Acids Res 41:D524–D529. https://doi.org/10.1093/nar/gks1169
19. Berman HM (2000) The Protein Data Bank. Nucleic Acids Res 28:235–242. https://doi.org/10.1093/nar/28.1.235
20. Pruitt KD, Tatusova T, Brown GR, Maglott DR (2012) NCBI reference sequences (RefSeq): current status, new features and genome annotation policy. Nucleic Acids Res 40:D130–D135. https://doi.org/10.1093/nar/gkr1079
21. UniProt Consortium (2014) Activities at the universal protein resource (UniProt). Nucleic Acids Res 42:D191–D198. https://doi.org/10.1093/nar/gkt1140
22. Nemoto W, Toh H (2012) Functional region prediction with a set of appropriate homologous sequences-an index for sequence selection by integrating structure and sequence information with spatial statistics. BMC Struct Biol 12:11. https://doi.org/10.1186/1472-6807-12-11

23. Valdar WSJ (2002) Scoring residue conservation. Proteins 48:227–241. https://doi.org/10.1002/prot.10146
24. Shi J, Blundell TL, Mizuguchi K (2001) FUGUE: sequence-structure homology recognition using environment-specific substitution tables and structure-dependent gap penalties. J Mol Biol 310:243–257. https://doi.org/10.1006/jmbi.2001.4762
25. Berman HM, Westbrook J, Feng Z et al (2000) The Protein Data Bank. Nucleic Acids Res 28:235–242. https://doi.org/10.1093/nar/28.1.235
26. Borroto-Escuela DO, Brito I, Romero-Fernandez W et al (2014) The G protein-coupled receptor heterodimer network (GPCR-HetNet) and its hub components. Int J Mol Sci 15:8570–8590. https://doi.org/10.3390/ijms15058570
27. Khelashvili G, Dorff K, Shan J et al (2010) GPCR-OKB: the G protein coupled receptor oligomer knowledge base. Bioinformatics 26:1804–1805. https://doi.org/10.1093/bioinformatics/btq264
28. Nemoto W, Fukui K, Toh H (2011) GRIPDB - G protein coupled receptor interaction partners DataBase. J Recept Signal Transduct Res 31:199–205. https://doi.org/10.3109/10799893.2011.563312
29. Nemoto W, Yamanishi Y, Limviphuvadh V et al (2016) GGIP: structure and sequence-based GPCR-GPCR interaction pair predictor. Proteins 84:1224–1233. https://doi.org/10.1002/prot.25071
30. Babu M, Vlasblom J, Pu S et al (2012) Interaction landscape of membrane-protein complexes in Saccharomyces cerevisiae. Nature 489:585–589. https://doi.org/10.1038/nature11354

Chapter 7

Memdock: An α-Helical Membrane Protein Docking Algorithm

Naama Hurwitz and Haim J. Wolfson

Abstract

Memdock is a tool for docking α-helical membrane proteins which takes into consideration the lipid bilayer environment. Given two α-helical membrane located protein molecules, the method outputs a list of potential complexes sorted by energy criteria. The program includes three steps: docking, refinement, and re-ranking of the results. All three docking steps have been customized to the membrane environment in order to improve performance and reduce program run-time. In this chapter, we describe the application of our web server, referred to as Memdock, for prediction of the docking complex for a pair of input membrane protein structures. Memdock is freely available for academic users without registration at http://bioinfo3d.cs.tau.ac.il/Memdock/index.html.

Key words Protein–protein docking, Membrane proteins, Lipid bilayer

1 Introduction

Analysis of the complete genomic sequences for several organisms indicates that 30% of all genes code for membrane proteins [1, 2]. Due to their central role in a wide range of fundamental biological processes, membrane proteins constitute around 60% of approved drug targets [3]. However, despite their large number and importance and even though there is an increasing number of solved membrane protein structures, they are still under-represented in the Protein Data Bank [4]. This is owing to the difficulties involved in expressing, purifying, and crystallizing these proteins [5]. In the light of this deficit of empirical information, it is particularly important to develop efficient computational methods for predicting the structure, interactions, and docking of membrane proteins.

Memdock is an integrated docking tool for α-helical membrane proteins. Given protein structures of two α-helical protein chains, Memdock docks the structures based on shape complementarity considerations, refines the docking predictions allowing side chain

Irina S. Moreira et al. (eds.), *Computational Design of Membrane Proteins*, Methods in Molecular Biology, vol. 2315, https://doi.org/10.1007/978-1-0716-1468-6_7, © Springer Science+Business Media, LLC, part of Springer Nature 2021

rearrangement and limited backbone flexibility, and re-ranks the candidate solutions by a score which is customized to the membrane environment.

In this chapter, we describe the use of Memdock for docking pairs of membrane proteins.

2 Materials

2.1 An Overview of the Memdock Algorithm

Memdock is a docking algorithm for membrane proteins. It has three major execution steps, which are (a) rigid docking; (b) docking refinement to accommodate full side chain flexibility and limited backbone flexibility of the docking partners; and (c) re-ranking of the top scoring solutions by a physics driven energy function. Memdock adopts our previously developed docking methods to all these steps, while adjusting them to the lipid bilayer environment. The first "rigid docking" step is based on the PatchDock algorithm [6]. PatchDock is an efficient rigid "unbound" soft docking algorithm, which ranks docking candidates mainly by geometric shape complementarity between the docked proteins. In the modified PatchDock, customized to the membrane (Mem-PatchDock), the hypothesized placement of the docked proteins is enforced to be in a membrane-consistent orientation. The guiding principle behind this enforcement is that proteins in the membrane are restricted in their movement between their apo (conformation before docking) and holo structures (conformation after docking). The movement restriction is demonstrated in Fig. 1. Given an apo structure, let a 3D Euclidean coordinate frame XYZ be attached to this structure, centered at its center of mass, and having the Z axis parallel to the membrane normal. After docking this structure will be transformed to the holo structure with the coordinate frame $X'Y'Z'$. Due to the membrane environment restrictions, the center of mass of the holo structure can have only a limited translation in the Z direction and the angle between the Z and Z' axes is limited as well.

In the second step, a flexible refinement of the docking candidates is done by the FiberDock algorithm [7], which remodels the side chain interactions and also allows limited backbone flexibility in the receptor. In the modified FiberDock, customized to the membrane (Mem-FiberDock), perturbations of the ligand are enforced to be in a membrane consistent orientation, similarly to the rigid docking step. Finally, the refined models are re-scored and re-ranked according to an energy function (Memscore) that was designed specifically for membrane proteins. For a comprehensive description of the Memdock algorithm, *see* [8].

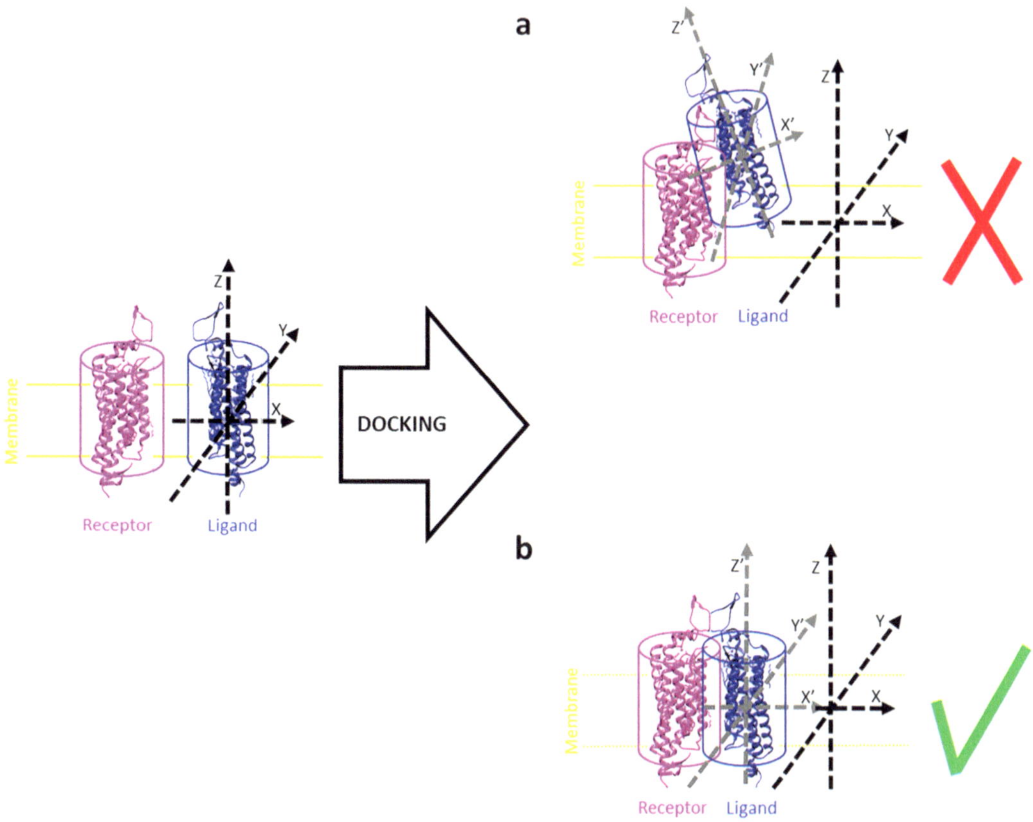

Fig. 1 The baseline of docking membrane proteins: A Euclidean coordinate frame XYZ is attached to the ligand structure (in black), centered at its center of mass and having the Z axis parallel to the membrane normal. The movement of the ligand is shown by drawing the gray Euclidean coordinates (X ′ Y ′ Z ′). (**a**) A *membrane-inconsistent* orientation docking and (**b**) A *membrane-consistent* orientation docking

2.2 The Memdock Website

Memdock can be accessed through http://bioinfo3d.cs.tau.ac.il/Memdock/index.html. The submission page is shown in Fig. 2. The details are described below.

2.3 Input

Memdock accepts as its input two membrane protein chains in the PDB format: a ligand and a receptor to which the user wants to dock the ligand to. Both the PDB coordinate files should be output from the OPM server [9]. Then, their coordinates are pre-calculated and positioned in the membrane bilayer.

3 Methods

3.1 Submitting Jobs to Memdock

1. Upload the receptor file (in PDB format, OPM output).
2. Write down the receptor chains (the chains must appear in the PDB file), separated by comas, spaces, or "All" if you wish to dock the whole structure.

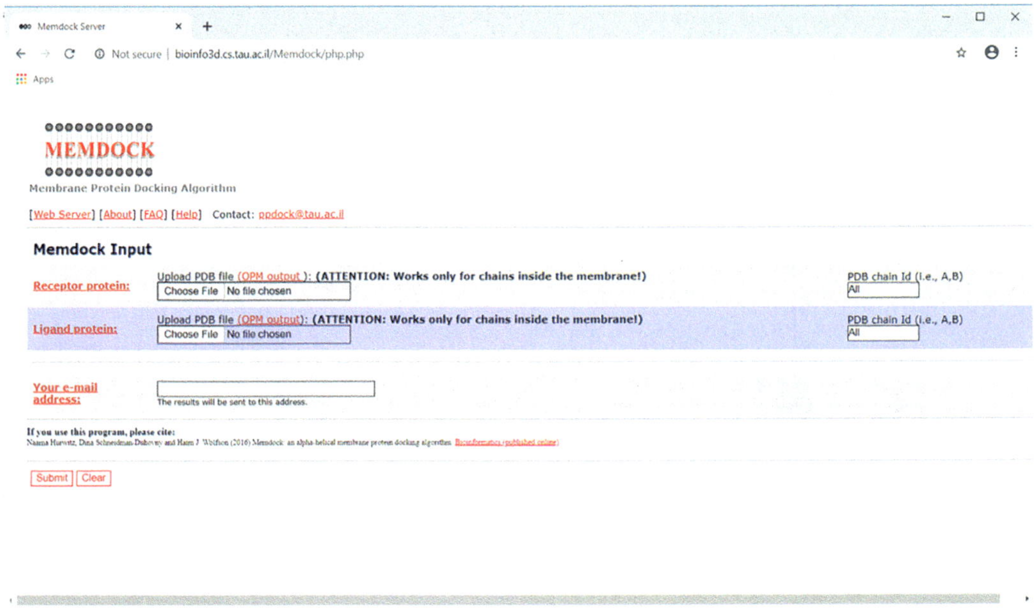

Fig. 2 The Memdock web server

3. Upload the ligand file (in PDB format, OPM output).
4. Write down the ligand chains (the chains must appear in the PDB file), separated by comas or "All" if you wish to dock the whole structure.

3.2 Output

The results will be sent via email (compressed file). The best 20 sorted docking candidate complexes will be sent in PDB file format. In addition, you will receive a summary output file, "mem_results.csv," which includes the table of the output docking candidates, ranked by the "Memscore."

4 A Case Study

We use "Multisubunit cytochrome b" (PDB; 1Q90) as an example. Multisubunit cytochrome b is interacting with the complex Photosystems I and II which reaction centers capture light energy to derive oxygenic photosynthesis.

4.1 Input Entry

PDB file 1Q90, a complex of a dimer, was input to Memdock server for both receptor and ligand, choosing B as the receptor and S as the ligand (Fig. 3).

4.2 Output

The user receives a mail with the following output zipped file—memdock.tar.gz. This file includes 20 PDB files (resX.pdb) which are the best output docking candidates. In addition, a file

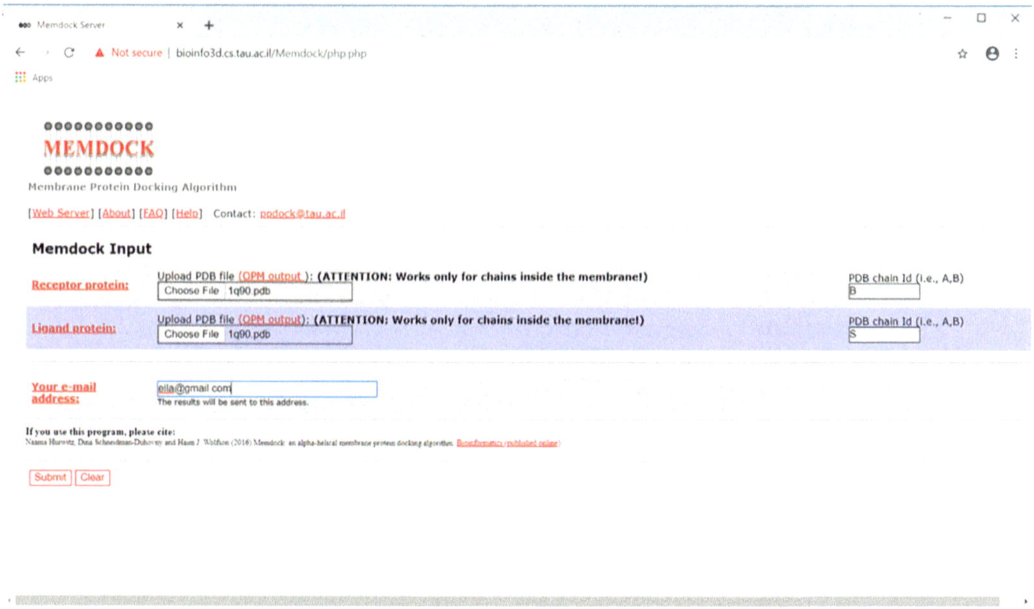

Fig. 3 Input screen of the Memdock web server

"mem_results.csv," includes a table of the output docking candidates, ranked by the "Memscore" (Fig. 4). Figure 5 shows the best 1Q90 docking candidate, namely res0.pdb.

5 Conclusions and Future Work

We introduced an algorithm with excellent performance for docking and refinement of α-helical membrane proteins. The improved performance of this algorithm compared to standard protein–protein docking algorithms is due to the adaptation of its search and scoring methods to the membrane protein environment. First, we enforce membrane-consistent orientation in both docking and refinement steps as an integral part of the algorithm rather than a post-processing step. Second, the refined docking candidates undergo re-ranking using a scoring function formulated from a combination of knowledge-based potentials derived from statistical analyses of known 3D structures of α-helical membrane proteins and the FiberDock scoring function terms which are based on the binding forces between the receptor and the ligand.

File mem_results.csv

```
Output_file          Memscore       Fiberdock_score
res0.pdb  -707.98575889        -86.4968
res1.pdb  -632.039453181       -58.2535
res2.pdb  -610.285111481       -49.4556
res3.pdb  -603.130184926       -36.6721
res4.pdb  -597.058908288       -95.3879
res5.pdb  -580.952862463       -96.6619
res6.pdb  -556.408916045       -60.0548
res7.pdb  -515.824975351       -53.6264
res8.pdb  -500.280579927       -54.9106
res9.pdb  -498.050334865       -54.399
res10.pdb -491.39897129        -29.4764
res11.pdb -488.587307314       -35.7046
res12.pdb -486.819582923       -50.6834
res13.pdb -486.052151761       42.2147
res14.pdb -484.13599362        -70.975
res15.pdb -483.81700437        6.7491
res16.pdb -481.596478254       -94.0621
res17.pdb -464.29665452        -17.423
res18.pdb -457.517969201       -85.747
res19.pdb -448.502923492       -81.9913
```

Fig. 4 Memdock output file, case study: file mem_results.csv—This file includes 20 PDB files (resX.pdb) which are the best output docking candidates

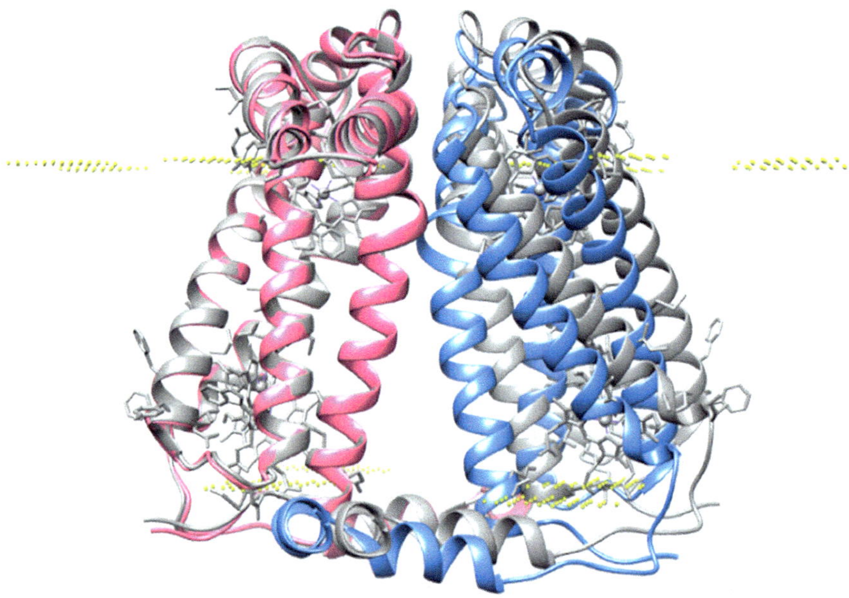

Fig. 5 Memdock best docking structure, case study: The structure of 1Q90, chains B and S in gray. The yellow spots represent the membrane location. Memdock best docking candidate (res0.pdb) is shown in color: chain B as the receptor in pink and chain S as the ligand in light blue

References

1. Jones DT (1998) Do transmembrane protein superfolds exist? FEBS Lett 423(3):281–285. https://doi.org/10.1016/S0014-5793(98)00095-7
2. Wallin E, Heijne GV (1998) Genome-wide analysis of integral membrane proteins from eubacterial, archaean, and eukaryotic organisms. Protein Sci 7(4):1029–1038. https://doi.org/10.1002/pro.5560070420
3. Yıldırım MA, Goh KI, Cusick ME et al (2007) Drug—target network. Nat Biotechnol 25 (10):1119–1126. https://doi.org/10.1038/nbt1338
4. Berman HM, Westbrook J, Feng Z et al (2000) The protein data bank. Nucleic Acids Res 28 (1):235–242. https://doi.org/10.1093/nar/28.1.235
5. Carpenter EP, Beis K, Cameron AD, Iwata S (2008) Overcoming the challenges of membrane protein crystallography. Curr Opin Struct Biol 18(5):581–586. https://doi.org/10.1016/j.sbi.2008.07.001
6. Duhovny D, Nussinov R, Wolfson HJ (2002) Efficient unbound docking of rigid molecules. In: Guigó R, Gusfield D (eds) Algorithms in bioinformatics. WABI 2002, Lecture notes in computer science, vol 2452. Springer, Berlin, pp 185–200. https://doi.org/10.1007/3-540-45784-4_14
7. Mashiach E, Nussinov R, Wolfson HJ (2010) FiberDock: flexible induced-fit backbone refinement in molecular docking. Proteins 78 (6):1503–1519. https://doi.org/10.1002/prot.22668
8. Hurwitz N, Schneidman-Duhovny D, Wolfson HJ (2016) Memdock: an α-helical membrane protein docking algorithm. Bioinformatics 32 (16):2444–2450. https://doi.org/10.1093/bioinformatics/btw184
9. Lomize MA, Pogozheva ID, Joo H et al (2012) OPM database and PPM web server: resources for positioning of proteins in membranes. Nucleic Acids Res 40(Database issue):D370–D376. https://doi.org/10.1093/nar/gkr703

Part IV

Coarse-Grained and Atomistic MD Simulations

Chapter 8

Identification and Characterization of Specific Protein–Lipid Interactions Using Molecular Simulation

Robin A. Corey, Mark S. P. Sansom, and Phillip J. Stansfeld

Abstract

Interactions with lipids can dramatically shape and define the activity of membrane proteins. Here, we describe tools that allow the identification of these interactions using molecular dynamics simulation. Additionally, we provide the details of how to use different methods to probe the affinity of these interactions.

Key words Molecular dynamics, Membrane proteins, Lipids, Protein–lipid interactions, Free energy calculations

1 Introduction

Integral membrane proteins are embedded in a lipid bilayer, meaning they are constantly in contact with a wide number of different lipids from their surroundings. These interactions can be either general or specific. General interactions will be important for anchoring the protein within the membrane and ensuring the proteins is localized to—and possibly only active in—certain regions of the membrane, e.g., rafts [1].

Specific protein–lipid interactions have been shown to have a number of profound effects on the function of the protein. Such roles include contributing to protein oligomerization or stabilizing conformational states, such as outward facing states in transporter proteins or open states in channel proteins.

While the identification of lipids bound to proteins in structural densities is on the rise (e.g., *see* refs. [2, 3]), particularly owing to recent developments in both cryo-EM and nanodisc technology, most structural analyses require solubilization of the protein in detergent, which often removes the bulk of the bound lipids.

Irina S. Moreira et al. (eds.), *Computational Design of Membrane Proteins*, Methods in Molecular Biology, vol. 2315, https://doi.org/10.1007/978-1-0716-1468-6_8,

Additionally, even when present, assignment of the molecular identity of the lipid species can be non-trivial. Finally, it is unclear whether the interactions seen in the purified protein–lipid complex will necessarily reflect those seen in the native membrane environment. To address some of these problems, additional information can be obtained by applying molecular dynamics (MD) simulations of the protein reconstituted in native-like membranes. Here, a molecular system can be described with accuracy at a high spatial and temporal resolution. By following the motions of the surrounding lipids in relation to the protein over time, it is possible to identify any interactions between specific residues on the protein and a given lipid species. This can provide an unambiguous assignment of molecular species for lipids identified in structural density, as well as identifying new sites not identified experimentally.

Traditionally, atomistic MD has been limited to studying only very short timescales, on the order of ns, owing to prohibitive amounts of computer resource being needed. This makes analysis of protein–lipid interactions difficult, as these typically occur on the μs timescale. Recent increases in computational efficiency has now made these timescales more attainable, but even now the modeling of a single protein–lipid interaction requires considerable investment of computational resource, making it slow and somewhat limited in scope.

This restrictive cost can be ameliorated through use of a coarse-grained (CG) biomolecular force field, such as Martini (see http://cgmartini.nl and [4, 5]). Here, groups of approximately four heavy atoms and associated hydrogens are modeled as a single bead. This both reduces the number of MD calculations required per step by a factor of ~10 and allows the use of timesteps of 20–40 fs (instead of 2 fs). Both of these features greatly enhance the speed of sampling compared to fully atomistic simulations.

Here, we outline in details the steps needed to build and run Martini CG simulations of integral membrane proteins in membranes, using the *Gromacs* simulation package (http://www.gromacs.org [6, 7]). We then provide a practical basis for identifying and determining the affinity of any specific protein–lipid interactions which occur throughout the simulation. We primarily discuss integral membrane proteins; however, each step of the protocol can be easily adapted for peripheral membrane proteins.

2 Protocol

Our protocol follows the diagram in Fig. 1. Each step of the process is described in detail below.

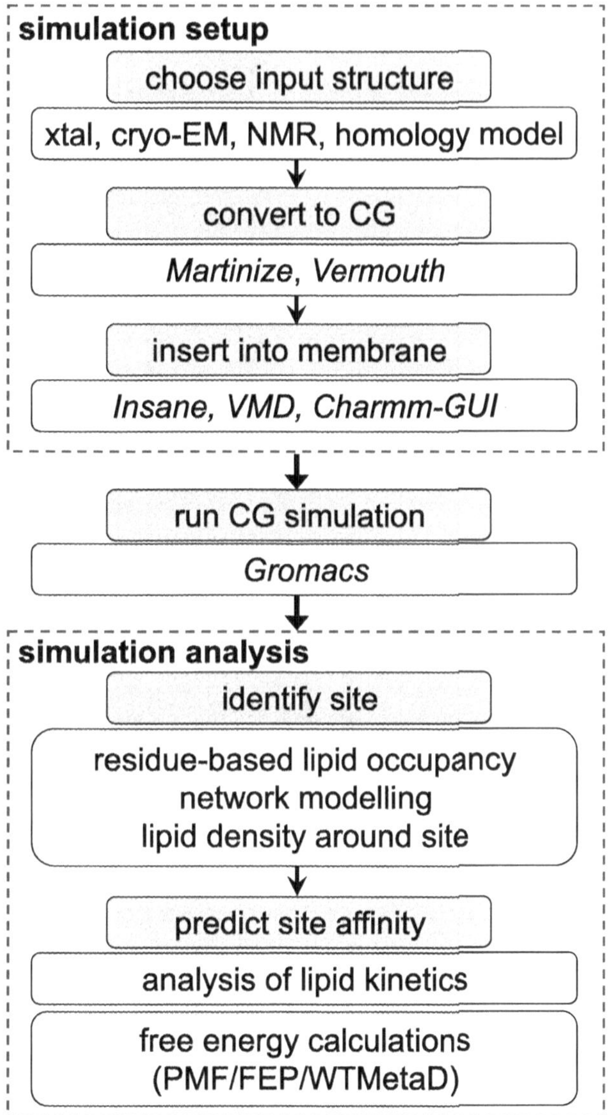

Fig. 1 Flowchart for modeling protein-lipid interactions using CG-MD. Grey boxes represent the primary steps in building, running, and analyzing the simulations. White boxes represent the programs or techniques used for each step

2.1 Simulation Setup

2.1.1 Choice of Input Structure

The choice of input coordinates for the simulation is very important. Ideally, a structure produced by experimental data—such as cryo-EM or X-ray crystallography—will be used. Alternatively, a homology model may be used.

If a protein exists in multiple states, such as seen with transporters or channels, then the conformation used in the simulations will be of importance. For instance, if a specific interaction is thought to be state-dependent, then it is most pertinent to simulate protein–lipid interactions in this configuration.

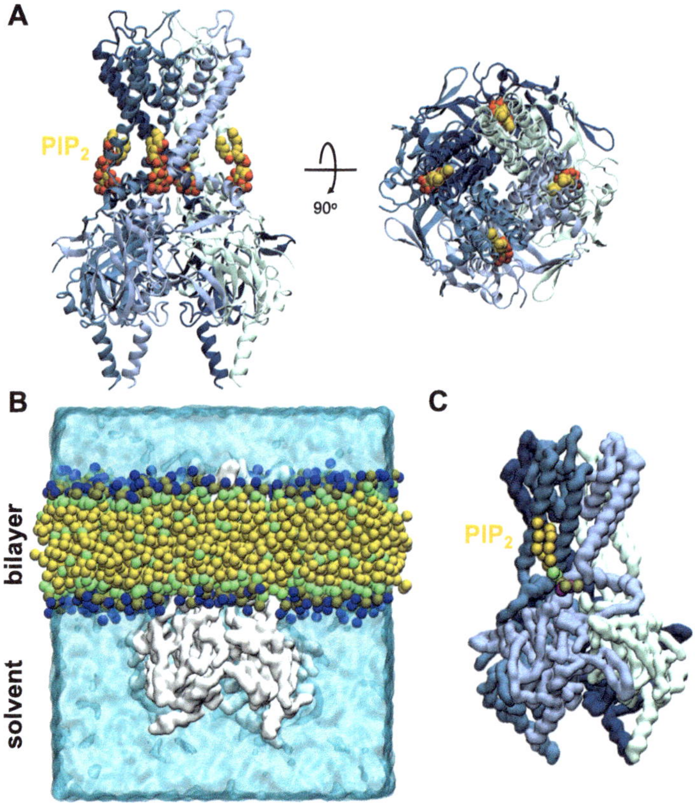

Fig. 2 Overview of protein–lipid interactions, focusing on Kir2.2 and PIP_2. (**a**) View of Kir2.2 with bound PIP_2-like molecules from structural data (PDB: 3SPI). The four separate chains are colored different shades of blue and the lipids are shown in colored spheres. (**b**) View of 3SPI converted to CG, built into a solvated membrane and relaxed with MD simulation. The protein is shown as white surface, the water as cyan surface and the lipids (POPC and PIP_2) as colored spheres. (**c**) View of a Kir2.2–PIP_2 interaction as sampled in a simulation like that shown in panel B. The binding site is similar to that seen in panel A

An example of this can be seen with the potassium channel Kir2.2, a member of the inwardly rectifying ion channel family. Upon binding of the phospholipid PIP_2 to Kir2.2, the channel is activated [8]. A structure-based prediction of a PIP_2 lipid bound to a Kir channel was first proposed using CG simulations [9], before being confirmed via X-ray crystallography [10] (Fig. 2a). This interaction will be used as an example throughout this chapter.

2.1.2 Conversion to CG

Once an input structure has been decided upon and prepared for simulation, it needs to be converted to a CG representation, inserted into a periodic unit cell and immersed in solvent and lipids (Fig. 2b).

For CG conversion, this depends on which CG force field is being used. For simplicity, this chapter only discusses simulations run in Martini 2.2 [4, 5], although it should be noted that other comparable CG force fields are available, notably the Sirah force field [12], which has recently been extended to include lipids [13, 14].

In Martini, proteins are typically described with each bead representing a group of four C, N, O, P, or S atoms. Backbones are modeled as "BB" beads, and side chains as between 0 and 5 "SC" beads. More complete formal descriptions of the forcefield are provided in reference [15]. *See* **Note 1** for a discussion on the different options available when using Martini.

To convert a protein to a Martini 2.2 description, this is typically done using the *Martinize* program (https://github.com/cgmartini/martinize.py [16]) or the newer *Vermouth* (https://github.com/marrink-lab/vermouth-martinize). Here, the user can decide how the system is built, including if additional elastic networks are applied to the protein. This step is particularly important, as membrane proteins in CG tend to undergo unnatural dynamics without additional restraints. These are applied as harmonic bonds between BB beads within a defined cut-off, with a set force constant of e.g. 500 kJ mol^{-1} nm^{-2}. Additional accuracy may be obtained by comparing backbone RMSDs or residue RMSFs of Martini simulations at different force constants with atomistic simulations and choosing the force constant which best reproduces the atomistic data.

An example *Martinize* command is as follows:

```
python martinize.py -v -f input.pdb -merge all -name Protein
-dssp /path/to/dssp -ff martini22 -v -p backbone -elastic -ef
500 -el 0.5 -eu 1 -ea 0 -ep 0 -x protein-cg.pdb -o protein-cg.
top
```

This takes an input atomistic pdb file and converts it to Martini 2.2 beads, applying an elastic network of 500 kJ mol^{-1} nm^{-2} bonds (`-ef`) between BB atoms within 0.5–1 nm of each other (`-el -eu`) with the decay factor and power defined by `-ea -ep`. A CG coordinate and topology file are produced (`-x -o`). Note that either a DSSP binary [17, 18] (`-dssp`) or web server produced DSSP file for the protein of interest needs to be supplied (`-ss`) to assign BB types. We recommend installing DSSP using *Anaconda* (https://www.anaconda.com), with the following command: `conda install -c salilab dssp`.

2.1.3 Insertion into a Membrane

The next thing is to insert the protein into a membrane. For this, careful consideration needs to be taken of which lipids are used. The simplest option is to build a membrane from a single, representative, lipid such as 1-palmitoyl-2-oleoyl-sn-glycero-3-

phosphocholine (POPC) or 1-palmatoyl-2-oleoyl-sn-glycero-3-phosphoethanolamine (POPE). This has the benefit of being simple, making it most suitable for systems where only generic interactions with lipids are of interest.

If it is already known which lipid is expected to bind, then a common tactic is to build a binary lipid system comprising the lipid of interest and a generic lipid. For instance, for modeling Kir2.2-PIP_2 interactions, a membrane comprising POPC and approximately eight copies of PIP_2 in the appropriate (cytoplasmic) leaflet could be used.

Alternatively, far more complex lipid compositions can be used, to better reflect the high complexity of the biological membrane. Commonly, this diversity is expressed in the headgroups of the lipid, with model membranes comprising multiple headgroup types (e.g., PC, PE, GM3, PS, and PIP_2) but often with the same acyl tails. An example of Kir2.2 probed in this way can be found in [11]. Otherwise, the headgroups may remain constant and the lipid tails varied, for instance, modeling how lipid tail saturation [19] or length [20] affects interaction with proteins.

Note that with increased complexity of the lipid membrane, the time required to reach convergence will likely also increase, so that the level of complexity should be tailored to the biological question at hand.

To build the membrane, a number of different methods are available. Perhaps the simplest is the Python-based *insane* approach (https://github.com/Tsjerk/Insane [21]), although other powerful tools are available, including *VMD* [22] and *Charmm-GUI* [23, 24]. With *insane* you are able to fully describe the desired membrane, giving the lipid types and ratios for either the whole membrane or each leaflet separately. An example is given below:

```
python insane.py -f protein-cg.pdb -o all-cg.gro -x 14 -y 14 -z
13 -l POPC:80 -l POPE:10 -l CHOL:10 -u POPC:100 -sol W -p temp.
top
```

Here, a membrane is built comprising 80% POPC, 10% POPE, and 10% cholesterol in the lower leaflet (`-l`) and just POPC in the upper leaflet (`-u`). The system has also been solvated with Martini waters (`-sol`). The box size is manually set using the `-x -y -z` flags. It is recommended to leave plenty of space around the protein for the lipids to diffuse, at least 2–3 nm (Fig. 2b).

Note that it can often be useful to first orient the protein before inserting into the membrane. There are a number of useful tools for this including *memembed* (https://github.com/timnugent/memembed [25]), and the *Orientations of Proteins in Membranes* (OPM) database (https://opm.phar.umich.edu [26]).

2.1.4 Running Simulations

Typically, the system will next be energy minimized using the steepest descent algorithm, and then equilibrated and simulated using a 20–30 fs time step. Example mdp input files can be found on the Martini website (http://cgmartini.nl/index.php/force-field-parameters/input-parameters). If unfamiliar with *Gromacs*, we strongly recommend following a tutorial, such as http://www.mdtutorials.com/gmx/. The length of the simulation will depend on the system and what the desired observation is, but will typically be in excess of 4–5 μs. *See* **Note 2** for a discussion on convergence.

2.1.5 Conversion Back to AT

Once a pose has been chosen which provides a useful state of the system, for instance with a lipid of interest bound the protein (see below for more details on this), the system can be converted back to an atomistic description for additional analyses. This is more complicated than the initial atomistic to CG conversion, but fortunately there are a number of programs available for this purpose, *see* ref. [27] or [28], or CG2AT2, a new program built for this purpose at https://github.com/owenvickery/cg2at and https://zenodo.org/record/3890164.

3 Simulation Analysis

3.1 Identifying Specific Lipid-Binding Sites

Over the course of a simulation, the lipids will likely equilibrate into annular shells around the embedded protein [29]. Additionally, individual lipid molecules might bind to a specific "site" on the surface of the protein, defined here as a group of residues that interact simultaneously with a given lipid [11], usually focusing on the lipid headgroup for phospholipids or the whole molecule for sterols.

These interactions are typically far more long-lasting than non-specific interactions and can be of substantial functional importance to the protein. As a result, the identification of specific protein–lipid interactions will likely be one of the primary analyses performed on the system (Fig. 2c). Here we describe a few different methods to analyze lipid-binding sites on the protein surface.

3.1.1 Occupancy

A simple initial analysis is to model the occupancy of each residue in the protein for a given lipid species. Here, the occupancy can be defined as:

$\mathrm{Occ} = \frac{t_{\mathrm{bound}}}{t_{\mathrm{free}}}$ where t_{bound} is the amount of time (i.e., number of frames) where the residue is in contact with a lipid type, and t_{free} is the amount of time it is not. Contact is often simply defined as the two groups being within a certain distance, e.g., 0.6 nm for a CG system. Note that, as lipid tails are highly flexible and dynamic, it is

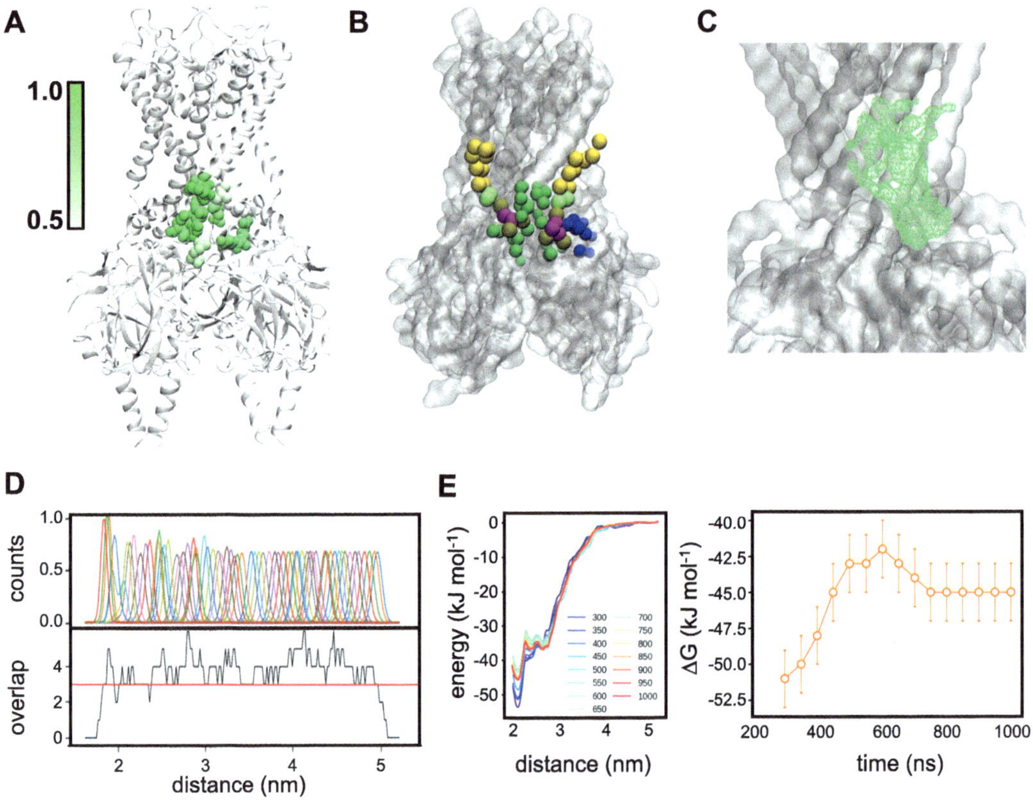

Fig. 3 Protein–lipid site identification. (**a**) The occupancy (between 0.5 and 1.0) of PIP_2 headgroups for each residue of Kir2.2 was calculated over many s of CG simulation and mapped onto the atomistic Kir2.2 coordinates (PDB 3SPI). (**b**) View of two discrete sites identified using network analyses. The green site has a residence time of >50 s, the blue site of approximately 10–20 s (Adapted from [11]). (**c**) Computed density from CG simulation of an entire PIP_2 molecule in relation to Kir2.2. (**d**) Coverage analysis of a PMF calculation, showing the histograms of each US window (top) and the number of overlapping histograms at each point of the reaction coordinate (bottom). Plots generated using https:/zenodo.org/record/3890063 and adapted for use here. (**e**) Convergence analyze of a PMF calculation. On the left is the PMF as calculated using simulation times from 300 ns (purple) to 1000 ns (red). On the right is the computed binding energy from each of these traces

typical to focus on residue interactions with the headgroup. Projecting this information onto the protein structure is often sufficient to highlight discrete binding regions for a given lipid, e.g., Fig. 3a.

This approach has the benefit that it is easy to implement (e.g., using *gmx distance* or *gmx mindist* which comes as part of the *Gromacs* package or *MDAnalysis* [30, 31]) and provides a straightforward, protein-oriented output, identifying residues that are important for lipid binding. These can then be mutated experimentally for validation of the simulation results, using techniques such as native mass spectrometry [32]. In addition, if sufficient binding and unbinding events occur during the simulation, it may be possible to predict a free energy for the interaction from the occupancy (e.g., see [33]).

As a downside, it does not directly provide much information on the nature of the interaction and can result in considerable ambiguity with regard to what exactly constitutes a lipid-binding site, as no information is present on whether nearby residues are binding the lipid concurrently. For example, the occupancy analysis of Kir2.2–PIP_2 interactions shown in Fig. 3a might suggest a single binding site, a notion challenged by the next section.

3.1.2 Network Modeling

Another approach is to build an interaction profile for each individual residue–lipid pair over the course of the simulation, and cluster residues which have a high incidence of simultaneously binding the same lipid or lipid headgroup. For this, one can apply graph theory, where the residues form the graph nodes, with the edges linking the nodes weighted according to the frequency of the residues interacting with the same lipid concurrently. Residues can then be grouped into communities [34], allowing the identification of specific sites. This approach was recently applied to study the interactions of Kir2.2 with both cholesterol [35] and a more complex mixture of lipids [11].

This technique has the potential to be very powerful, as it provides a detailed map of distinct sites across the surface of the protein. Once a site has been identified, it is then possible to model the length of time a given lipid will spend in contact with the site, giving a residence time and k_{off} (see below for more details), allowing a prediction of relative affinity of each site for a given lipid. It is, however, more difficult to implement than the occupancy analysis. To assist with this process, we have produced a Python program, *PyLipID*, which allows for the automatic detection and kinetic analysis of sites (https://github.com/wlsong/PyLipID). This approach has been successfully applied in a number of studies, including the identification of >700 cardiolipin binding sites from CG MD simulations of 42 different membrane proteins [64].

3.1.3 Lipid Density

Finally, the density of the lipid during the entire simulation can be computed around the surface of the protein, using such tools as the *Gromacs* tool *gmx density*, *MDAnalysis*, or the *VolMap* tool of *VMD* [22]. This has the benefit of being extremely quick and providing more information on the behavior of the lipids around the binding site, including the existence of lipid-stacking as seen in certain cholesterol-binding sites [36], *see* [37] for a recent application of this in CG. In addition, it is conceptually straightforward to compare to densities obtained by structural analysis, for instance, Fig. 7 panels A and B from [38].

On the other hand, it can be nontrivial to distinguish discrete binding sites in regions of generally high occupancy, and the approach lacks an easy ability to quantify the interaction probability. It is, however, very powerful for looking at more general structural changes imposed on the membrane by the protein [39] or lipid "fingerprinting" around the protein [40, 41]. As well, it is useful for modeling more dynamic protein–lipid interactions, as might be seen during phospholipid scrambling [42].

3.2 Characterizing the Affinity of the Interaction

Given the high diversity [43] and cross-membrane asymmetry [44] of lipids present in a typical biological membrane, hundreds of different lipid species might be jostling for position around a given protein at any one time. Therefore, it is important to not only know if a specific protein–lipid interaction can occur, but what the relative affinity of this interaction is in relation to other potential interactions. Here we outline a few methods that can be used for this purpose.

3.2.1 Kinetics of Binding

As mentioned above, the residence time (equal to $1/k_{\mathrm{off}}$) of a lipid can be computed from simulation data, with residence time defined as the average time that a single lipid (j) spends continuously interacting with a given site before exchange with the annular shell around the protein [45]. This is commonly [11, 46] calculated from a normalized survival time correlation function $\sigma(t)$,

$$\sigma(t) = \frac{1}{N_j} \frac{1}{T - t} \sum_{j=1}^{N_j} \sum_{\nu=0}^{T} \rho_j(\nu, \nu + t).$$

where N_j is the total number of the lipid type with a nonzero site interaction time, T is the total simulation time, and $\rho_j(\nu, \nu + t)$ is a function which has a value of 1 if lipid j interacts with the site continuously from time ν to $\nu + t$, and 0 if it does not. By modeling the data using a double exponential, both fast (likely failed binding attempts) and slow (successful binding events) interactions can be extracted. The residence times for different sites and lipids can then be compared to provide insight into the strength of different interactions.

As mentioned before, implementation of this analysis can be fairly time consuming and require considerable input to ensure accuracy in the data. Therefore, we recommend using the *PyLipID* method (https://github.com/wlsong/PyLipID) which automatically detects and provides residence times for all lipid sites on the protein, requiring only an input trajectory and coordinate file.

3.2.2 Free Energy Calculations: PMF

Alternatively, the protein–lipid interaction can be more directly examined using free energy calculations [47]. The best established for this particular biological question is potential of mean force (PMF) calculations [46, 48], which allow the construction of a typically one-dimensional reaction coordinate through the free energy landscape of a system, for instance, the distance between the headgroup of a lipid molecule and the center of mass of the membrane protein. This can be achieved via a number of methods; however, it is common to use steered MD to "pull" or "push" the lipid onto or off of the binding site, and then run umbrella sampling (US) simulations seeded using snapshots from the steered MD. These US simulations allow calculation of the force acting on the lipid at each position along the coordinate.

For the steered MD simulations, an umbrella pulling force (e.g., of 1000 kJ mol^{-1} nm^{-2}) can be applied along the desired coordinate (e.g., along the x-axis), with a defined rate, e.g., 0.1 nm ns^{-1}. This can be defined in the *Gromacs* mdp file as:

```
pull = yes
pull_ngroups = 2
pull_ncoords = 1
pull_group1_name = Lipid
pull_group2_name = Protein
pull_coord1_type = umbrella
pull_coord1_groups = 1 2
pull_coord1_rate = 0.0001
pull_coord1_k = 1000
pull_coord1_start = yes
pull-nstfout = 50
pull_coord1_geometry = direction
pull-coord1-vec = 1 0 0
```

(NB the exact syntax can vary between *Gromacs* versions). Note that an index file with a group name "Lipid" containing the specific beads of the target lipid headgroup needs to be supplied to the *grompp* command. For the steered MD, the writing out of the forces (`pull-nstfout`) should match the writing of coordinates to the trajectory (`nstxout-compressed`—not shown in the example).

Once the steered MD has been run (e.g., the lipid is >3 nm further along the coordinate than its starting point), frames can be extracted at specific points along the coordinate. This can be achieved manually or can be done in an automated fashion using a script written specifically for the problem, such as: https://zenodo.org/record/3890063.

Then, individual US can be run, using exactly the same mdp settings as for the steered MD above, but with `pull_coord1_rate` set to 0 and `nstxout-compressed` to 10,000 or similar. To prevent any rotational movement of the proteins in the PMF calculations, mild *xy* positional restraints (e.g., of

100 kJ mol^{-1} nm^{-2}) can be applied to 3–4 backbone beads in the protein.

Once the simulations have completed to convergence (*see* Fig. 3e and **Note 3**), the energy landscape can be constructed using the produced force xvg file, using a technique such as the weighted histogram analysis methods (WHAM; [49]), as implemented in a program such as *gmx wham* [50].

3.2.3 Free Energy Calculations: FEP/ABFE

Another way of estimating the free energy of a protein–lipid interaction is through either free energy perturbation (FEP) or absolute binding free energy (ABFE) calculations. Here, the target lipid molecule is "alchemically" perturbed *in silico* into a generic lipid (FEP) or fully decoupled from the system (ABFE). Both approaches require construction of a coordinate termed λ. When $\lambda = 0$, the system contains the lipid of interest and when $\lambda = 1$, a generic lipid in its place. When $\lambda = 0.5$ the lipid is at a halfway point between the two—this state is likely non-physiological. For this specific application, FEP and ABFE estimate the same coordinate, as ABFE decoupling the lipid will allow a generic lipid to bind in its place (*see* [47] for a more detailed discussion).

By perturbing the lipid while both bound to the protein and free in the membrane, the difference between these states can be calculated to provide the free energy associated with lipid binding to the protein. FEP/ABFE appears to give equivalent values to PMF calculations, while typically using less computational resource [51].

There are many different ways of setting up and running FEP calculations [52]; here we focus on the dual topology method. As an example, for the system described in Fig. 1, it might make sense to perturb a PIP_2 lipid into the generic POPC lipid. This can be done in Martini 2.2 by the "removal" of beads PO1, PO2, RP1, and RP2 and a conversion of bead RP3 from Martini type SP1 to Q0, including changing the charge from 0 to +1. To achieve this, an itp needs to be constructed with type, charge, and mass columns representing PIP_2, and typeb, chargeb, and massb columns representing POPC, as below:

```
[atoms]
; id type resnr resid atom cgnr charge mass typeb chargeb massb
 1 Qa 1 P2PC PO1 1 -2.0 72 Dum 0 72
 2 Qa 1 P2PC PO2 2 -2.0 72 Dum 0 72
 3 SP1 1 P2PC RP1 3 0 72 Dum 0 72
 4 SP1 1 P2PC RP2 4 0 72 Dum 0 72
 5 SP1 1 P2PC RP3 5 0 72 Q0 1.0 72
 6 Qa 1 P2PC PO3 6 -1.0 72 Qa -1.0 72
 7 Na 1 P2PC GL1 7 0 72 Na 0 72
 8 Na 1 P2PC GL2 8 0 72 Na 0 72
 9 C1 1 P2PC C1A 9 0 72 C1 0 72
```

```
10 C1 1 P2PC C2A 10 0 72 C1 0 72
11 C1 1 P2PC C3A 11 0 72 C1 0 72
12 C1 1 P2PC C4A 12 0 72 C1 0 72
13 C1 1 P2PC C1B 13 0 72 C1 0 72
14 C1 1 P2PC C2B 14 0 72 C1 0 72
15 C1 1 P2PC C3B 15 0 72 C1 0 72
16 C1 1 P2PC C4B 16 0 72 C1 0 72
17 C1 1 P2PC C5B 17 0 72 C1 0 72
```

Note that we typically use the input bonded parameters for all λ states, as these will contribute little to the interaction and usually cancel out between the bound and free simulations. A new bead type "Dum" will need to be added to the root martini itp file and set to have no Lennard–Jones interactions with itself or other beads, i.e., add

```
Dum 72 0 A 0.0 0.0
```

under [atomtypes] and

```
Dum Dum 1 0 0
```

under [nonbond_params].

To keep the overall charge of the box neutral at each λ state (which is not strictly necessary as Martini uses reaction field electrostatics), the charge can be removed from five sodium beads, as defined below.

```
[atoms]
;id type resnr residu atom cgnr charge mass typeb chargeb massb
 1 Qd 1 ION NAP 1 1.0 72 Qd 0 72
```

Once the topology has been built, simulations can be run at different values of λ. Normally, Coulombic (coul_lambdas) and Lennard Jones (LJ; vdw_lambdas) terms are decoupled separately, to prevent charge–charge interactions causing beads at low LJ to overlap. An example of the mdp file settings used to perturb PIP_2 to POPC is given below:

```
free_energy = yes
init_lambda_state = ${NUM} ; change for each window
delta_lambda = 0
calc-lambda-neighbors = -1 ; Need -1 for MBAR
;init_lambda_state 0 1 2 3 4 5 6 7 8 9 10
vdw_lambdas = 0.0 0.0 0.0 0.0 0.0 0.0 0.2 0.4 0.6 0.8 1.0
coul_lambdas = 0.0 0.2 0.4 0.6 0.8 1.0 1.0 1.0 1.0 1.0 1.0
sc-alpha = 0.5 ; soft-core LJ potentials are on
sc-coul = no
sc-power = 1
sc-sigma = 0.3
```

```
nstdhdl = 100
```

We have shown the mdp settings for 11 states; however, more states may be desirable, for example, 21 steps of magnitude 0.1 for both Coulombic then LJ [51].

The user should write separate mdp files where the ${NUM} selection is changed to an integer representing a specific λ state. These can then be used to seed production simulations (using *gmx grompp*), from which a dhdl.xvg file will be written. The dhdl.xvg files for all the windows can be combined using a technique such as thermodynamic integration or multistate Bennett acceptance ratio to get the value of perturbing along λ. We strongly recommend the use of specialist software to do this, such as *alchemlyb* [53] (https://github.com/alchemistry/alchemlyb). The analysis also allows the user to make sure the simulations are long enough to be converged (*see* **Note 4**), typically for a protein–lipid interaction we run 100–250 ns simulation windows, and run multiple repeats to allow statistical reporting on our free energies. It may be that shorter or longer windows are more suitable to different systems.

ABFE calculations follow a similar process but require the imposition of restraints keeping the lipid in place for low LJ values. We recommend the restraint scheme constructed by Salari et al. [54], adapted for CG in [51].

3.2.4 Free Energy Calculations: WTMetaD

In addition, 2D free energy landscapes can be built using well-tempered meta-dynamics (WTMetaD) [55, 56] using, for example, the *x* and *y* position of the lipid headgroup in relation to the protein as the collective variable (CV). This has the advantage of providing a more complete landscape of the protein–lipid interaction and not needing to identify the precise binding pose before running the calculations.

On the other hand, their running is slightly more complicated, currently requiring the use of *Gromacs* patched with *Plumed* [57, 58]. In addition, the calculations can be costlier than PMF or FEP, and more difficult to determine convergence for. For details on how to set up these calculations, we recommend reference [51].

4 Summary

In summary, CG MD offers a powerful tool to identify protein–lipid interactions, with the strength of the interactions quantifiable in the form of an affinity from either kinetic analysis or free energy calculation. In this chapter, we have outlined the necessary steps to achieve these goals, along with guidance to ensure that the analyses are robust. In time, we expect that there will be an increase in the use of atomistic force fields, which require far more computational time to sample a given system. These atomic-level simulations pose additional complications, but permit a more accurate description of the molecular system. We end this chapter, however, by stating that the real challenge is, and will likely continue to be, relating the information obtainable from the MD simulation to the biological activity of the protein.

5 Notes

1. Martini force field decisions

   Several decisions are to be made when choosing the version of Martini force field used. Typically, we use Martini 2.2 with non-polarizable beads, Na^+ and Cl^- ions and elastic networks between nearby BB beads. Other choices exist, including polarizable Martini [59] and "dry" Martini using implicit solvent [60], as well as the elastic networks being handled by the ElNeDyn force field [61]. Also of note is the upcoming release of Martini 3, which should represent a substantial increase in accuracy of the force field [62].

2. Convergence of equilibrium simulations

   Given Martini's high speed in sampling, convergence is not as difficult as it can be for atomistic simulation. Typically, the length of the simulation should be sufficient to sample multiple binding and unbinding events for a given protein–lipid interaction. This could be as short at 5 μs or take longer than 50 μs [11]. Having a rigorous method to resolve this can be of great use—we recommend our Python program *PyLipID* which uses kinetic modeling to determine whether convergence has been reached (https://github.com/wlsong/PyLipID). Alternatively, for a quick analysis, the user can count the interactions with each type of lipid (such as with *gmx select* in *Gromacs*) and run the simulations until this value stabilizes.

3. Convergence of PMF calculations

   There are two considerations for ensuring convergence in PMF calculations. First, is the reaction coordinate sufficiently sampled? For this, we recommend taking snapshots at intervals of at least 0.1 nm along the coordinate, although 0.05 nm is preferable (see Supporting Fig. 13B of ref. [51]). To check that the coordinate is covered, the histograms outputted from the WHAM analysis should be plotted. Ideally each histogram should overlap with the two preceding and proceeding histograms, with the higher number of overlapping histograms the better. To check the overlap between histograms and to fill in any missing umbrella windows, we recommend the following Python script: https://zenodo.org/record/3890063.

Second, is each US simulation long enough? This will be when all possible configurations within each window have been sampled. This can be tested by comparing calculated PMF values from either increasing lengths of the simulation (such as in ref. [51]) or through subsampling the PMF profile and evaluating with respect to the final profile [63]. We have found that at least 1 μs per umbrella window is necessary for most protein–lipid interactions.

4. Convergence of FEP calculations

As with PMF calculations, both the coordinate spacing and simulation length are of importance for checking convergence. Ideally, the coordinate spacing (i.e., how many λ states are sampled) and simulation length should be chosen to keep the errors low and consistent across the coordinate. We typically use 21 windows (10 to perturb Coulombic interactions, 10 to perturb LJ) of 100–250 ns each, although overlapping Coulombic and LJ windows can also be used. We also run 3–5 replicates to allow calculation of the mean and standard deviation. However, being familiar with the convergence analyses of *alchemlyb* [53] (https://github.com/alchemistry/alchemlyb) is very highly recommended, as is running very short simulations to check λ spacing before committing to longer simulations.

Acknowledgments

The authors thank Drs Anna Duncan, Wanling Song, and Owen Vickery for critical reading and useful discussions.

References

1. Lingwood D, Simons K (2010) Lipid rafts as a membrane-organizing principle. Science 327:46–50. https://doi.org/10.1126/science.1174621
2. Whorton MR, MacKinnon R (2011) Crystal structure of the mammalian GIRK2 K + channel and gating regulation by G proteins, PIP 2, and sodium. Cell 147:199–208. https://doi.org/10.1016/j.cell.2011.07.046
3. Norimatsu Y, Hasegawa K, Shimizu N, Toyoshima C (2017) Protein–phospholipid interplay revealed with crystals of a calcium pump. Nature 545:193. https://doi.org/10.1038/nature22357
4. Marrink SJ, Risselada HJ, Yefimov S et al (2007) The MARTINI force field: coarse grained model for biomolecular simulations. J Phys Chem B 111:7812–7824. https://doi.org/10.1021/jp071097f
5. Monticelli L, Kandasamy SK, Periole X et al (2008) The MARTINI coarse-grained force field: extension to proteins. J Chem Theory Comput 4:819–834. https://doi.org/10.1021/ct700324x
6. Berendsen HJC, van der Spoel D, van Drunen R (1995) GROMACS: a message-passing parallel molecular dynamics implementation. Comput Phys Commun 91:43–56. https://doi.org/10.1016/0010-4655(95)00042-E
7. Van Der Spoel D, Lindahl E, Hess B et al (2005) GROMACS: fast, flexible, and free. J Comput Chem 26:1701–1718. https://doi.org/10.1002/jcc.20291
8. Hilgemann DW, Feng S, Nasuhoglu C (2001) The complex and intriguing lives of PIP2 with

ion channels and transporters. Sci STKE 2001: re19. https://doi.org/10.1126/stke.2001.111.re19
9. Stansfeld PJ, Hopkinson R, Ashcroft FM, Sansom MSP (2009) PIP2-binding site in Kir channels: definition by multiscale biomolecular simulations. Biochemistry 48:10926–10933. https://doi.org/10.1021/bi9013193
10. Hansen SB, Tao X, MacKinnon R (2011) Structural basis of PIP 2 activation of the classical inward rectifier K+ channel Kir2.2. Nature 477:495. https://doi.org/10.1038/nature10370
11. Duncan AL, Corey RA, Sansom MSP (2020) Defining how multiple lipid species interact with inward rectifier potassium (Kir2) channels. Proc Natl Acad Sci 117:7803–7813. https://doi.org/10.1073/pnas.1918387117
12. Darré L, Machado MR, Brandner AF et al (2015) SIRAH: a structurally unbiased coarse-grained force field for proteins with aqueous solvation and long-range electrostatics. J Chem Theory Comput 11:723–739. https://doi.org/10.1021/ct5007746
13. Barrera EE, Frigini EN, Porasso RD, Pantano S (2017) Modeling DMPC lipid membranes with SIRAH force-field. J Mol Model 23:259. https://doi.org/10.1007/s00894-017-3426-5
14. Barrera EE, Machado MR, Pantano S (2019) Fat SIRAH: coarse-grained phospholipids to explore membrane–protein dynamics. J Chem Theory Comput 15:5674–5688. https://doi.org/10.1021/acs.jctc.9b00435
15. Periole X, Marrink SJ (2013) The martini coarse-grained force field. Methods Mol Biol 924:533–565. https://doi.org/10.1007/978-1-62703-17-5_20
16. de Jong DH, Singh G, Bennett WFDD et al (2012) Improved parameters for the martini coarse-grained protein force field. J Chem Theory Comput 9:687–697. https://doi.org/10.1021/ct300646g
17. Kabsch W, Sander C (1983) Dictionary of protein secondary structure: pattern recognition of hydrogen-bonded and geometrical features. Biopolymers 22:2577–2637. https://doi.org/10.1002/bip.360221211
18. Joosten RP, Te Beek TAH, Krieger E et al (2010) A series of PDB related databases for everyday needs. Nucleic Acids Res 39:D411–D419. https://doi.org/10.1093/nar/gkq1105
19. Javanainen M, Enkavi G, Guixà-Gonzaléz R et al (2019) Reduced level of docosahexaenoic acid shifts GPCR neuroreceptors to less ordered membrane regions. PLoS Comput Biol 15:e1007033. https://doi.org/10.1371/journal.pcbi.1007033
20. Parker JL, Corey RA, Stansfeld PJ, Newstead S (2019) Structural basis for substrate specificity and regulation of nucleotide sugar transporters in the lipid bilayer. Nat Commun 10:1–10. https://doi.org/10.1038/s41467-019-12673-w
21. Wassenaar TA, Ingólfsson HI, Böckmann RA et al (2015) Computational lipidomics with insane: a versatile tool for generating custom membranes for molecular simulations. J Chem Theory Comput 11:2144–2155. https://doi.org/10.1021/acs.jctc.5b00209
22. Humphrey W, Dalke A, Schulten K (1996) VMD: visual molecular dynamics. J Mol Graph 14:33–38. https://doi.org/10.1016/0263-7855(96)00018-5
23. Jo S, Kim T, Iyer VG, Im W (2008) CHARMM-GUI: a web-based graphical user interface for CHARMM. J Comput Chem 29:1859–1865. https://doi.org/10.1002/jcc.20945
24. Qi Y, Ingólfsson HI, Cheng X et al (2015) CHARMM-GUI martini maker for coarse-grained simulations with the martini force field. J Chem Theory Comput 11:4486–4494. https://doi.org/10.1021/acs.jctc.5b00513
25. Nugent T, Jones DT (2013) Membrane protein orientation and refinement using a knowledge-based statistical potential. BMC Bioinformatics 14:276. https://doi.org/10.1186/1471-2105-14-276
26. Lomize MA, Pogozheva ID, Joo H et al (2012) OPM database and PPM web server: resources for positioning of proteins in membranes. Nucleic Acids Res 40:D370–D376. https://doi.org/10.1093/nar/gkr703
27. Stansfeld PJ, Sansom MSP (2011) From coarse grained to atomistic: a serial multiscale approach to membrane protein simulations. J Chem Theory Comput 7:1157–1166. https://doi.org/10.1021/ct100569y
28. Wassenaar TA, Pluhackova K, Böckmann RA et al (2014) Going backward: a flexible geometric approach to reverse transformation from coarse grained to atomistic models. J Chem Theory Comput 10:676–690. https://doi.org/10.1021/ct400617g
29. Stansfeld PJ, Jefferys EE, Sansom MSP (2013) Multiscale simulations reveal conserved patterns of lipid interactions with aquaporins. Structure 21:810–819. https://doi.org/10.1016/j.str.2013.03.005
30. Michaud-Agrawal N, Denning EJ, Woolf TB, Beckstein O (2011) MDAnalysis: a toolkit for

the analysis of molecular dynamics simulations. J Comput Chem 32:2319–2327. https://doi.org/10.1002/jcc.21787

31. Gowers RJ, Linke M, Barnoud J et al (2019) MDAnalysis: a python package for the rapid analysis of molecular dynamics simulations. Los Alamos National Lab.(LANL), Los Alamos, NM
32. Corey RA, Pyle E, Allen WJ et al (2018) Specific cardiolipin–SecY interactions are required for proton-motive force stimulation of protein secretion. Proc Natl Acad Sci U S A 115:7967–7972. https://doi.org/10.1073/pnas.1721536115
33. Lee JY, Lyman E (2012) Predictions for cholesterol interaction sites on the A2Aadenosine receptor. J Am Chem Soc 134:16512–16515. https://doi.org/10.1021/ja307532d
34. Fortunato S (2010) Community detection in graphs. Phys Rep 486:75–174. https://doi.org/10.1016/j.physrep.2009.11.002
35. Barbera N, Ayee MAA, Akpa BS, Levitan I (2018) Molecular dynamics simulations of Kir2. 2 interactions with an ensemble of cholesterol molecules. Biophys J 115:1264–1280. https://doi.org/10.1016/j.bpj.2018.07.041
36. Fantini J, Barrantes FJ (2013) How cholesterol interacts with membrane proteins: an exploration of cholesterol-binding sites including CRAC, CARC, and tilted domains. Front Physiol 4:31. https://doi.org/10.3389/fphys.2013.00031
37. Hedger G, Koldsø H, Chavent M et al (2019) Cholesterol interaction sites on the transmembrane domain of the hedgehog signal transducer and class F G protein-coupled receptor smoothened. Structure 27:549–559.e2. https://doi.org/10.1016/j.str.2018.11.003
38. Wang Q, Corey RA, Hedger G et al (2020) Lipid interactions of a ciliary membrane TRP channel: simulation and structural studies of polycystin-2. Structure 28:169–184. https://doi.org/10.1016/j.str.2019.11.005
39. Zhou W, Fiorin G, Anselmi C et al (2019) Large-scale state-dependent membrane remodeling by a transporter protein. elife 8:e50576. https://doi.org/10.7554/eLife.50576
40. Corradi V, Mendez-Villuendas E, Ingólfsson HI et al (2018) Lipid–protein interactions are unique fingerprints for membrane proteins. ACS Cent Sci 4:709–717. https://doi.org/10.1021/acscentsci.8b00143
41. Shearer J, Jefferies D, Khalid S (2019) Outer membrane proteins OmpA, FhuA, OmpF, EstA, BtuB, and OmpX have unique lipopolysaccharide fingerprints. J Chem Theory Comput 15:2608–2619. https://doi.org/10.1021/acs.jctc.8b01059
42. Bushell SR, Pike ACW, Falzone ME et al (2019) The structural basis of lipid scrambling and inactivation in the endoplasmic reticulum scramblase TMEM16K. Nat Commun 10:1–16. https://doi.org/10.1038/s41467-019-11753-1
43. Harayama T, Riezman H (2018) Understanding the diversity of membrane lipid composition. Nat Rev Mol Cell Biol 19(5):281–296. https://doi.org/10.1038/nrm.2017.138
44. Lorent JH, Levental KR, Ganesan L et al (2020) Plasma membranes are asymmetric in lipid unsaturation, packing and protein shape. Nat Chem Biol 16:644–652. https://doi.org/10.1038/s41589-020-0529-6
45. Bolla JR, Corey RA, Sahin C et al (2019) A mass spectrometry-based approach to distinguish annular and specific lipid binding to membrane proteins. Angew Chemie Int Ed Engl 59:3523–3528. https://doi.org/10.1002/anie.201914411
46. Arnarez C, Marrink SJ, Periole X (2013) Identification of cardiolipin binding sites on cytochrome c oxidase at the entrance of proton channels. Sci Rep 3:1–9. https://doi.org/10.1038/srep01263
47. Corey RA, Stansfeld PJ, Sansom MSP (2019) The energetics of protein–lipid interactions as viewed by molecular simulations. Biochem Soc Trans 48:25–37. https://doi.org/10.1042/BST20190149
48. Hedger G, Rouse SL, Domański J et al (2016) Lipid-loving ANTs: molecular simulations of cardiolipin interactions and the organization of the adenine nucleotide translocase in model mitochondrial membranes. Biochemistry 55:6238–6249. https://doi.org/10.1021/acs.biochem.6b00751
49. Souaille M, Roux B (2001) Extension to the weighted histogram analysis method: combining umbrella sampling with free energy calculations. Comput Phys Commun 135:40–57. https://doi.org/10.1016/S0010-4655(00)00215-0
50. Hub JS, De Groot BL, Van Der Spoel D (2010) g_wham — a free weighted histogram analysis implementation including robust error and autocorrelation estimates. J Chem Theory Comput 6:3713–3720. https://doi.org/10.1021/ct100494z
51. Corey RA, Vickery ON, Sansom MSP, Stansfeld PJ (2019) Insights into membrane protein–lipid interactions from free energy calculations. J Chem Theory Comput

15:5727–5736. https://doi.org/10.1021/acs.jctc.9b00548

52. Shirts MR, Mobley DL (2013) An introduction to best practices in free energy calculations. In: Biomolecular simulations. Springer, pp 271–311
53. Klimovich PV, Shirts MR, Mobley DL (2015) Guidelines for the analysis of free energy calculations. J Comput Aided Mol Des 29:397–411. https://doi.org/10.1007/s10822-015-9840-9
54. Salari R, Joseph T, Lohia R et al (2018) A streamlined, general approach for computing ligand binding free energies and its application to GPCR-bound cholesterol. J Chem Theory Comput 14:6560–6573. https://doi.org/10.1021/acs.jctc.8b00447
55. Bussi G, Laio A, Parrinello M (2006) Equilibrium free energies from nonequilibrium metadynamics. Phys Rev Lett 96:10–13. https://doi.org/10.1103/PhysRevLett.96.090601
56. Barducci A, Bussi G, Parrinello M (2008) Well-tempered metadynamics: a smoothly converging and tunable free-energy method. Phys Rev Lett 100:20603. https://doi.org/10.1103/PhysRevLett.100.020603
57. Tribello GA, Bonomi M, Branduardi D et al (2014) PLUMED 2: new feathers for an old bird. Comput Phys Commun 185:604–613. https://doi.org/10.1016/j.cpc.2013.09.018
58. Bonomi M, Branduardi D, Bussi G et al (2009) PLUMED: a portable plugin for free-energy calculations with molecular dynamics. Comput Phys Commun 180:1961–1972. https://doi.org/10.1016/j.cpc.2009.05.011
59. Yesylevskyy SO, Schäfer LV, Sengupta D, Marrink SJ (2010) Polarizable water model for the coarse-grained MARTINI force field. PLoS Comput Biol 6:1–17. https://doi.org/10.1371/journal.pcbi.1000810
60. Arnarez C, Uusitalo JJ, Masman MF et al (2015) Dry martini, a coarse-grained force field for lipid membrane simulations with implicit solvent. J Chem Theory Comput 11:260–275. https://doi.org/10.1021/ct500477k
61. Periole X, Cavalli M, Marrink S-J, Ceruso MA (2009) Combining an elastic network with a coarse-grained molecular force field: structure, dynamics, and intermolecular recognition. J Chem Theory Comput 5:2531–2543. https://doi.org/10.1021/ct9002114
62. Alessandri R, Souza PCT, Thallmair S et al (2019) Pitfalls of the martini model. J Chem Theory Comput 15:5448–5460. https://doi.org/10.1021/acs.jctc.9b00473
63. Klesse G, Rao S, Tucker SJ, Sansom MSP (2020) Induced polarization in molecular dynamics simulations of the 5-HT3 receptor channel. J Am Chem Soc 142:9415–9427. https://doi.org/10.1021/jacs.0c02394
64. Corey R, Song W, Duncan A et al (2021) Identification and assessment of cardiolipin interactions with E. coli inner membrane proteins. BioRxiv. https://doi.org/10.1101/2021.03.19.436130

Chapter 9

Molecular Dynamics Simulation of Lipid-Modified Signaling Proteins

Vinay V. Nair and Alemayehu A. Gorfe

Abstract

In this chapter, we provide a practical guide on how to plan, execute, and interpret atomistic and coarse-grained molecular dynamics (MD) simulations of lipid-modified proteins in model membranes. After outlining some key practical considerations when planning such simulations, we survey resources and techniques to obtain force field parameters for nonconventional amino acids, such as posttranslationally lipid-modified amino acids that are unique to this class of proteins. We then describe the protocols to build, setup, and run the simulations, followed by a brief comment on the analysis and interpretation of the simulations. Finally, examples of insights that could be gained from atomistic and coarse-grained MD simulations of lipidated proteins will be provided, using RAS proteins as illustrative examples. Throughout the chapter, we highlight the main advantages and limitations of simulating RAS and related lipid-modified G-proteins in biomimetic membranes.

Key words Lipidated proteins, Membrane, Lipid bilayer, Molecular dynamics simulation, G-proteins

1 Introduction

Biological membranes are the fundamental structural frameworks of the cell and have many diverse functions. They serve as a selectively permeable protective envelope of the cell, compartmentalize various organelles that have specialized functions, and provide a two-dimensional structural framework to organize signaling proteins and lipids. A wide variety of signaling proteins, such as the guanine triphosphate hydrolyzing family of enzymes (GTPases or G-proteins), are targeted to the plasma membrane (PM) or various internal membranes via lipid-based membrane-targeting motifs. Among these motifs, the most common are generated by posttranslational modification of glycine and cysteine amino acids by acyl and prenyl groups. The quintessential example of acetylated and prenylated G-proteins is the RAS family of small GTPases.

Irina S. Moreira et al. (eds.), *Computational Design of Membrane Proteins*, Methods in Molecular Biology, vol. 2315, https://doi.org/10.1007/978-1-0716-1468-6_9,

RAS proteins function as intracellular molecular switches to control a wide variety of signal transduction pathways. All RAS proteins, including HRAS, NRAS, and KRAS that represent the three most common RAS isoforms in humans, cycle between active and inactive states by binding to GTP and GDP. Active RAS interacts with multiple downstream effectors to drive cell growth and proliferation. Somatic mutations that impair hydrolysis of GTP to GDP stabilize RAS in its activated form, causing sustained proliferative signaling and oncogenesis. In fact, *RAS* genes are found mutated in nearly 30% of all cancers and are the primary drivers of many lethal cancers such as pancreatic, colon, and lung cancer [1]. Despite years of studies, however, pharmacologically targeting RAS has proved extremely tricky. A discussion of the challenges in drugging RAS is beyond the scope of this chapter but can be found in our recent review [2].

An aspect of the biology of RAS and related G-proteins that is gaining increasing attention for its therapeutic prospect is their dynamic localization to and fluctuations at membrane surfaces. Unfortunately, both the process of membrane targeting and the dynamics of the complex are difficult to study experimentally due to the resolution limit of current techniques. In contrast, the exponential growth in computational power over the last decade has begun to enable a detailed characterization of these phenomena using atomic-level molecular simulations. In particular, invaluable insights have been gained into the dynamics of RAS in complex with biomimetic membranes using molecular dynamics (MD) simulations. The results were instrumental in bringing RAS back to the forefront of anticancer drug discovery by, for example, showing how interactions of RAS with lipids, partner proteins, or ligands can be modulated by allosteric transitions and conformational selection [3–11].

In this chapter, we provide a practical guide on how to plan, setup, and conduct MD simulations of lipidated proteins in various lipid bilayers. Our discussion is largely based on our own experiences over the last decade, but we also highlight key contributions from other laboratories. We use RAS proteins for illustration because they are the best characterized and most representative example of proteins that are targeted to membranes by lipid-based motifs. The chapter is organized as follows: we first provide an overview of practical considerations to keep in mind during the planning stage of a simulation, followed by a survey of resources and approaches to obtain force field parameters for nonconventional amino acids. We then describe protocols for system construction and setup, and for running and analyzing MD simulations of lipid-modified proteins. This is followed by examples of insights that could be gained from atomistic and coarse-grained MD simulation of lipidated proteins using RAS proteins as an example. Strengths and limitations of MD simulations to study RAS and related proteins are highlighted throughout the chapter.

2 Methods

2.1 During Planning: Practical Considerations About System Size and Model Resolution

In a typical MD simulation, the position and momenta of each particle in the system is computed by integrating Newtonian equations of motion at discrete time intervals. This results in a series of snapshots that can be used to track the state of the system throughout the simulation and to characterize its dynamical or equilibrium properties. These capabilities and its high (atomic or semi-atomic) resolution made MD a powerful tool to study complex systems in great detail, such as monomeric or multimeric RAS in the membrane. A detailed discussion of MD is beyond the scope of this chapter and can be found elsewhere (e.g., [12, 13]). For our purposes, it suffices to note that there are two important issues to carefully consider during the planning stage of MD: system size and resolution. Because MD calculates the pairwise interactions between nearly all the particles in the system, it has an $O(N^2)$ complexity (actual number could be slightly lower due to the use of cutoff schemes for long-range vdW and electrostatic interactions). Therefore, performing MD on large systems at atomic resolution can be resource intensive.

1. One way to mitigate this challenge is to choose a model system with minimal complexity but sufficient to address a given set of specific questions, and simulating it for a sufficiently long duration to ensure convergences of the results. For example, in the case of RAS, a typical consideration for tradeoff between complexity and sampling efficiency would be to ask whether one should simulate the minimal membrane-targeting motif (the lipid anchor), the hypervariable region (HVR), or the full-length protein (Fig. 1). The answer almost always depends on the questions to be addressed by the simulation. The smaller system size made possible by the use of the isolated lipid anchor or the HVR allows for conducting multiple relatively long simulations with minimal computational cost. The goal of such a simulation could be to study the role of individual residues on these segments for membrane insertion. Depending on the simulation length and the resolution of the model (see below), another goal of simulating the isolated lipid anchor or the HVR could be to study their lateral dynamics and self-assembly in bilayers. Simulating the larger system associated with the full-length RAS comes at a higher computational cost, but would allow one to additionally study the external as well as internal dynamics and interactions of the catalytic domain with partner proteins or lipids.
2. A related issue is compositional complexity, which is especially important in membrane simulations because the more diverse the lipid species in the simulation box the longer it takes to

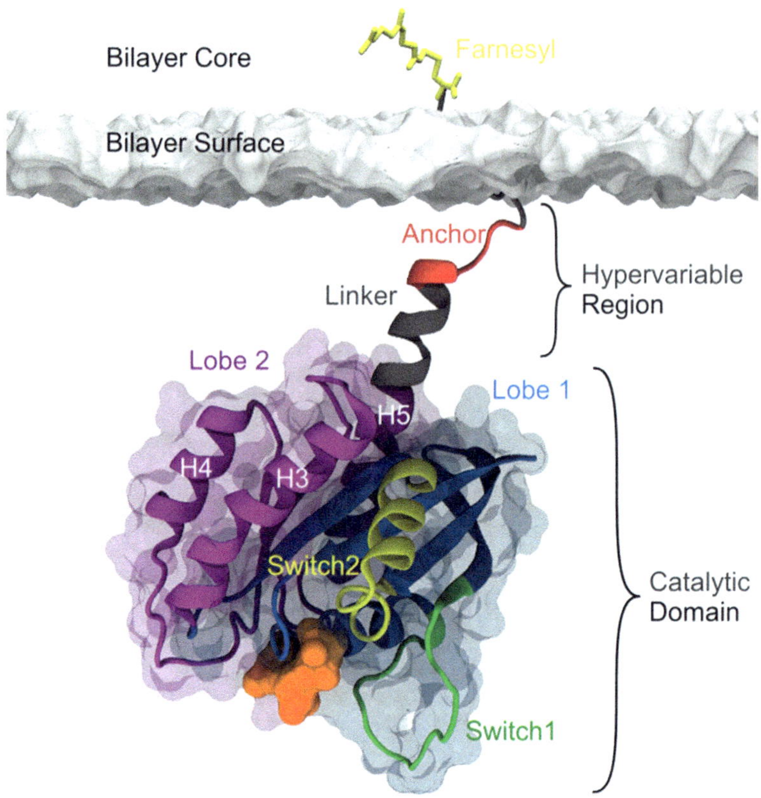

Fig. 1 Domain architecture and membrane targeting of lipid-modified small GTPases using the KRAS protein as example. The conserved catalytic domain (residues 1–166) has two lobes: Lobe 1 (residues 1–86) and Lobe 2 (residues 87–166) and harbors switch 1 (residues 25–40) and switch 2 (residues 60–75) regions that are involved in nucleotide (shown in orange), and effector and regulator binding. The hypervariable C-terminus (HVR, residues 167–185) consists of a flexible linker region and a farnesylated (yellow) polybasic (red) lipid anchor. A portion of hypothetical membrane is shown as a gray surface to illustrate tethering of lipid-modified proteins to one leaflet of a membrane. Actual organization of the catalytic domain and the HVR on membrane surfaces is more complicated, highly dynamic, and will depend on sequence and structural details

achieve equilibration. Here, too, one needs to carefully balance complexity with resource and time requirements to achieve the desired goal. A rule of thumb may be to choose—guided by prior knowledge of the system—the minimum number of lipid species required to address the question at hand. We will reiterate this issue later in this chapter in reference to simulating RAS proteins.

3. Another way of reducing computational cost is to use a coarse-grained (CG) model, wherein a group of physically connected non-hydrogen atoms (typically about four) are represented by a

single-reaction center. This reduces the degrees of freedom in the molecules as well as the number of particles to be simulated, and averages out the high-frequency fluctuations. As a result, CG models allow one to simulate larger systems for longer durations [12]. Coarse-grained MD simulation is particularly relevant to study long timescale processes such as RAS lateral dynamics and clustering in domain-forming bilayers.

2.2 Before Getting Started: Force Field Parameters for Posttranslationally Modified Amino Acids

In this section, we use RAS proteins as an example to provide a brief description of force field-related issues relevant for simulating lipidated proteins in both the all-atom (AA) and coarse-grained (CG) models. Force field parameters for AA-MD simulation of standard amino acids, lipids, and nucleic acids have been optimized over the decades and included in all of the popular force field distributions including CHARMM [14], AMBER [15–17], and GROMOS [18]. Similarly, parameters for the CG-MD simulation of standard amino acids and common lipids are available from various sources. Therefore, our discussion below focuses on the non-standard, post-translationally modified amino acids. Moreover, for brevity, we limit our discussion to the CHARMM force field [14] for AA-MD and the MARTINI [19] parameter set for CG-MD simulations, but the basic concepts discussed here also apply to other popular force fields.

1. AA force field parameters for the most common lipidations, such as palmitoylation, myristoylation, and farnesylation, are now available in the CHARMM36 force field [20] (http://mackerell.umaryland.edu/charmm_ff.shtml). CHARMM-GUI (http://www.charmm-gui.org) may represent the easiest pipeline to set up AA-MD simulations using these parameters. However, there may be situations where simulations need to be setup from scratch or the desired lipidated amino acid is not yet parameterized. Parameters compatible with an earlier version of the CHARMM force field (CHARMM27) have been developed for farnesylated and palmitoylated cysteine using analogy to phospholipid and cysteine parameters [21, 22]. Missing parameters for lipidated amino acids that are compatible with the latest force fields can be prepared using either an *ab initio*-based or a semiempirical approach. The *ab initio* approach entails geometry optimization using quantum mechanical (QM) calculations to obtain partial charges, bond lengths, and angles. Such an approach has been used to obtain parameters for farnesyl [23] using the QM software Gaussian [24], and a more recent study described how to do so using NWCHEM [25] in a manner that can serve as a useful guide to parameterize other lipidated amino acids [26]. The semiempirical approach uses already parameterized analogous molecules as a reference to assign parameters to new molecules,

followed by iterative optimization steps to reproduce experimental observations. This technique is most successfully implemented in the CHARMM Generalized Force Field (CGenFF) toolkit. CGenFF is primarily designed to generate parameters for small molecule ligands [27], but with little extra effort it could be used to generate parameters for farnesylated [28, 29] or other lipidated amino acids. Also, the force field ToolKit (ffTK) plugin in VMD (Visual Molecular Dynamics: https://www.ks.uiuc.edu/Research/vmd/) employs the same basic principle to generate parameters for small molecules and may be used to parameterize lipidated amino acids [30]. A detailed tutorial for ffTK can be found at https://www.ks.uiuc.edu/Research/vmd/plugins/fftk/.

2. In addition to lipid modification, mono- or poly-ubiquitination is common in many proteins including RAS, which has been shown to be ubiquitinated at multiple lysine residues such as Lys104, Lys147, and Lys170 [31]. These modifications may affect the dynamics and/or interactions of RAS with lipids, but there are no ready-to-use parameters for ubiquitinated lysine. Lysine ubiquitination entails formation of an isopeptide bond between the C-terminal glycine of ubiquitin and the lysine side chain. We recently used analogy to parameterize ubiquitin-modified Lys [84]. Briefly, a modified lysine residue, Lyq, was defined with the backbone NH2 atom type of Lys replaced by NH and the positive charge on the HZ2 atom redistributed to the neighboring ε-CH2 atoms. A covalent bond was then defined between the carboxy-terminal carbon of a modified glycine, Glq, and the side chain NH atom of Lyq. The bond length and corresponding angles were defined by analogy to backbone peptide bonds.
3. As in the atomistic models, efforts have been made to develop parameters for the CG-MD simulation of proteins harboring lipidated amino acids. A notable example in this regard is the MARTINI model. In this CG model, on average four covalently linked heavy atoms and associated hydrogens are represented by a single-interaction center or bead. Depending on the polarity of the atoms involved, the bead may be polar, intermediate, apolar, or charged (see http://cgmartini.nl/ for more detail). Parameters for palmitoylated and farnesylated cysteines compatible with an earlier version of MARTINI were developed using analogy and used to simulate relatively large RAS/membrane complexes for up to 40 μs, yielding invaluable insights into the mechanisms of RAS clustering and domain partitioning [32–34]. Following an update of the force field (MARTINI 2.2) to more accurately describe, for example, membrane partitioning of peptides [35], newer parameters for lipidated amino acids including prenylated cysteine

have been developed [36]. The protocol described in ref. [36] can be used to parameterize those not already parameterized.

2.3 Getting Started: System Construction and Simulation Setup

With the appropriate force field parameters in hand, one can begin to build an initial configuration of the system to setup the simulation.

1. *Preparing the protein*: For most lipidated proteins including RAS, the structure of the highly dynamic lipid anchor and a number of residues in its vicinity is unavailable in the Protein Data Bank (PDB). In fact, almost all available RAS structures consist only of the catalytic G-domain. Therefore, the structure of the lipid anchor and the linker connecting it to the catalytic domain need to be modeled, typically as an extended structure. This can be done using any of the many different structure prediction algorithms such as MODELLER [37], the Molefacture plugin in VMD [38], the Build Structure functionality in UCSF-Chimera [39], or PyMol [40]. If needed, the resulting structure can be energy-minimized and subjected to a short MD simulation using standard protocols. Then, it is ligated to the structure of the catalytic domain downloaded from the PDB to build the full-length protein structure (see for example ref. [22]); missing residues in flexible regions of the PDB structure, if any, can be modeled using the tools mentioned above. Alternatively, the full-length protein structure may be directly built using homology modeling programs such as MODELLER (https://salilab.org/modeller/tutorial/). The isolated lipid anchor or the full-length protein can then be prepared for simulation after appropriate patching for disulfide bonds (if any) or chain termini (negative, positive, or neutral). For example, RAS has an oxymethylated C-terminus, which can be added to the structure by selecting CT1 as a C-terminal patch. The resulting structure may then be subjected to energy-minimization and a short equilibration to ensure integrity.
2. *Preparing the bilayer*: In parallel, a bilayer patch of the desired size and lipid composition is prepared either from a pre-equilibrated solvated bilayer or from scratch (e.g., using CHARMM-GUI (http://www.charmm-gui.org/?doc=input/membrane.bilayer) [41, 42] or similar other resources). The size of the bilayer patch or the number of lipids is determined by the nature (e.g., isolated lipid anchor versus full-length) and the number of the protein to be simulated. In general, the bilayer should be large enough to accommodate the full extent of anticipated conformational fluctuations of the protein in a manner that avoids boundary effects. The nature of the protein to be simulated and its target membrane in the cell also dictate the choice of the lipid composition in the bilayer.

For example, the PM has a different lipid composition from the generally more curved internal membranes such as the endoplasmic reticulum. However, the amino acid composition is of the protein that interacts with lipids is perhaps more critical for the choice of lipids in the model membrane. For example, while all three most common human RAS proteins are targeted to the inner leaflet of the PM, the prenylated polybasic lipid anchor of KRAS, but not the palmitoylated and prenylated HRAS or NRAS, preferentially interacts with anionic phospholipids such as POPS (1-palmitoyl-2-oleoyl-sn-glycero-3-phospho-L-serine), which is enriched at the inner leaflet of the PM. Therefore, KRAS is typically simulated in an anionic membrane such as a POPC/POPS (20–30%) bilayer while NRAS and HRAS can be simulated in a pure POPC bilayer. Once the appropriate bilayer model is constructed, it can be subjected to standard energy-minimization steps to relieve bad atomic contacts and equilibrated through a series of restrained and unrestrained MD runs to obtain an equilibrated bilayer model satisfying the expected measures of bilayer structural integrity, such as bilayer thickness and area per lipid.

3. *Building and solvating the protein–membrane system*: The protein (or peptide) and membrane models prepared as described above can be assembled into a single system using, for example, the protein structure file generator (PSFGEN) program (https://www.ks.uiuc.edu/Research/vmd/plugins/psfgen/) and VMD. This is typically followed by a series of manual or automated translations and rotations of the protein relative to the bilayer so that the soluble portion remains away from the bilayer, and the lipid anchor is placed near the bilayer surface oriented in such a way as to increase the likelihood of its membrane insertion. At this stage, the system can be solvated (e.g., using the *Solvate* plugin in VMD and the TIP3P water model) and ions and, as needed, salt are added to obtain a charge-neutralized system having the desired ionic strength (such as the physiologic ionic strength of 150 mM). In principle, the resulting system can now be energy-minimized, equilibrated, and simulated following standard protocols. In practice, however, the available search space and the barrier at the lipid–water interface make the spontaneous insertion of the lipid anchor into the bilayer costly in terms of both resource and time. A quicker work around is to slowly pull one or more of the lipid chains on the protein (e.g., using a soft harmonic potential applied to the terminal carbon and the bilayer center) during an MD run until a portion of the hydrocarbon chain (about five carbon atoms) reaches the hydrophobic core of the bilayer [21]. This would allow for the hydrophobic lipid anchor to cross the energy barrier at the lipid–water interface so that, once released, the rest of the anchor inserts spontaneously. This

protocol has proven highly effective in a number of previous simulations. Alternatively, a portion of the lipid anchor can be manually inserted into the bilayer using a series of translations and rotations, coupled with deletion or displacement of water and/or lipid molecules that overlap with protein atoms, followed by multiple steps of energy-minimization and equilibration runs. Some of these procedures can nowadays be done more easily using CHARMM-GUI [43]. These include building a bilayer of the desired lipid composition, adding a lipid modification to an otherwise complete protein structure (using the Add Lipid Tail option), and placing the protein in proximity to or in the membrane. One could also orient a desired surface of the protein toward the bilayer or position it at a desired distance from the membrane. Finally, if the protein were to be tethered only to one leaflet, the surface area of that leaflet may become larger than the opposing leaflet, causing membrane bending. This can be avoided fairly easily by adding the appropriate number of extra lipids in the protein-free leaflet or by removing from the leaflet housing the protein until the monolayer areas are equalized.

4. *Converting to a CG- model*: A fully atomistic protein–membrane–water system constructed as described above and, preferably, subjected to an AA-MD simulation can serve as a starting structure for a CG-MD simulation. For the MARTINI CG model, this can be achieved through AA-to-CG mapping using the *martinize.py* tool available in the MARTINI website (http://cgmartini.nl/). The CG-MD simulation can directly commence from this step if the target is the minimal membrane anchor, which in most lipidated proteins is devoid of a secondary structure. For the full-length protein, elastic network-based restraints need to be applied on the secondary structure elements to maintain the integrity of the tertiary structure. This can be done using the *ElNeDyn* utility [44]. Because a CG approach is primarily chosen to gain access to more complicated processes such as lipid domain formation, the bilayer lipid composition is typically more complex than in an AA model. In fact, the bilayer may be made up of two to tens of lipid species [45, 46] and can be simulated alone for reference or with the protein of interest bound. Moreover, it is often desirable to have multiple copies of the same protein (e.g., to increase sampling or to study self-assembly), or to have different proteins of varying numbers (e.g., to study the formation of signaling complexes). Construction of such a complicated system can be achieved using the *insane* utility, which allows the user to generate bilayers of essentially any lipid composition (provided that the parameters exist) and to insert the protein (s) of interest to any desired depth [47].

2.4 Running the Simulation and Analyzing the Data

Once the system is properly setup, the production phase of the simulations can be conducted following standard protocols. The AA-MD simulation can be run with NAMD [48], AMBER [49], or GROMACS [50] just to name a few, while the MARTINI-based CG-MD is typically run with GROMACS. In each case, no significant system-specific adjustments are required during the production phase of simulating lipidated proteins, with the possible exception of the need to regularly monitor insertion and equilibration of the lipid anchor in the bilayer. Similarly, most of the protocols and tools used to analyze MD trajectories of other protein–membrane complexes are directly applicable to lipidated proteins. Additional analysis techniques that are somewhat specific to surface-bound proteins, such as defining appropriate reaction coordinates to study orientations of the catalytic domain on membrane surfaces, can be found in several recent reports [51–54]. Limitations associated with force fields or the sampling of phase space that are typical of other protein–lipid complexes also apply to RAS and related proteins. A possible exception here may be the need to run multiple copies for as long as possible. This is because, first, the simulations are often started from structures that lack the flexible region connecting the catalytic domain with the lipid anchor. This requires extensive sampling to achieve convergence and internal consistency of the results, for example, in terms of conformations sampled and distributions of residue-lipid interactions. Second, this class of proteins is highly dynamic and tends to adopt many diverse transient conformations that may not be captured in a single-short simulation. Third, there are accumulating evidences suggesting that many lipidated proteins, especially the RAS family of small GTPases, may form transient oligomers for which experimental data for benchmarking is scarce or entirely unavailable. Therefore, simulation of the oligomerization process of lipidated proteins on membrane surfaces, often conducted using CG-MD for sampling efficiency, should ideally be coupled with new experiments however limited or indirect, and the results interpreted with care.

2.5 What Can We Learn from the Simulations? Some Illustrative Examples

In this section, we briefly review some of the important and likely broadly applicable observations from previous atomistic or coarse-grained MD simulations of RAS proteins conducted in our laboratory and those of others. Although not discussed here to save space, we note that MD simulations have been applied to a number of other RAS-like lipidated small GTPases [55–58].

1. *AA-MD simulation of the RAS lipid anchors:* RAS localizes to the PM via an S-farnesyl cysteine carboxylmethyl ester (common to all RAS proteins), complemented by palmitoylation of proximal cysteine(s) (in NRAS and HRAS) or a polybasic domain of six contiguous lysine residues (KRAS). Prior et al.

[59] identified residues 175–185 as the minimal membrane-targeting motif of KRAS and residues 180–186 as those of HRAS or NRAS. In the cell, these peptides (referred to as tK, tH, and tN) are targeted to the PM in a manner that generally recapitulates the full-length cognate [60]. These motifs can therefore be used as model systems in MD simulations to study the structural and physiochemical factors underlying RAS insertion into and dynamics at membranes. In one of the earliest such studies, AA-MD simulation was conducted on tN in a 1,2-dimyristoyl-sn-glycero-3-phosphocholine (DMPC) bilayer [21]. It was found that the insertion of about five terminal carbon atoms of either the palmitoyl at Cys181 or the farnesyl at Cys186 (modeled by a hexadecyl) into the bilayer core is sufficient to cross the barrier at the bilayer–water interface. Once the barrier is crossed, the rest of the peptide spontaneously partitioned into the bilayer, with tN lipids immersing into the hydrophobic core dragging with them the hydrophobic side chains of Met182 and Leu184; hydrogen bonds between the peptide backbone and lipid head groups provided additional stabilization [21]. These observations were consistent with and helped explain data from a combined FTIR, solid-state NMR, and neutron diffraction spectroscopy analysis [61]. A similar study of tH yielded the same results: the farnesylated (Cys186) and palmitoylated (Cys181 and Cys184) cysteines plus Met182 are engaged in vdW interactions with lipid acyl chains while the peptide backbone and polar side chains form hydrogen bonds with lipid head groups [62]. Similarly, an AA-MD study of tK in a POPC/POPG (2-oleoyl-1-palmitoyl-sn-glycero-3-phosphoglycerol) bilayer showed that insertion of the farnesyl tail and electrostatic interactions of the lysine residues with the anionic POPG lipids together stably tether tK to the membrane [63]. This observation was supported by more recent extended simulations of tK in a physiologically more relevant POPC/POPS bilayer [64] and experiments in cells [64]. The key role of electrostatics in tK-membrane association was further demonstrated by the observation that its phosphorylation of Ser181 decreases membrane affinity [65].

AA-MD simulations were also used to examine the energetics of tH-membrane interaction [66]. For example, potential of mean force calculations of tH insertion into a DMPC lipid bilayer using the adaptive biasing force (ABF) technique indicated a steep decline in the free energy profile after insertion of a few carbon atoms of the lipid anchor, corroborating an earlier observation [21, 62]. Follow up studies to investigate the relative role of individual lipid modifications and the thermodynamics of tH-DMPC interactions arrived at two important conclusions: (a) the palmitoyl modification at Cys181 contributes more to the insertion free energy than that at Cys184 [67], and (b) tH-DMPC binding is dominated by an enthalpy-driven hydrophobic effect resulting primarily from the vdW interactions between tH and membrane lipids [68].

2. *AA-MD simulation of full-length RAS proteins:* The first AA-MD simulation of full-length RAS was performed on HRAS in complex with a DMPC bilayer, and the results were compared with simulations of tH and the HVR (the hypervariable region, residues 167–186) [22]. It was found that although the residues at the lipid anchor are the primary determinants of membrane binding, the rest of the protein is not a passive spectator. In particular, in the context of the full-length protein, a pair of positively charged residues at the HVR (Arg169 and Lys170) or the catalytic domain (Arg128 and Arg135) interacts with lipids directly but not simultaneously. The catalytic domain adopting two distinct orientations with respect to the membrane plane made this possible. During similar simulations of full-length KRAS in a DMPC bilayer, the catalytic domain was highly dynamic and did not adopt a stable orientation [51, 69]. This was because KRAS requires anionic lipids for a stable membrane binding, as demonstrated by more recent extended simulations of KRAS in anionic lipid bilayers [51, 54]. One of these studies, conducted in the more physiologically relevant POPC/POPS bilayer and involving over 10 MD runs of up to 1 μs each, identified multiple conformational states in which the catalytic domain interacts with the membrane [51]. These included orientation state 1 (OS1) in which the three C-terminal α-helices face the bilayer and orientation state 2 (OS2) in which β-strands 1–3 and α-helix 2 are near the bilayer. In the latter, the effector binding switch loops are mostly occluded from solvent (and hence partner proteins) by the membrane [51]. Subsequent 20 μs-long AA-MD simulations on three different mutants of KRAS showed that [52, 53]: (a) transitions between OS1 and OS2 occur through an intermediate orientation OS0, (b) the signaling-competent (non-occluded) S0 and OS1 orientations dominate the overall population of orientation states, and (c) the orientational dynamics is underpinned largely by intrinsic conformational fluctuations.

While most AA-MD simulations including those previously conducted on RAS and discussed above were limited to tens of microseconds at best, a recent simulation of KRAS in a POPC and POPC/POPS bilayers achieved a millisecond aggregate time by running 290 copies of 5 μs-long simulations [54]. The overall conclusions of the study regarding protein–lipid interactions and KRAS membrane reorientation were consistent with the previous observations. However, the enormous data generated by the study enabled the quantification of the extent to which POPS slows down the dynamics of KRAS on the bilayer. It was found that, compared with a pure POPC bilayer, the presence of 30% POPS slowed down the translational diffusion of KRAS by a factor of ~2 and the likelihood of its disengagement from the membrane by a factor of 8. Other AA-MD simulations of membrane-bound RAS focused on characterizing protein–protein interaction interfaces (PPIs) involved in RAS dimerization [70–73] or its interaction with downstream effectors Raf [74] and PI3K [75].

3. *CG-MD simulation of Ras proteins:* Imaging and other cell biology experiments had shown that RAS proteins form spatially segregated nanoclusters on the membrane, but the structural details and physicochemical basis of this process remained elusive [60]. In one of the earliest CG-MD studies of RAS, 64 tH molecules were arrayed in a domain-forming bilayer made up of dipalmitoylphosphatidylcholine (DPPC) and dilinoleoylphosphatidylcholine (DLiPC) plus varying concentrations of cholesterol, and simulated at different temperatures for up to 40 μs effective time [34]. It was found that tH spontaneously assembles into clusters of 4–10 molecules. The clusters preferentially localized to the boundary between the liquid ordered (lo) and liquid disordered (ld) phases driven primarily by the affinity of the saturated palmitoyl and poly-unsaturated farnesyl moieties for DMPC and DLiPC, respectively. Using the same CG-MD model, subsequent studies investigated the effect of tH concentration and lipid composition on the formation [33] and stability [76] of tH nanoclusters. These simulations predicted a minimal concentration of tH below which nanoclustering may not occur and a saturating concentration beyond which increasing tH concentration may not result in changes in cluster size. Changes in cholesterol concentration did not affect the size of the nanoclusters but rather their stability through its impact on the stability of membrane domains. This is because, as noted above, tH nanoclusters tend to accumulate at lipid domain boundaries whose stability and size is a function of cholesterol content. Moreover, it was found that the accumulation of asymmetrically bound tH [32] and full-length HRAS [32, 77] nanoclusters at domain boundaries reduces the boundary line tension and stabilizes membrane curvature, largely through monolayer area expansion. This has been further demonstrated using the mesoscale dissipative particle dynamics simulation technique applied to monomeric and variously cross-linked simplified model systems that partitioned to the domain boundary of a small unilamellar vesicle [78].

4. *MD simulations for the identification of interaction hot spots:* The examples described in the previous subsections illustrate the most widely used applications of MD in the study of lipid-modified proteins. Another appealing application is probe-based MD (pMD), a solvent mapping technique that uses small organic molecules as probes to identify cryptic ligand binding sites for therapeutic targeting. When this technique was applied to the soluble catalytic domain of KRAS [79], it was found to be effective in identifying all previously characterized allosteric pockets [6]. However, most of the commonly used probes, such as isopropyl alcohol, tend to partition into membranes, limiting the application of pMD only to soluble proteins. To overcome this, a modified version of pMD called *pMD-membrane* has been proposed [80] wherein the affinity of probe molecules for the bilayer core is reduced by modifying selected probe–lipid pairwise vdW interactions. This can be done fairly easily using the correction term NBFIX for simulations with the CHARMM force field. Typically, the well depth of the Lennard-Jones potential is reduced to about 0.01 kcal/mol, and the minimum interparticle distance increased to about 7 Å. This prevents the probes from penetrating the hydrophobic core of the bilayer while providing access to the side of the protein facing the bilayer. A pMD-membrane simulation starts by solvating the bilayer-bound protein with water containing ~20% probe and adding K^+, Na^+, or Cl^- ions to neutralize the system and to achieve a desired ionic strength. The system is energy-minimized and the probes homogenized in the bulk water using simulated annealing with the protein heavy atoms harmonically restrained to prevent unfolding. The rest of the simulation follows standard protocols and can be run in multiple copies for several hundred nanoseconds to achieve convergence. The combined trajectories can then be analyzed using various techniques, including a recently introduced technique based on surface topography maps [81], to determine probe occupancy in putative binding pockets. Application of pMD-membrane on KRAS bound to a POPC/POPS bilayer in various orientations [80, 81] suggested that the way in which the catalytic domain interacts with lipids significantly affects the probe accessibility of druggable pockets, suggesting a similar potential impact on drug binding.

3 Summary and Future Prospects

Elucidating the interaction of surface-bound lipid-modified proteins with membranes presents multiple difficulties to experimental techniques due to the complex dynamics of these systems in space and time. Molecular dynamics simulations conducted at different levels of detail can help fill this gap, as demonstrated in this chapter using RAS proteins as an example. Advances in computer power and the continuing development of force field parameters and analysis tools mean that MD simulations are poised to address a wide range of questions whose answer may lie in visualizing how lipidated proteins engage membrane lipids. We hope that the protocols for parameterization, system construction, system setup, and simulation described in this chapter will facilitate the application of MD in its various flavors to any lipid-modified protein monomer or multimer tethered to a model membrane whose lipid composition mimics the protein's biological host membrane. Note that it is possible to draw some general principles by studying a few representatives from the different families of lipidated proteins, as we have attempted to show using RAS proteins. However, the substantial diversity of lipid-modified signaling proteins in sequence/structure, function, subcellular localization, posttranslational modification, and pathophysiology suggests that we have barely scratched the surface. Therefore, MD simulation will continue to play an indispensable role in shining light on the many hidden mysteries behind the seemingly simple domain architecture of lipidated signaling proteins. The following are two examples of interesting future applications: The PM is characterized by an asymmetric distribution of lipids [82]; in particular, the intracellular side is enriched by POPS and therefore negatively charged. Further, the differential distribution of ions across the PM generates a transmembrane potential. Both of these may modulate interaction of surface-bound proteins with lipids and/or the orientation, lateral dynamics, and assembly of the proteins on membrane surfaces. Therefore, simulating lipidated proteins tethered to an asymmetric bilayer using classical AA and CG models on the one hand and a polarizable force field (see ref. [83] for a review) on the other would be an appealing future work.

Acknowledgments

This work was supported in part by the National Institutes of Health grant R01GM124233 and the Cancer Prevention and Research Institute of Texas (CPRIT) grant RP190366. V.N. is supported by UTHealth Innovation for Cancer Prevention Research Training Program Pre-Doctoral Fellowship (Cancer Prevention and Research Institute of Texas grant RP160015).

References

1. Prior IA, Lewis PD, Mattos C (2012) A comprehensive survey of Ras mutations in cancer. Cancer Res 72(10):2457–2467. https://doi.org/10.1158/0008-5472.CAN-11-2612
2. Gorfe AA, Cho KJ (2019) Approaches to inhibiting oncogenic K-Ras. Small GTPases 12 (2):96-105. https://doi.org/10.1080/21541248.2019.1655883
3. Prakash P, Gorfe AA (2013) Lessons from computer simulations of Ras proteins in solution and in membrane. Biochim Biophys Acta 1830(11):5211–5218. https://doi.org/10.1016/j.bbagen.2013.07.024
4. Parker JA, Mattos C (2018) The K-Ras, N-Ras, and H-Ras isoforms: unique conformational preferences and implications for targeting oncogenic mutants. Cold Spring Harb Perspect Med 8(8):a031427. https://doi.org/10.1101/cshperspect.a031427
5. Zhou Y, Prakash P, Gorfe AA, Hancock JF (2018) Ras and the plasma membrane: a complicated relationship. Cold Spring Harb Perspect Med 8(10):a031831. https://doi.org/10.1101/cshperspect.a031831
6. Grant BJ, Lukman S, Hocker HJ, Sayyah J, Brown JH, McCammon JA, Gorfe AA (2011) Novel allosteric sites on Ras for lead generation. PLoS One 6(10):e25711. https://doi.org/10.1371/journal.pone.0025711
7. Grant BJ, McCammon JA, Gorfe AA (2010) Conformational selection in G-proteins: lessons from Ras and Rho. Biophys J 99(11): L87–L89. https://doi.org/10.1016/j.bpj.2010.10.020
8. Grant BJ, Gorfe AA, McCammon JA (2009) Ras conformational switching: simulating nucleotide-dependent conformational transitions with accelerated molecular dynamics. PLoS Comput Biol 5(3):e1000325. https://doi.org/10.1371/journal.pcbi.1000325
9. Gorfe AA, Grant BJ, McCammon JA (2008) Mapping the nucleotide and isoform-dependent structural and dynamical features of Ras proteins. Structure 16(6):885–896. https://doi.org/10.1016/j.str.2008.03.009
10. Grant BJ, Gorfe AA, McCammon JA (2010) Large conformational changes in proteins: signaling and other functions. Curr Opin Struct Biol 20(2):142–147. https://doi.org/10.1016/j.sbi.2009.12.004
11. Hocker HJ, Cho KJ, Chen CY et al (2013) Andrographolide derivatives inhibit guanine nucleotide exchange and abrogate oncogenic Ras function. Proc Natl Acad Sci U S A 110 (25):10201–10206. https://doi.org/10.1073/pnas.1300016110
12. Jefferies D, Khalid S (2021) Atomistic and coarse-grained simulations of membrane proteins: a practical guide. Methods 185:15–27. https://doi.org/10.1016/j.ymeth.2020.02.007
13. Hug S (2013) Classical molecular dynamics in a nutshell. Methods Mol Biol 924:127–152. https://doi.org/10.1007/978-1-62703-017-5_6
14. Zhu X, Lopes PE, Mackerell AD Jr (2012) Recent developments and applications of the CHARMM force fields. Wiley Interdiscip Rev Comput Mol Sci 2(1):167–185. https://doi.org/10.1002/wcms.74
15. Cheatham TE 3rd, Case DA (2013) Twenty-five years of nucleic acid simulations. Biopolymers 99(12):969–977. https://doi.org/10.1002/bip.22331
16. Dickson CJ, Madej BD, Skjevik AA et al (2014) Lipid14: the Amber lipid force field. J Chem Theory Comput 10(2):865–879. https://doi.org/10.1021/ct4010307
17. Maier JA, Martinez C, Kasavajhala K et al (2015) ff14SB: improving the accuracy of protein side chain and backbone parameters from ff99SB. J Chem Theory Comput 11 (8):3696–3713. https://doi.org/10.1021/acs.jctc.5b00255
18. Schmid N, Eichenberger AP, Choutko A et al (2011) Definition and testing of the GROMOS force-field versions 54A7 and 54B7. Eur Biophys J 40(7):843–856. https://doi.org/10.1007/s00249-011-0700-9
19. Marrink SJ, Risselada HJ, Yefimov S et al (2007) The MARTINI force field: coarse grained model for biomolecular simulations. J Phys Chem B 111(27):7812–7824. https://doi.org/10.1021/jp071097f
20. Huang J, AD MK Jr (2013) CHARMM36 all-atom additive protein force field: validation based on comparison to NMR data. J Comput Chem 34(25):2135–2145. https://doi.org/10.1002/jcc.23354
21. Gorfe AA, Pellarin R, Caflisch A (2004) Membrane localization and flexibility of a lipidated Ras peptide studied by molecular dynamics simulations. J Am Chem Soc 126 (46):15277–15286. https://doi.org/10.1021/ja046607n
22. Gorfe AA, Hanzal-Bayer M, Abankwa D et al (2007) Structure and dynamics of the full-length lipid-modified H-Ras protein in a

1,2-dimyristoylglycero-3-phosphocholine bilayer. J Med Chem 50(4):674–684. https://doi.org/10.1021/jm061053f

23. Jang H, Abraham SJ, Chavan TS et al (2015) Mechanisms of membrane binding of small GTPase K-Ras4B farnesylated hypervariable region. J Biol Chem 290(15):9465–9477. https://doi.org/10.1074/jbc.M114.620724
24. Frisch MJ, Trucks GW, Schlegel HB et al (2016) Gaussian 16 Rev. C.01. Gaussian, Inc., Wallingford, CT
25. Valiev M, Bylaska EJ, Govind N et al (2010) NWChem: a comprehensive and scalable open-source solution for large scale molecular simulations. Comput Phys Commun 181 (9):1477–1489. https://doi.org/10.1016/j.cpc.2010.04.018
26. Neale C, Garcia AE (2018) Methionine 170 is an environmentally sensitive membrane anchor in the disordered HVR of K-Ras4B. J Phys Chem B 122(44):10086–10096. https://doi.org/10.1021/acs.jpcb.8b07919
27. Vanommeslaeghe K, Hatcher E, Acharya C et al (2010) CHARMM general force field: a force field for drug-like molecules compatible with the CHARMM all-atom additive biological force fields. J Comput Chem 31 (4):671–690. https://doi.org/10.1002/jcc.21367
28. Li Z, Buck M (2020) Computational design of myristoylated cell-penetrating peptides targeting oncogenic K-Ras.G12D at the effector-binding membrane interface. J Chem Inf Model 60(1):306–315. https://doi.org/10.1021/acs.jcim.9b00690
29. Li Z-L, Buck M (2017) Computational modeling reveals that signaling lipids modulate the orientation of K-Ras4A at the membrane reflecting protein topology. Structure 25 (4):679–689.e672. https://doi.org/10.1016/j.str.2017.02.007
30. Mayne CG, Saam J, Schulten K, Tajkhorshid E et al (2013) Rapid parameterization of small molecules using the force field toolkit. J Comput Chem 34(32):2757–2770. https://doi.org/10.1002/jcc.23422
31. Sasaki AT, Carracedo A, Locasale JW et al (2011) Ubiquitination of K-Ras enhances activation and facilitates binding to select downstream effectors. Sci Signal 4(163):ra13. https://doi.org/10.1126/scisignal.2001518
32. Li Z, Gorfe AA (2013) Deformation of a two-domain lipid bilayer due to asymmetric insertion of lipid-modified Ras peptides. Soft Matter 9(47):11249–11256. https://doi.org/10.1039/C3SM51388B
33. Li Z, Janosi L, Gorfe AA (2012) Formation and domain partitioning of H-ras peptide nanoclusters: effects of peptide concentration and lipid composition. J Am Chem Soc 134 (41):17278–17285. https://doi.org/10.1021/ja307716z
34. Janosi L, Li Z, Hancock JF, Gorfe AA (2012) Organization, dynamics, and segregation of Ras nanoclusters in membrane domains. Proc Natl Acad Sci U S A 109(21):8097–8102. https://doi.org/10.1073/pnas.1200773109
35. de Jong DH, Singh G, Bennett WF et al (2013) Improved parameters for the Martini coarse-grained protein force field. J Chem Theory Comput 9(1):687–697. https://doi.org/10.1021/ct300646g
36. Atsmon-Raz Y, Tieleman DP (2017) Parameterization of palmitoylated cysteine, farnesylated cysteine, geranylgeranylated cysteine, and myristoylated glycine for the Martini force field. J Phys Chem B 121 (49):11132–11143. https://doi.org/10.1021/acs.jpcb.7b10175
37. Webb B, Sali A (2016) Comparative protein structure modeling using MODELLER. Curr Protoc Protein Sci 86:2 9 1–2 9 37. https://doi.org/10.1002/cpps.20
38. Humphrey W, Dalke A, Schulten K (1996) VMD: visual molecular dynamics. J Mol Graph 14(1):33–38, 27-38. https://doi.org/10.1016/0263-7855(96)00018-5
39. Pettersen EF, Goddard TD, Huang CC et al (2004) UCSF chimera—a visualization system for exploratory research and analysis. J Comput Chem 25(13):1605–1612. https://doi.org/10.1002/jcc.20084
40. The PyMOL Molecular Graphics System, Version 1.2r3pre, Schrödinger, LLC
41. Lee J, Cheng X, Swails JM, Yeom MS et al (2016) CHARMM-GUI input generator for NAMD, GROMACS, AMBER, OpenMM, and CHARMM/OpenMM simulations using the CHARMM36 additive force field. J Chem Theory Comput 12(1):405–413. https://doi.org/10.1021/acs.jctc.5b00935
42. Jo S, Kim T, Iyer VG et al (2008) CHARMM-GUI: a web-based graphical user interface for CHARMM. J Comput Chem 29 (11):1859–1865. https://doi.org/10.1002/jcc.20945
43. Wu EL, Cheng X, Jo S et al (2014) CHARMM-GUI membrane builder toward realistic biological membrane simulations. J Comput Chem 35(27):1997–2004. https://doi.org/10.1002/jcc.23702

44. Periole X, Cavalli M, Marrink SJ et al (2009) Combining an elastic network with a coarse-grained molecular force field: structure, dynamics, and intermolecular recognition. J Chem Theory Comput 5(9):2531–2543. https://doi.org/10.1021/ct9002114
45. Ingolfsson HI, Carpenter TS, Bhatia H et al (2017) Computational lipidomics of the neuronal plasma membrane. Biophys J 113 (10):2271–2280. https://doi.org/10.1016/j.bpj.2017.10.017
46. Ingolfsson HI, Melo MN, van Eerden FJ et al (2014) Lipid organization of the plasma membrane. J Am Chem Soc 136 (41):14554–14559. https://doi.org/10.1021/ja507832e
47. Wassenaar TA, Ingolfsson HI, Bockmann RA et al (2015) Computational lipidomics with *Insane*: a versatile tool for generating custom membranes for molecular simulations. J Chem Theory Comput 11(5):2144–2155. https://doi.org/10.1021/acs.jctc.5b00209
48. Phillips JC, Braun R, Wang W et al (2005) Scalable molecular dynamics with NAMD. J Comput Chem 26(16):1781–1802. https://doi.org/10.1002/jcc.20289
49. Case DA, Cheatham TE 3rd, Darden T et al (2005) The Amber biomolecular simulation programs. J Comput Chem 26 (16):1668–1688. https://doi.org/10.1002/jcc.20290
50. Van Der Spoel D, Lindahl E, Hess B et al (2005) GROMACS: fast, flexible, and free. J Comput Chem 26(16):1701–1718. https://doi.org/10.1002/jcc.20291
51. Prakash P, Zhou Y, Liang H et al (2016) Oncogenic K-Ras binds to an anionic membrane in two distinct orientations: a molecular dynamics analysis. Biophys J 110(5):1125–1138. https://doi.org/10.1016/j.bpj.2016.01.019
52. Prakash P, Gorfe AA (2019) Probing the conformational and energy landscapes of KRAS membrane orientation. J Phys Chem B 123 (41):8644–8652. https://doi.org/10.1021/acs.jpcb.9b05796
53. Prakash P, Litwin D, Liang H et al (2019) Dynamics of membrane-bound G12V-KRAS from simulations and single-molecule FRET in native nanodiscs. Biophys J 116 (2):179–183. https://doi.org/10.1016/j.bpj.2018.12.011
54. Neale C, Garcia AE (2020) The plasma membrane as a competitive inhibitor and positive allosteric modulator of KRas4B signaling. Biophys J 118(5):1129–1141. https://doi.org/10.1016/j.bpj.2019.12.039
55. Edler E, Stein M (2017) Probing the druggability of membrane-bound Rab5 by molecular dynamics simulations. J Enzyme Inhib Med Chem 32(1):434–443. https://doi.org/10.1080/14756366.2016.1260564
56. Prakash P, Gorfe AA (2017) Membrane orientation dynamics of lipid-modified small GTPases. Small GTPases 8(3):129–138. https://doi.org/10.1080/21541248.2016.1211067
57. Karandur D, Nawrotek A, Kuriyan J et al (2017) Multiple interactions between an Arf/-GEF complex and charged lipids determine activation kinetics on the membrane. Proc Natl Acad Sci U S A 114(43):11416–11421. https://doi.org/10.1073/pnas.1707970114
58. Munzberg E, Stein M (2019) Structure and dynamics of mono- vs. doubly lipidated Rab5 in membranes. Int J Mol Sci 20(19):4773. https://doi.org/10.3390/ijms20194773
59. Prior IA, Harding A, Yan J et al (2001) GTP-dependent segregation of H-ras from lipid rafts is required for biological activity. Nat Cell Biol 3(4):368–375. https://doi.org/10.1038/35070050
60. Prior IA, Muncke C, Parton RG et al (2003) Direct visualization of Ras proteins in spatially distinct cell surface microdomains. J Cell Biol 160(2):165–170. https://doi.org/10.1083/jcb.200209091
61. Huster D, Vogel A, Katzka C et al (2003) Membrane insertion of a lipidated Ras peptide studied by FTIR, solid-state NMR, and neutron diffraction spectroscopy. J Am Chem Soc 125(14):4070–4079. https://doi.org/10.1021/ja0289245
62. Gorfe AA, Babakhani A, McCammon JA (2007) H-ras protein in a bilayer: interaction and structure perturbation. J Am Chem Soc 129(40):12280–12286. https://doi.org/10.1021/ja073949v
63. Janosi L, Gorfe AA (2010) Segregation of negatively charged phospholipids by the polycationic and farnesylated membrane anchor of Kras. Biophys J 99(11):3666–3674. https://doi.org/10.1016/j.bpj.2010.10.031
64. Zhou Y, Prakash P, Liang H et al (2017) Lipid-sorting specificity encoded in K-Ras membrane anchor regulates signal output. Cell 168 (1–2):239–251. e216. https://doi.org/10.1016/j.cell.2016.11.059
65. Cho KJ, Casteel DE, Prakash P et al (2016) AMPK and endothelial nitric oxide synthase signaling regulates K-Ras plasma membrane interactions via cyclic GMP-dependent protein

kinase 2. Mol Cell Biol 36(24):3086–3099. https://doi.org/10.1128/MCB.00365-16

66. Gorfe AA, Babakhani A, McCammon JA (2007) Free energy profile of H-ras membrane anchor upon membrane insertion. Angew Chem Int Ed Engl 46(43):8234–8237. https://doi.org/10.1002/anie.200702379
67. Gorfe AA, McCammon JA (2008) Similar membrane affinity of mono- and Di-S-acylated ras membrane anchors: a new twist in the role of protein lipidation. J Am Chem Soc 130 (38):12624–12625. https://doi.org/10.1021/ja805110q
68. Gorfe AA, Baron R, McCammon JA (2008) Water-membrane partition thermodynamics of an amphiphilic lipopeptide: an enthalpy-driven hydrophobic effect. Biophys J 95 (7):3269–3277. https://doi.org/10.1529/biophysj.108.136481
69. Abankwa D, Gorfe AA, Inder K et al (2010) Ras membrane orientation and nanodomain localization generate isoform diversity. Proc Natl Acad Sci U S A 107(3):1130–1135. https://doi.org/10.1073/pnas.0903907107
70. Sarkar-Banerjee S, Sayyed-Ahmad A, Prakash P et al (2017) Spatiotemporal analysis of K-Ras plasma membrane interactions reveals multiple high order homo-oligomeric complexes. J Am Chem Soc 139(38):13466–13475. https://doi.org/10.1021/jacs.7b06292
71. Prakash P, Sayyed-Ahmad A, Cho KJ et al (2017) Computational and biochemical characterization of two partially overlapping interfaces and multiple weak-affinity K-Ras dimers. Sci Rep 7:40109. https://doi.org/10.1038/srep40109
72. Jang H, Muratcioglu S, Gursoy A et al (2016) Membrane-associated Ras dimers are isoform-specific: K-Ras dimers differ from H-Ras dimers. Biochem J 473(12):1719–1732. https://doi.org/10.1042/BCJ20160031
73. Sayyed-Ahmad A, Cho KJ, Hancock JF et al (2016) Computational equilibrium thermodynamic and kinetic analysis of K-Ras dimerization through an effector binding surface suggests limited functional role. J Phys Chem B 120(33):8547–8556. https://doi.org/10.1021/acs.jpcb.6b02403
74. Travers T, Lopez CA, Van QN et al (2018) Molecular recognition of RAS/RAF complex at the membrane: role of RAF cysteine-rich domain. Sci Rep 8(1):8461. https://doi.org/10.1038/s41598-018-26832-4
75. Zhang M, Jang H, Nussinov R (2019) The structural basis for Ras activation of PI3Kalpha lipid kinase. Phys Chem Chem Phys 21 (22):12021–12028. https://doi.org/10.1039/c9cp00101h
76. Lin X, Li Z, Gorfe AA (2015) Reversible effects of peptide concentration and lipid composition on H-Ras lipid anchor clustering. Biophys J 109(12):2467–2470. https://doi.org/10.1016/j.bpj.2015.11.009
77. Li H, Gorfe AA (2014) Membrane remodeling by surface-bound protein aggregates: insights from coarse-grained molecular dynamics simulation. J Phys Chem Lett 5(8):1457–1462. https://doi.org/10.1021/jz500451a
78. Li Z, Gorfe AA (2014) Modulation of a small two-domain lipid vesicle by linactants. J Phys Chem B 118(30):9028–9036. https://doi.org/10.1021/jp5042525
79. Prakash P, Hancock JF, Gorfe AA (2015) Binding hotspots on K-ras: consensus ligand binding sites and other reactive regions from probe-based molecular dynamics analysis. Proteins 83 (5):898–909. https://doi.org/10.1002/prot.24786
80. Prakash P, Sayyed-Ahmad A, Gorfe AA (2015) pMD-membrane: a method for ligand binding site identification in membrane-bound proteins. PLoS Comput Biol 11(10):e1004469. https://doi.org/10.1371/journal.pcbi.1004469
81. Sayyed-Ahmad A, Gorfe AA (2017) Mixed-probe simulation and probe-derived surface topography map analysis for ligand binding site identification. J Chem Theory Comput 13(4):1851–1861. https://doi.org/10.1021/acs.jctc.7b00130
82. Lorent J, Levental K, Ganesan L et al (2020) The mammalian plasma membrane is defined by transmembrane asymmetries in lipid unsaturation, leaflet packing, and protein shape. bioRxiv:698837. https://doi.org/10.1101/698837
83. Jing Z, Liu C, Cheng SY et al (2019) Polarizable force fields for biomolecular simulations: recent advances and applications. Annu Rev Biophys 48:371–394. https://doi.org/10.1146/annurev-biophys-070317-033349
84. Yin G, Zhang J, Nair V, Truong V, Chaia A, Petela J, Harrison J, Gorfe AA, Campbell SL (2020) KRAS ubiquitination at lysine 104 retains exchange factor regulation by dynamically modulating the conformation of the interface. Iscience 23(9):101448. https://doi.org/10.1016/j.isci.2020.101448

Chapter 10

In Silico Prediction of the Binding, Folding, Insertion, and Overall Stability of Membrane-Active Peptides

Nicolas Frazee, Violeta Burns, Chitrak Gupta, and Blake Mertz

Abstract

Membrane-active peptides (MAPs) are short-length peptides used for potential biomedical applications in diagnostic imaging of tissues, targeted drug delivery, gene delivery, and antimicrobials and antibiotics. The broad appeal of MAPs is that they are infinitely variable, relatively low cost, and biocompatible. However, experimentally characterizing the specific properties of a MAP or its many variants is a low-resolution and potentially time-consuming endeavor; molecular dynamics (MD) simulations have emerged as an invaluable tool in identifying the biophysical interactions that are fundamental to the function of MAPs. In this chapter, a step-by-step approach to discreetly model the binding, folding, and insertion of a membrane-active peptide to a model lipid bilayer using MD simulations is described. Detailed discussion is devoted to the critical aspects of running these types of simulations: prior knowledge of the system, understanding the strengths and weaknesses of molecular mechanics force fields, proper construction and equilibration of the system, realistically estimating both experimental and computational timescales, and leveraging analysis to make direct comparisons to experimental results as often as possible.

Key words Membrane-active peptides (MAPs), Molecular dynamics (MD) simulation, Lipid bilayer, Force field, Time-step, Partitioning–folding coupling

1 Introduction

Predicting the behavior of membrane-active peptides (MAPs) based on their sequence is an inherently difficult task. A multitude of pathways exist by which MAPs interact with the membrane, such as monomeric partitioning–folding coupling [1], surface-bound aggregation to induce membrane thinning and pore formation [2], or cytosolic delivery via endocytosis [3]. The propensity for a MAP to utilize one pathway over another is dependent on many factors beyond the peptide sequence. Specifically, a certain degree of coupling exists between membrane biophysics and the function (and ultimately the mechanism) of a MAP; however, fundamental understanding of this relationship requires a molecular level of detail. Both experimental and computational techniques are useful

Irina S. Moreira et al. (eds.), *Computational Design of Membrane Proteins*, Methods in Molecular Biology, vol. 2315, https://doi.org/10.1007/978-1-0716-1468-6_10,

in this endeavor. Solid-state NMR spectroscopy can determine the equilibrium dynamics of a lipid bilayer as well as quantifying interatomic distances on the sub-nanometer scale [1, 4–8]. Electron paramagnetic resonance (EPR) spectroscopy is well-suited to mapping out distances between transmembrane helices in integral membrane proteins [9] as well as the orientation of amphiphilic helical peptides [10, 11]. In contrast to experimental approaches, molecular dynamics (MD) simulations are able to produce a detailed picture of the molecular fluctuations that drive biological function, albeit on a smaller time and length scale than is accessible with experimental approaches.

Significant advances have been made in the past 25 years in our ability to use MD simulations to probe membrane-based systems [12–20]. Atomistic MD simulations capture many of the critical phenomena that explain interactions of membrane proteins with the membrane environment, such as binding, folding, and dimerization of proteins [19, 21] and changes in the lateral pressure profile, chemical potential, and dipole moment of the bilayer [22]. Coarse-grained MD simulations allow for longer time and length scales (nm and μs-ms), enabling massive sampling of interactions between MAPs and membranes (with the caveat that peptide conformation is restricted by an elastic network) [23].

In particular, MD simulations have provided invaluable insights into the partitioning, folding, and insertion of MAPs. Equilibrium MD simulations have been used to characterize the formation of pores that are characteristic of cell-penetrating peptides [24, 25] as well as translocation of a peptide across the entire membrane [26]. Modified models of the membrane environment that accelerate the timescales for lateral diffusion along the membrane surface have been extremely useful in identifying key interactions that stabilize the bound state of peptides like the amphipathic lipid packing sensor (ALPS) motif and in the oligomeric interfaces of single-span transmembrane helical peptides [14, 27, 28]. In addition, enhanced sampling techniques such as temperature-based replica exchange, umbrella sampling, metadynamics, and replica exchange with solute scaling have all been useful in characterizing the free-energy landscape of these processes [20, 29–31].

The immense diversity of membrane-active peptides, potential mechanisms of action, and numerous MD approaches employed to study these systems can be overwhelming. This guide is designed to provide the reader with a general workflow for using MD simulations to discreetly model the mechanistic steps of a membrane-active peptide (Fig. 1), with detailed notes to inform about best practices and potential pitfalls to avoid. Subheading 2 covers general information that should be gathered before setting up simulations. Subheading 3 provides step-by-step processes to set up, run, and analyze MD simulations of a membrane-active peptide during the (1) binding, (2) inserted, and (3) insertion/exit processes.

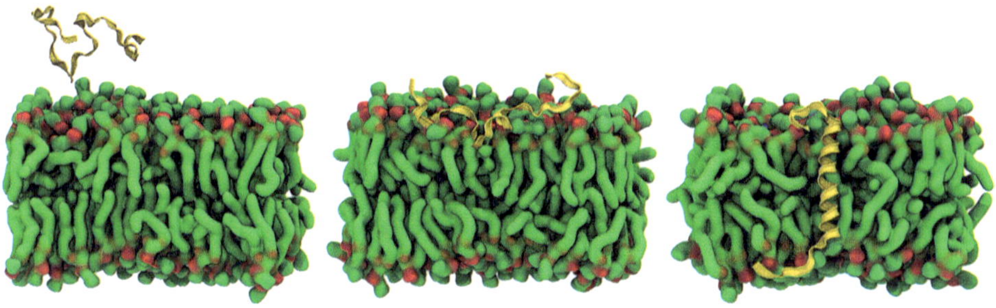

Fig. 1 General mechanism of a membrane-active peptide (MAP). *Left*: A MAP diffuses through bulk solvent in a coiled conformation. *Middle*: Upon binding to the bilayer surface, the peptide partitions into the headgroup of the outer leaflet. This partitioning is often accompanied by folding into an alpha helix, a phenomenon known as partitioning–folding coupling. *Right*: Upon folding into an alpha helix, it is common for a MAP to insert into the membrane. A transmembrane span can be achieved, depending on the length and tilt angle of the peptide. If the peptide is too short to completely span the membrane, it can associate with other peptides to form a pore

Subheading 4 is a compilation of notes that provide common approaches and oft-overlooked details that will help the reader maximize their ability to carry out impactful MD simulations of MAP-based systems.

2 Materials

The first rule of thumb with designing an MD simulation to model a MAP-based system is to have a *thorough understanding of the governing biophysical characteristics of (1) the peptide and (2) the membrane*. This detail cannot be emphasized enough: the combination of longer timescales, heterogeneity of the membrane environment, and variability in peptide mechanisms creates an enormous phase space that cannot be explored in a completely unbiased manner. It is the responsibility of the simulator to utilize as much existing data as possible to inform the design of the project.

1. The very first considerations for running MD simulations are to select:
   - An MD engine.
   - A force field (*see* **Note 1**).
2. To properly build the simulation system, consideration must be given to the following:
   - Lipid:peptide ratio.
   - Hydration.
   - Salts and concentrations.
   - Membrane patch size.
   - Temperature and pressure settings (*see* **Note 2**).

3. To properly run an MD simulation, the following parameters must be considered:
   - Thermostat.
   - Barostat.
   - Time-step.
   - Force cutoffs.
   - Water model.
   - The rate to write to trajectory (*see* **Note 3**).

3 Methods

In order to complete the following steps, we assume that the reader has a working knowledge of the tools described below.

3.1 Binding of MAPs to Model Bilayers

3.1.1 Generating a Peptide for Binding Simulations (See Note 4)

1. Find or generate a .pdb of the peptide.
2. Mutate residues and patch termini if necessary.
3. Make a structure file.
4. Solvate the peptide in a water box of sufficient size. If necessary, add ions to the system to counterbalance the peptide charge.
5. Using the selected engine, perform minimization on the system. Run a 100 ps *NVT* equilibration with the selected barostat, thermostat, time-step, and other parameters to optimize the packing of waters in the box around the peptide.
6. If the peptide inherently adopts a coiled conformation in solution, a good rule of thumb is to apply a high-temperature MD simulation to generate a series of randomized starting structures. Initiate an equilibrium MD simulation with the selected barostat, thermostat, and time-step at 300 K. Over 20 ps increase the temperature linearly to 700 K and run at 700 K for 1 ns or until the peptide has visibly lost any preexisting conformation. Save starting structures from the trajectory, ensuring the ω backbone dihedral angles lie at $\pm 180°$.
7. Equilibrate the peptide for 10 ns in the *NPT* ensemble at the chosen temperature with the same thermostat and time-step to allow the peptide to adopt a natural conformation in solution.

3.1.2 Generating a Membrane for Binding Simulations (See Note 5)

1. *Create a membrane with water and ions.* Generate a membrane with the proper lipid:peptide and lipid:water ratios and type of lipids. Add ions to obtain a salt concentration consistent with experiment or physiological conditions.
2. *Minimize and equilibrate membrane.* Minimize and equilibrate the membrane, typically in the *NPT* ensemble for at least 50 ns. Save the final frame of the equilibrated membrane.

3.1.3 Running a Binding/Unbinding Simulation (See ***Note 6****)*

1. *Combine membrane and peptide.* Place the peptide at least 10–12 Å above the bilayer surface to avoid biasing of peptide binding via nonbonded attractions between the peptide and bilayer. Remove any water molecules that overlap with the peptide. Create a .pdb and a structure file for the combined system. Repeat this step with several peptide starting structures to remove potential bias in obtaining bound proteolipid complexes.
2. *Equilibration.* Run a simulation in the *NVT* ensemble for 100 ps to equilibrate to the proper system density.
3. *Production run.* Run a long equilibrium simulation in the *NPT* ensemble for each of the prepared systems. It is important to occasionally visualize the trajectories to ensure the systems are behaving properly.

3.1.4 Analysis of a Binding/Unbinding Simulation

1. *Possible analyses for lipids* (*see* **Note 7**):
   - Area per lipid.
   - Bilayer thickness.
   - Order parameters.
   - Bilayer defects.
   - Electron/mass densities.
   - Lateral pressure of the bilayer.
   - Mean-squared displacement (MSD).
2. *Possible analyses for peptides* (*see* **Note 8**):
   - Secondary structure.
   - Ramachandran plot.
   - Lipid–peptide contacts.
   - Angle of the peptide with respect to its principal axis and the membrane normal.
   - Per-residue distance from the bilayer.
   - Per-residue sidechain orientation with respect to the bilayer normal.
   - Rotation of the principal axis of the peptide with respect to the membrane surface.
   - Peptide radius of gyration.
   - Binding rates.
   - Autocorrelation time.
   - Analyses that could be relevant to the peptide in question.

3.2 Modeling Inserted States of MAPs

3.2.1 Generating Systems with a MAP Inserted into a Model Bilayer (See Notes 4 and 9)

1. Find or generate a .pdb of the peptide.
2. Mutate residues and patch termini as needed.
3. *Create the proteolipid complex.* Generate a membrane surrounding the peptide with the proper lipid:peptide and lipid:water ratio. Add ions to obtain a salt concentration consistent with experiment or physiological conditions.

3.2.2 Running Simulations of Inserted MAPs (See Note 10)

1. *Minimize and equilibrate the proteolipid complex.* Minimize and equilibrate the membrane, typically in the *NPT* ensemble for at least 50 ns. For systems with oligomeric peptides, consider applying constraints to maintain stability of the oligomeric complex.
2. *Production run.* Conduct a production run (typically in the *NPT* ensemble). Visually inspect the trajectories to ensure the systems are behaving properly. Anticipate longer timescales for formation of oligomeric complexes/pores.

3.2.3 Analysis of Simulations with an Inserted MAP

1. *General analysis for membrane* (*see* **Notes 7** and **11**):
   - Order parameters.
   - Radial distribution function (RDF).
   - Mass/electron density profile of lipid components along the membrane normal.
   - Mean square displacement (MSD) of lipids in the lateral plane of the bilayer.
2. *General analysis for peptides* (*see* **Notes 8** and **12**):
   - Determination of per-residue secondary structure.
   - Ramachandran plots of backbone phi/psi angles.
   - Tilt angle of helix with respect to the membrane normal.
   - Radius of gyration.
   - Dihedral angles of amino acid sidechains.
   - Per-residue distance with respect to bilayer normal.
3. *Analysis for combined system* (*see* **Note 13**):
   - Measurement of the kinetic rates of water molecules entering into the hydrophobic region of the bilayer and residence times of water molecules within the bilayer.
   - RDF of waters with respect to the inserted peptide.
   - Time-dependent and averaged water densities.

3.3 Mapping out the Insertion and Exit Pathways of MAPs

Modeling partitioning–folding coupling in combination with insertion in a single simulation is a significant endeavor and should not be taken lightly. Many of the steps outlined in Subheadings 3.1 and 3.2 apply to mapping out the entire insertion/exit pathway. Refer to our statement at the beginning of Methods for a refresher on intimate working knowledge of a system.

3.3.1 System Considerations and Setup

1. It is common to utilize an enhanced sampling (ES) technique, due to the fact that the timescales for insertion/exit are long (μs-ms).
2. Define a collective variable (CV) that is appropriate to the system and the ES technique (*see* **Note 14**).
3. Set up systems according to the ES technique being used. At minimum, it is desirable to have two endpoints (i.e., surface-bound and inserted). Refer to **Notes 4–6** for general advice on setting up proteolipid systems.

3.3.2 Simulating Insertion and Exit Pathways for Peptides

1. Run the simulation (*see* **Note 15**). This is typically carried out in the *NPT* ensemble.
2. After preliminary analysis, extend the simulation if necessary.

3.3.3 Analysis of Simulations Characterizing the Pathway of MAP Insertion

1. Analyze the simulation (*see* **Note 16**).
2. Determine if the CV space has been sufficiently sampled (*see* **Note 17**).

4 Notes

1. *How to pick an MD engine.* NAMD, GROMACS, OpenMM, AMBER, GENESIS. MD engines are generally similar across different codes. Know the strengths and weaknesses of each code with respect to (a) the computer hardware available and (b) the techniques (outside of equilibrium MD) that can be utilized.

   Choosing a force field. Each lipid and protein force field have a characteristic strength and weakness. The predominant force fields are CHARMM36, AMBER, and GROMOS [32–35]. In most cases, it is best practice to use the most recent version of a given force field for proteins, lipids, and solvent/ions. However, caveat emptor! It is ultimately the user's responsibility to consult the primary literature and understand the strengths and weaknesses of a particular force field. For example, the Amber force fields use a noticeably different cutoff from the CHARMM force field (*see* **Note 3** below).

2. *Hydration.* It is imperative that the number of waters per lipid matches or exceeds the experimental value; too low a hydration number will lead to abnormal behavior of the bilayer. There are several approaches to determining this quantity experimentally, so be prepared to find multiple values. The work of John Nagle, Stephanie Tristram-Nagle, John Katsaras, and Michael Brown, to name a few, are highly informative [36–38].

   Bilayer patch size. The main factor constraining the size of the bilayer is the relationship between the size of the peptide and periodic boundary conditions (PBCs). To avoid interaction of the peptide with its image (which can introduce artifacts into the simulation), the bilayer patch must be approximately 15 Å larger than the size of the peptide. Smaller patch sizes lead to greater computational efficiency but can also suppress undulations and deformations that naturally occur in the cell membrane.

3. *Selecting a thermostat.* Langevin or stochastic rescaling [39] is the most common. Care needs to be taken that the thermostat does not violate the equipartition theorem via its approach to velocity rescaling (the so-called "flying ice cube" effect) [40, 41].

   Selecting a barostat. Nosé-Hoover is generally best, as it combines the advantages of using the MTK and Langevin barostats [42, 43]. More recently, the Monte Carlo barostat has been implemented in OpenMM and AMBER [44]. In addition, the Berendsen barostat is used in AMBER and GROMACS. Furthermore, check the force field being used to see if it can be run in a tensionless configuration or if a surface tension needs to be applied. With any membrane simulation, it is important to be able to decouple the direction of the barostat in the *xy*-plane from the *z*-axis; this allows the bilayer patch to undergo compression in the *z* dimension while allowing the lateral plane of the bilayer to remain tensionless. Anisotropic pressure coupling (i.e., independent in all directions) is normally sufficient to achieve this result. With longer timescale simulations (μs and longer), it is not uncommon for the lipid patch to adopt a rectangular shape. If this occurs, switch to a semi-isotropic pressure scheme that maintains the ratio of the *xy* dimensions.

   Selecting a time-step. A 2 fs time-step is usually sufficient for maintaining system energy in equilibrium MD simulations. However, some force fields will require a 1 fs time-step and some options in the MD engine (e.g., hydrogen mass repartitioning [45]) will allow for longer time-steps.

   Force cutoffs. Cutoffs for nonbonded interactions (van der Waals and electrostatic forces) are very specific to force fields, and a critical detail that should not go overlooked in membrane simulations [46]. Two main cutoffs exist: force switching and hard cutoff. Force switching is traditionally used to prevent errors due to truncation of long-range forces and is most commonly used in the CHARMM force field. A hard cutoff is much easier to implement and typically gives about a 15–20% speedup in performance. It is most commonly used in the Amber force field.

Selection of water model. Water models are generally categorized by the number of sites (points of interaction of the water molecule) and their flexibility. The most commonly used model is the TIP3P (3-site) model due to its computational efficiency; however, the OPC (4-site) model has been quite successful in reproducing experimental results [47, 48]. It is critical to utilize the water model that was used for parameterization of the other force field components (protein and lipid) in a system in order to reproduce expected protein and lipid behavior.

Write rates to trajectories. The rate to write to the trajectory must strike a balance between being frequent enough that important motions are not missed but large enough that the trajectory is not massive in file size.

4. *Finding/generating a peptide.* Search the Protein Data Bank (www.rcsb.org) for the peptide in question and if available, download the coordinates (.pdb file format). If coordinates are unavailable, they can be generated based on the primary amino acid sequence using one of the following programs: Molefacture plugin in VMD [49], PyMOL [50], Avogadro [51], or tleap in AMBER.

   Mutating and patching the peptide. If the .pdb file exists but the sequence does not completely match the target peptide, apply point mutations. Additionally, ensure the termini of the peptide have the proper atoms on the N and C termini (i.e., "patching"). Mutation and patching can be accomplished with the Mutate Residue plugin in VMD, tleap in AMBER, psfgen in VMD, or the CHARMM-GUI input generator [52].

   Creating a structure file. Most engines have a unique nomenclature for a structure file. Consult their documentation for detailed walkthroughs for generation of a structure file. This file describes the bonds, angles, dihedral angles, and other possible interactions between atoms in the system.

   Tools for solvation. Suggested tools for solvation are the solvate plugin in VMD, Packmol [53], OpenMM [54], and tleap.

   Water box size. To heat the peptide in solution, measure the length of the peptide along its principal axis and add 10 Å of water on all sides of the peptide. This should be sufficient to prevent the peptide interacting with its periodic image.

Methods for adding ions. The ionize plugin in VMD, CHARMM-GUI, and the optimal membrane generator (OMG) in LOOS [55] are all well-suited to this task. It is also highly recommended to consult the SLTCAP method [56] for determination of the proper number of ions necessary for a specific salt concentration, as the method for calculation used in tools listed above may over- or underestimate the actual number of ions.

Minimization. If the system is well-constructed, minimization should take 5000–10,000 steps to properly minimize the system.

5. *Creation of the lipid bilayer, minimization, and equilibration.* Several options are available to generate a lipid bilayer: CHARMM-GUI [52], Packmol [53], the OMG in LOOS [55], and the inflate-gro tool in GROMACS [57]. Generate a lipid bilayer, solvate and ionize the system (using the same tools as above). After this, perform a minimization (5000–10,000 steps as above), and follow equilibration steps (*NVT* ensemble to the *NPT* ensemble) similar to that in **Note 4**. Each equilibration step (*NVT* and *NPT* ensembles) will typically take a minimum of 50 ns. It is highly recommended to monitor the stability of the system during equilibration by calculating the area per lipid of the bilayer.
6. *Timescales for binding/unbinding simulations.* Effectively simulating the binding/unbinding of a peptide to a membrane is challenging for several reasons: (1) random diffusion of the peptide through solution leads to inconsistent binding, (2) some force fields are more "sticky," making dissociation from the membrane unlikely, and (3) the number of necessary binding/unbinding events required for an accurate estimate of binding constants makes this type of simulation computationally expensive.

   Examples of common errors in simulations. If a system has not reached the proper density, a void (or multiple voids) can form to compensate for this discrepancy (this usually simply requires running the simulation in the *NPT* ensemble). If the system size is too small (either size of the lipid patch or not enough water molecules), then the peptide will interact with its periodic image or with the opposite leaflet in the neighboring periodic cell.
7. *Possible analyses for lipids.*
   - Area per lipid can usually be measured simply by dividing the x-y area of the simulation box by the number of lipids in the system (Fig. 2). A more advanced method of determination is through a Voronoi decomposition which allows a layered area per lipid as a function of the depth of the bilayer.

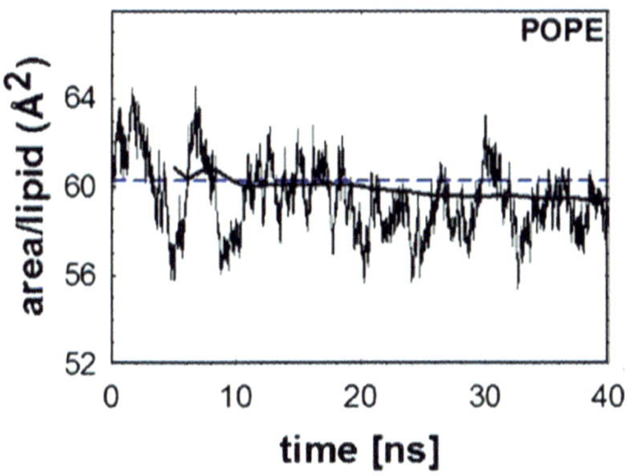

Fig. 2 Calculating area per lipid for a bilayer. Time-dependent area per lipid (in angstroms squared) of a 1-palmitoyl-2-oleoyl-*sn*-3-glycero-phosphoethanolamine (POPE) bilayer. The trend line shows that even over a 40 ns timescale, the area per lipid has not converged. Most lipid systems equilibrate within 50–100 ns [33]

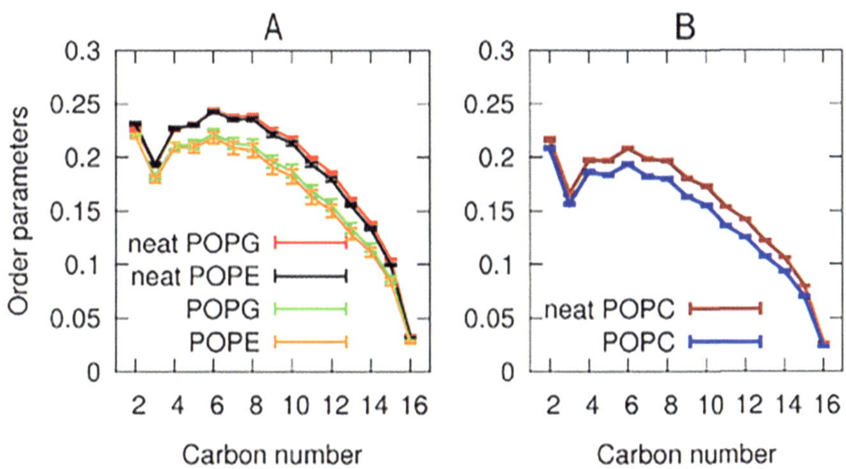

Fig. 3 Calculation of ^{2}H order parameters reveals the degree of ordering within a bilayer. Plots of ^{2}H order parameters for a series of three different lipids (1-palmitoyl-2-oleoyl-*sn*-3-glycero- phospho- glycerol (POPG), 1-palmitoyl-2-oleoyl-*sn*-3- glycero-phosphoethanolamine (POPE), and 1-palmitoyl- 2-oleoyl-*sn*-3-glycero-phosphocholine (POPC)) without ("neat") and with a lipidic fungicide bound to the bilayer surface. The order parameter decreases with increasing distance from the glycerol backbone, indicating a higher degree of flexibility towards the midplane of the bilayer [58]

- Bilayer thickness is typically measured as the interleaflet distance between P atoms. This distance can be calculated by measuring the mass/electron density.
- ^{2}H order parameters can be calculated by tools included in most MD analysis packages (cpptraj, GROMACS, LOOS). It is a quantitative measure of the disorder of the acyl chains in a lipid bilayer and is one of the few ways to make a direct comparison between MD simulations and experiment (Fig. 3).

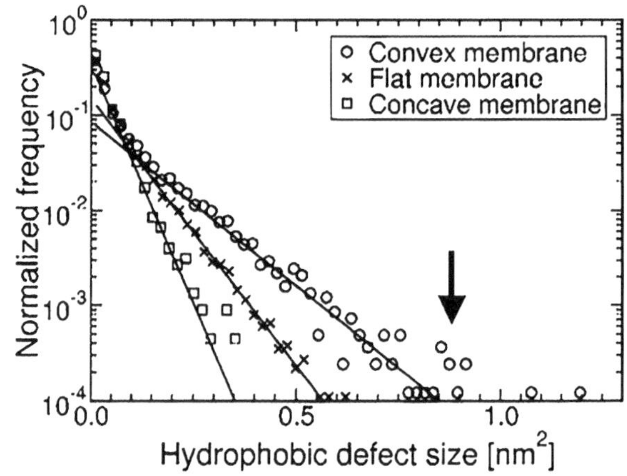

Fig. 4 Calculation of membrane defects in a bilayer surface. Defects were calculated using the SASA of the lipid bilayer. A correlation exists between the number of defects and size of the defects, with an overall increase in defect size in a convex membrane and an overall decrease in defect size in a concave membrane [59]

- Bilayer defects are a measurement of voids in the bilayer surface (potential binding sites). They are often a hallmark of stable binding interactions between a MAP and the bilayer surface, characterized by shape complementarity. Several methods exist for calculating defects, with the two most common techniques being a calculation based on solvent-accessible surface area (SASA) [59] and a combinatorial calculation (Fig. 4) [60].
- Mass or electron densities can be used to characterize the position of peptide residues, headgroups, acyl chains, and changes in bulk water along the membrane normal (Fig. 5). This is especially useful in determining differences between the distributions of specific membrane components when comparing lipid systems.
- Lateral pressure of a membrane (Fig. 5) can only be calculated via simulation; an experimental method to determine lateral pressure profile does not currently exist. It is a useful metric for identifying large-scale perturbations to the water–bilayer interface.
- Mean square displacement (MSD) provides information on the movement of the selection over the course of the trajectory. It is most often used as a way to gauge diffusion in the lateral plane of a bilayer of the species of interest. Extreme care must be taken to avoid comparing estimates of MSD between lipid systems of different sizes, as periodic boundary conditions can skew MSD values above and below experimentally determined ones [62].

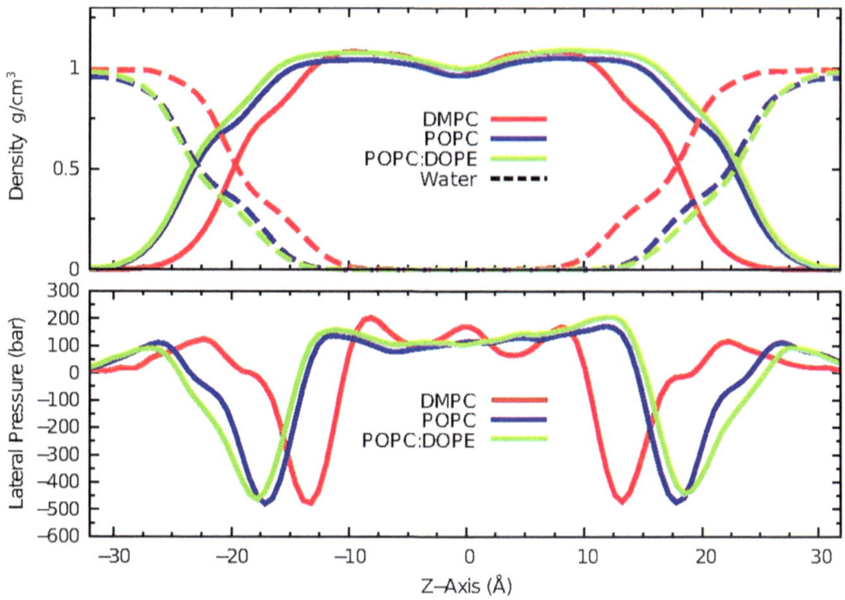

Fig. 5 Using density distributions and lateral pressure profiles to identify differences between lipid systems. *Top*: Density profiles of lipids (solid) and waters (dashed) in MD simulations of DMPC, POPC, and POPC:DOPE bilayers. *Bottom*: Lateral pressure profiles of the different lipid systems corresponding to the densities calculated in the first plot. Each plot aids in discerning differences between the lipid systems, both from the perspective of membrane thickness (*top*) and the water–lipid interface that characterizes the location of the glycerol backbone just below the headgroup region [61]

8. *Possible analyses for peptides.*
   - Secondary structure can be analyzed as an indicator of partitioning–folding coupling. Multiple tools are available for quantifying this interaction, such as the Timeline plugin in VMD, the secstruct tool in AmberTools, or a simple calculation-based algorithm like STRIDE [63] or the Lifson–Roig definition of secondary structure (Fig. 6) [65].
   - Ramachandran plots are a more generalized approach to characterize sampling of secondary structures in a simulation. These plots show the overall distribution of the phi/psi backbone dihedral angles of a peptide, providing a condensed dataset about the global sampling of peptide secondary structure.
   - Lipid–peptide contacts can act as a measurement of the extent and/or the quality of binding of the peptide to the bilayer surface.
   - Measuring the angle of the principal axis of the peptide with respect to the bilayer normal gives insight into preferred binding poses with the bilayer. This detail can be especially important for cell-penetrating peptides, in particular when they form pores via a carpet mechanism.

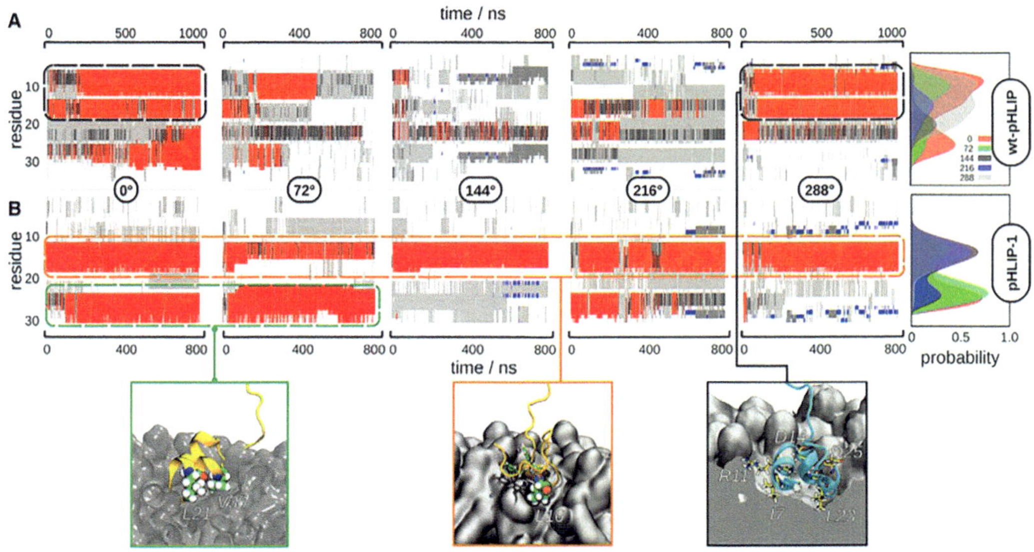

Fig. 6 Using calculation of peptide secondary structure to characterize conserved interactions in formation of peptide–bilayer complexes. Per-residue secondary structure map as a function of time for the pH-Low Insertion Peptide binding to a POPC bilayer. Analysis shows that a conserved segment of the peptide folds into an α-helix, facilitating binding to the bilayer [64]

- The rotation along the principal axis of the peptide with respect to the bilayer can be particularly helpful if the peptide has amphipathic faces. This allows for unambiguous determination of preferred orientation to form a bound complex.
- To identify residue-specific interactions with the bilayer, it is useful to measure both the sidechain orientation with respect to the membrane normal and the projection on the membrane normal of the distance between the residue and the membrane surface. This can provide insights into residues that are critical for stabilization of the proteolipid complex.
- Radius of gyration is a very general tool that provides a measurement of compactness; the larger the radius of gyration, the more globular the peptide. The distribution of the radius of gyration can also support arguments for peptides adopting a more broad or narrow range of conformations.
- Binding rates can be challenging to determine in equilibrium MD because of the number of binding events to properly determine the rate. The weighted ensemble method allows for calculation of binding rates, although it is extremely helpful to have a parallel resource to carry out these types of simulations [66].

- Autocorrelation time represents the lifetime for the observation of an independent event; this can be applied to determine the autocorrelation of peptide binding.
- Finally, use the information determined by other experiments to develop new analyses to corroborate the experimental results.

9. *Finding/generating a peptide.* The Orientations of Proteins in Membranes (OPM) database is a great resource for proper orientation of the inserted peptide, if available [67]. If an experimentally derived structure is unavailable, it is possible to use programs such as VMD, PyMOL, or AmberTools to generate a proteolipid complex. For oligomeric systems, verify the correct number of monomers required to form the complex. To model spontaneous assembly of inserted peptides in an oligomeric complex/pore, randomly distribute and rotationally orient the peptides in the bilayer.

   Mutate residues and patch termini (*see* **Note 4**).

   Create a membrane with water, ions, and the peptide. The most commonly used membrane generators are the CHARMM-GUI server, the Optimal Membrane Generator (OMG) in LOOS, and the inflate_gro utility in GROMACS.

   System size (*see* **Notes 2** and **5**).

   Addition of ions (*see* **Note 4**).

   Hydration (*see* **Note 2**).

   Insertion methods. There are two main procedures to generate a bilayer with an inserted peptide: insertion or replacement. The insertion process inserts the peptide into a hole about the size of the protein, in a membrane that has been previously equilibrated. The replacement method creates a dummy lipid-like bilayer around the peptide and then replaces the dummy lipids with the lipids chosen for the system [68].

10. *Minimization and equilibration of the proteolipid complex. See* **Note 5** for details. Unlike simulations modeling binding of MAPs to the membrane surface, it is common practice to initially constrain the peptide during minimization and equilibration. The timescales for equilibration of the bilayer are much longer than a transmembrane peptide.

    Examples of simulation errors (*see* **Note 6**).

    Timescales for formation of oligomeric complexes. Lateral diffusion and complete mixing of lipids and peptides typically occurs on the μs-ms timescale [5]. When conducting simulations of inserted MAPs to observe formation of oligomeric complexes, it is advisable to obtain access to special-purpose machines like Anton, which can routinely run μs-timescale MD simulations [25], or to utilize enhanced sampling techniques [69].

11. *Possible analyses of lipids.* (*See* **Note 7** for additional ideas.)
    - In proteolipid systems, it is useful to determine the degree of disorder in the bilayer with respect to distance from the inserted peptide. One way to do this is with the DIBMOPS tool in LOOS. DIBMOPS defines a principal axis along the acyl chain of each lipid in the bilayer along with secondary and tertiary axes (based on dummy particles associated with each lipid), which is then used to calculate a molecular order parameter.
12. *Possible analysis for peptides/proteins.* (*See* **Note 8** for additional ideas.)
    - Tilt angles are a common experimental observable of inserted MAPs (usually measured by solid-state NMR or oriented circular dichroism spectroscopy) [70]. In MD simulations, a MAP will reach its equilibrium tilt angle with tens to hundreds of ns, allowing for direct comparison to experiment. Calculation of tilt angles is accomplished by defining the principal axis of the MAP and measuring the angle between that vector and the membrane normal.
    - Calculations of the dihedral angles of sidechain rotamers are particularly important in identifying specific orientations that help contribute to the function of a MAP. Often an amino acid sidechain will adopt a particular orientation that facilitates nonbonded interactions with other peptides in the bilayer or with the headgroup region in each leaflet. Tools exist in multiple packages (VMD, LOOS, cpptraj) to carry out this calculation.
13. *Possible analyses for combined systems.*
    - Calculating kinetic rates of water diffusion through the interior of the bilayer requires a nuanced approach: one must account for the number of times that a water molecule crosses an arbitrarily defined "threshold" of the hydrophobic bilayer interior (n_{crossing}) as well as the lifetime of a given water molecule within the bilayer interior ($t_{\text{residence}}$). The easiest solution is to write in-house codes to carry out these calculations, but LOOS also has a tool ("crossing waters") that can also do this.
    - Quantifying the degree to which water invades the hydrophobic interior is also important, as this calculation can capture global behaviors between an inserted MAP and the bilayer. This involves calculation of water density profiles, which can be carried out with tools such as the Volmap plugin in VMD, the water-hist tool in LOOS, dxTuber [71], or writing in-house code.

14. *Choosing a collective variable.* This is arguably the most important choice in an ES simulation. A well-chosen CV can lead to quick convergence; a poorly chosen CV can converge on long timescales or provide erroneous results [72, 73]. Options include: (1) a distance-based CV such as the center of mass distance between the peptide and the bilayer [74]; (2) endpoints on a path, such as a surface-bound and an inserted peptide (would require structural knowledge of these endpoints) [59]; (3) a higher dimension CV (distance and tilt angle, tilt angle and helicity, number of contacts between the peptide and lipid headgroups, and distance and orientation of the peptide with respect to the bilayer surface (for amphipathic helices)) [75]. Alternatively, a more brute-force route could involve temperature-based replica exchange [29] or accelerated MD, both of which do not require prior knowledge of the CV but can be computationally expensive.
15. *Running the ES simulation.* Do NOT expect success on the first attempt! It is not unusual for application of a bias to follow a particular CV will degrade the integrity of the bilayer, denature the peptide, or both. Optimization of running an ES technique is highly dependent on (1) properly defined restraints, (2) available computer hardware, and (3) the ES technique [76]. With respect to restraints, it may be necessary to apply them to the secondary structure of the peptide or the position of the bilayer. For hardware, steered MD and metadynamics are fairly versatile and can run on any architecture; umbrella sampling (US) is embarrassingly parallel, provided that a system will not require extensive resources per window; weighted ensemble (WE) is also extremely parallel. Replica exchange (REX) tends to be the most computationally expensive (and least efficient). However, less-sophisticated ES techniques like US and REX are often better, simply because they do not break the system.
16. *Analyzing ES simulations.* The type of information that can be extracted from an ES simulation varies with the technique. Steered MD will provide a rough estimate of the pathway [77]. Umbrella sampling will provide a free energy surface of the process, but not necessarily mechanistic information [30]. Metadynamics will also generate a free energy surface [31]. WE approaches can potentially provide kinetics and thermodynamics [78].
17. *Determining sufficient sampling.* This is very dependent on (1) the CV selected and (2) the technique in use. A critical aspect is determining if the CV is the slowest degree of freedom in converging to a particular state. This requires intimate knowledge of the biophysics of the system. For phenomena like peptide binding to a bilayer surface, this can take place in as

little as 10 ns. For folding and insertion into a bilayer, it can take on the order of μs [19, 26]. It is the reader's responsibility to be familiar with the literature, which will provide a general idea as to the appropriate timescales and techniques that can be used. It is also possible to check sampling based on the technique: in US, each window should have overlapping distributions of the sampled CV; in REX, the rate and number of exchanges between replicas can be calculated; in WE, the flux between states in the progress coordinate can be quantified.

Acknowledgments

The authors would like to thank Alan Grossfield, Tod Romo, and others from the Grossfield lab that have contributed to our collective knowledge on best practices in running membrane simulations and in fully utilizing LOOS for analysis of our work. Thanks also go out to Chris Neale, Ed Lyman, Alex Sodt, Dan Zuckerman, and Michael Brown for their support in helping us successfully marry experiment with simulation and theory, as well as for their infectious enthusiasm for membrane biophysics.

References

1. Gleason NJ, Vostrikov VV, Greathouse DV, Koeppe RE (2013) Buried lysine, but not arginine, titrates and alters transmembrane helix tilt. Proc Natl Acad Sci U S A 110 (5):1692–1695. https://doi.org/10.1073/pnas.1215400110
2. Almeida PF, Ladokhin AS, White SH (2012) Hydrogen-bond energetics drive helix formation in membrane interfaces. Biochim Biophys Acta 1818(2):178–182. https://doi.org/10.1016/j.bbamem.2011.07.019
3. Kim SY, Pittman AE, Zapata-Mercado E et al (2019) Mechanism of action of peptides that cause the pH-triggered macromolecular poration of lipid bilayers. J Am Chem Soc 141 (16):6706–6718. https://doi.org/10.1021/jacs.9b01970
4. Leftin A, Molugu TR, Job C et al (2014) Area per lipid and cholesterol interactions in membranes from separated local-field (13)c NMR spectroscopy. Biophys J 107(10):2274–2286. https://doi.org/10.1016/j.bpj.2014.07.044
5. Sodt AJ, Sandar ML, Gawrisch K et al (2014) The molecular structure of the liquid-ordered phase of lipid bilayers. J Am Chem Soc 136 (2):725–732. https://doi.org/10.1021/ja4105667
6. Wolf J, Aisenbrey C, Harmouche N et al (2017) pH-dependent membrane interactions of the histidine-rich cell-penetrating peptide LAH4-L1. Biophys J 113(6):1290–1300. https://doi.org/10.1016/j.bpj.2017.06.053
7. Mandala VS, Williams JK, Hong M (2016) Structure and dynamics of membrane proteins from solid-state NMR. Annu Rev Biophys 47:201–222. https://doi.org/10.1146/annurev-biophys-070816-033712
8. Hanz SZ, Shu NS, Qian J et al (2016) Protonation-driven membrane insertion of a pH-low insertion peptide. Angew Chem Int Ed Engl 55(40):12376–12381. https://doi.org/10.1002/anie.201605203
9. Altenbach C, Kusnetzow AK, Ernst OP et al (2008) High-resolution distance mapping in rhodopsin reveals the pattern of helix movement due to activation. Proc Natl Acad Sci U S A 105(21):7439–7444. https://doi.org/10.1073/pnas.0802515105
10. Jao CC, Hegde BG, Chen J et al (2008) Structure of membrane-bound α-synuclein from site-directed spin labeling and computational refinement. Proc Natl Acad Sci U S A 105 (50):19666–19671. https://doi.org/10.1073/pnas.0807826105

11. Bravo-Arredondo JM, Kegulian NC, Schmidt T et al (2018) The folding equilibrium of huntingtin exon 1 monomer depends on its polyglutamine tract. J Biol Chem 293 (51):19613–19623. https://doi.org/10.1074/jbc.RA118.004808
12. Feller SE, Pastor RW (1996) On simulating lipid bilayers with an applied surface tension: periodic boundary conditions and undulations. Biophys J 71(3):1350–1355. https://doi.org/10.1016/S0006-3495(96)79337-2
13. Feller SE, Pastor RW (1999) Constant surface tension simulations of lipid bilayers: the sensitivity of surface areas and compressibilities. J Chem Phys 111:1281–1287. https://doi.org/10.1063/1.479313
14. Im W, Brooks CL (2005) Interfacial folding and membrane insertion of designed peptides studied by molecular dynamics simulations. Proc Natl Acad Sci U S A 102 (19):6771–6776. https://doi.org/10.1073/pnas.0408135102
15. Grossfield A, Feller SE, Pitman MC (2006) A role for direct interactions in the modulation of rhodopsin by w-3 polyunsaturated lipids. Proc Natl Acad Sci U S A 103(13):4888–4893. https://doi.org/10.1073/pnas.0508352103
16. Dror RO, Arlow DH, Maragakis P et al (2011) Activation mechanism of the β2-adrenergic receptor. Proc Natl Acad Sci U S A 108 (46):18684–18689. https://doi.org/10.1073/pnas.1110499108
17. Ingólfsson HI, Melo MN, van Eerden FJ et al (2014) Lipid Organization of the Plasma Membrane. J Am Chem Soc 136 (41):14554–14559. https://doi.org/10.1021/ja507832e
18. Neale C, García AE (2018) Methionine 170 is an environmentally sensitive membrane anchor in the disordered HVR of K-Ras4B. J Phys Chem B 122(44):10086–10096. https://doi.org/10.1021/acs.jpcb.8b07919
19. Ulmschneider JP, Smith JC, White SH, Ulmschneider MB (2011) In Silico partitioning and transmembrane insertion of hydrophobic peptides under equilibrium conditions. J Am Chem Soc 133(39):15487–15495. https://doi.org/10.1021/ja204042f
20. Smith AK, Lockhart C, Klimov DK (2016) Does replica exchange with solute tempering efficiently sample Aβ peptide conformational ensembles? J Chem Theory Comput 12 (10):5201–5214. https://doi.org/10.1021/acs.jctc.6b00660
21. Vila-Viçosa D, Silva TFD, Slaybaugh G et al (2018) Membrane-induced pKa shifts in wt-pHLIP and its L16H variant. J Chem Theory Comput 14(6):3289–3297. https://doi.org/10.1021/acs.jctc.8b00102
22. Allender DW, Sodt AJ, Schick M (2019) Cholesterol-dependent bending energy is important in cholesterol distribution of the plasma membrane. Biophys J 116 (12):2356–2366. https://doi.org/10.1016/j.bpj.2019.03.028
23. Khan HM, Souza PCT, Thallmair S et al (2020) Capturing choline–aromatics Cation–π interactions in the MARTINI force field. J Chem Theory Comput 16(4):2550–2560. https://doi.org/10.1021/acs.jctc.9b01194
24. Sun D, Forsman J, Woodward CE (2017) Molecular simulations of Melittin-induced membrane pores. J Phys Chem B 121 (44):10209–10214. https://doi.org/10.1021/acs.jpcb.7b07126
25. Pino-Angeles A, Lazaridis T (2018) Effects of peptide charge, orientation, and concentration on melittin transmembrane pores. Biophys J 114(12):2865–2874. https://doi.org/10.1016/j.bpj.2018.05.006
26. Ulmschneider JP (2017) Charged antimicrobial peptides can translocate across membranes without forming channel-like pores. Biophys J 113(1):73–81. https://doi.org/10.1016/j.bpj.2017.04.056
27. Wildermuth KD, Monje-Galvan V, Warburton LM, Klauda JB (2019) Effect of membrane lipid packing on stable binding of the ALPS peptide. J Chem Theory Comput 15 (2):1418–1429. https://doi.org/10.1021/acs.jctc.8b00945
28. Mirjalili V, Feig M (2015) Interactions of amino acid side-chain analogs within membrane environments. J Phys Chem B 119 (7):2877–2885. https://doi.org/10.1021/jp511712u
29. Nymeyer H, Woolf TB, Garcia AE (2005) Folding is not required for bilayer insertion: replica exchange simulations of an α-helical peptide with an explicit lipid bilayer. Proteins 59(4):783–790. https://doi.org/10.1002/prot.20460
30. Neale C, Hsu JCY, Yip CM, Pomès R (2014) Indolicidin binding induces thinning of a lipid bilayer. Biophys J 106(8):L29–L31. https://doi.org/10.1016/j.bpj.2014.02.031
31. Dignon GL, Zerze GH, Mittal J (2017) Interplay between membrane composition and structural stability of membrane-bound hIAPP. J Phys Chem B 121(37):8661–8668. https://doi.org/10.1021/acs.jpcb.7b05689
32. Antila H, Buslaev P, Favela-Rosales F et al (2019) Headgroup structure and cation binding in phosphatidylserine lipid bilayers. J Phys

Chem B 123(43):9066–9079. https://doi.org/10.1021/acs.jpcb.9b06091
33. Klauda JB, Venable RM, Freites JA et al (2010) Update of the CHARMM all-atom additive force field for lipids: validation on six lipid types. J Phys Chem B 114(23):7830–7843. https://doi.org/10.1021/jp101759q
34. Dickson CJ, Madej BD, Skjevik AA et al (2014) Lipid14: the Amber lipid force field. J Chem Theory Comput 10(2):865–879. https://doi.org/10.1021/ct4010307
35. Poger D, Caron B, Mark AE (2016) Validating lipid force fields against experimental data: Progress, challenges and perspectives. Biochim Biophys Acta 1858(7 Pt B):1556–1565. https://doi.org/10.1016/j.bbamem.2016.01.029
36. Kučerka N, Nieh M-P, Katsaras J (2011) Fluid phase lipid areas and bilayer thicknesses of commonly used phosphatidylcholines as a function of temperature. Biochim Biophys Acta 1808(11):2761–2771. https://doi.org/10.1016/j.bbamem.2011.07.022
37. Nagle JF, Tristram-Nagle S (2000) Structure of lipid bilayers. Biochim Biophys Acta 1469(3):159–195. https://doi.org/10.1016/S0304-4157(00)00016-2
38. Mallikarjunaiah KJ, Leftin A, Kinnun JJ et al (2011) Solid-state ^{2}H NMR shows equivalence of dehydration and osmotic pressures in lipid membrane deformation. Biophys J 100(1):98–107. https://doi.org/10.1016/j.bpj.2010.11.010
39. Bussi G, Donadio D, Parrinello M (2007) Canonical sampling through velocity rescaling. J Chem Phys 126(1):014101. https://doi.org/10.1063/1.2408420
40. Harvey SC, Tan RK-Z, Cheatham TE (1998) The flying ice cube: velocity rescaling in molecular dynamics leads to violation of energy equipartition. J Comput Chem 19(7):726–740. https://doi.org/10.1002/(SICI)1096-987X(199805)19:7<726::AID-JCC4>3.0.CO;2-S
41. Braun E, Moosavi SM, Smit B (2018) Anomalous effects of velocity rescaling algorithms: the flying ice cube effect revisited. J Chem Theory Comput 14(10):5262–5272. https://doi.org/10.1021/acs.jctc.8b00446
42. Martyna GJ, Tobias DJ, Klein ML (1994) Constant pressure molecular dynamics algorithms. J Chem Phys 101:4177–4189. https://doi.org/10.1063/1.467468
43. Feller SE, Zhang Y, Pastor RW, Brooks BR (1995) Constant pressure molecular dynamics simulation: the Langevin piston method. J Chem Phys 103:4613–4621. https://doi.org/10.1063/1.470648
44. Åqvist J, Wennerström P, Nervall M et al (2004) Molecular dynamics simulations of water and biomolecules with a Monte Carlo constant pressure algorithm. Chem Phys Lett 384(4–6):288–294. https://doi.org/10.1016/j.cplett.2003.12.039
45. Balusek C, Hwang H, Lau CH et al (2019) Accelerating membrane simulations with hydrogen mass repartitioning. J Chem Theory Comput 15(8):4673–4686. https://doi.org/10.1021/acs.jctc.9b00160
46. Lee J, Cheng X, Swails JM et al (2016) CHARMM-GUI input generator for NAMD, GROMACS, AMBER, OpenMM, and CHARMM/OpenMM simulations using the CHARMM36 additive force field. J Chem Theory Comput 12(1):405–413. https://doi.org/10.1021/acs.jctc.5b00935
47. Izadi S, Anandakrishnan R, Onufriev AV (2014) Building water models: a different approach. J Phys Chem Lett 5(21):3863–3871. https://doi.org/10.1021/jz501780a
48. Izadi S, Onufriev AV (2016) Accuracy limit of rigid 3-point water models. J Chem Phys 145(7):074501. https://doi.org/10.1063/1.4960175
49. Humphrey W, Dalke A, Schulten K (1996) VMD: visual molecular dynamics. J Mol Graph 14(1):33–38. https://doi.org/10.1016/0263-7855(96)00018-5
50. The PyMOL Molecular Graphics System, Version 1.7.0. Schrödinger, LLC
51. Hanwell MD, Curtis DE, Lonie DC et al (2012) Avogadro: an advanced semantic chemical editor, visualization, and analysis platform. J Cheminformatics 4(1):1–17. https://doi.org/10.1186/1758-2946-4-17
52. Jo S, Kim T, Iyer VG, Im W (2008) CHARMM-GUI: a web-based graphical user interface for CHARMM. J Comput Chem 29(11):1859–1865. https://doi.org/10.1002/jcc.20945
53. Martínez L, Andrade R, Birgin EG, Martínez JM (2009) PACKMOL: a package for building initial configurations for molecular dynamics simulations. J Comput Chem 30(13):2157–2164. https://doi.org/10.1002/jcc.21224
54. Eastman P, Swails J, Chodera JD et al (2017) OpenMM 7: rapid development of high performance algorithms for molecular dynamics. PLoS Comput Biol 13(7):e1005659. https://doi.org/10.1371/journal.pcbi.1005659
55. Romo TD, Leioatts N, Grossfield A (2014) Lightweight object oriented structure analysis: tools for building tools to analyze molecular

dynamics simulations. J Comput Chem 35 (32):2305–2318. https://doi.org/10.1002/jcc.23753
56. Schmit JD, Kariyawasam NL, Needham V, Smith PE (2018) SLTCAP: a simple method for calculating the number of ions needed for MD simulation. J Chem Theory Comput 14 (4):1823–1827. https://doi.org/10.1021/acs.jctc.7b01254
57. Abraham MJ, Murtola T, Schulz R et al (2015) GROMACS: high performance molecular simulations through multi-level parallelism from laptops to supercomputers. SoftwareX 1–2:19–25. https://doi.org/10.1016/j.softx.2015.06.001
58. Sur S, Romo TD, Grossfield A (2018) Selectivity and mechanism of fengycin, an antimicrobial lipopeptide, from molecular dynamics. J Phys Chem B 122(8):2219–2226. https://doi.org/10.1021/acs.jpcb.7b11889
59. Cui H, Lyman E, Voth GA (2011) Mechanism of membrane curvature sensing by amphipathic helix containing proteins. Biophys J 100 (5):1271–1279. https://doi.org/10.1016/j.bpj.2011.01.036
60. Gautier R, Bacle A, Tiberti ML et al (2018) PackMem: a versatile tool to compute and visualize interfacial packing defects in lipid bilayers. Biophys J 115(3):436–444. https://doi.org/10.1016/j.bpj.2018.06.025
61. Dominguez L, Foster L, Straub JE, Thirumalai D (2016) Impact of membrane lipid composition on the structure and stability of the transmembrane domain of amyloid precursor protein. Proc Natl Acad Sci U S A 113(36): E5281–E5287. https://doi.org/10.1073/pnas.1606482113
62. Camley BA, Lerner MG, Pastor RW, Brown FLH (2015) Strong influence of periodic boundary conditions on lateral diffusion in lipid bilayer membranes. J Chem Phys 143 (24):243113. https://doi.org/10.1063/1.4932980
63. Frishman D, Argos P (1995) Knowledge-based protein secondary structure assignment. Proteins 23(4):566–579. https://doi.org/10.1002/prot.340230412
64. Gupta C, Ren Y, Mertz B (2018) Cooperative nonbonded forces control membrane binding of the pH-low insertion peptide pHLIP. Biophys J 115(12):2403–2412. https://doi.org/10.1016/j.bpj.2018.11.002
65. Sorin EJ, Pande VS (2005) Exploring the helix-coil transition via all-atom equilibrium ensemble simulations. Biophys J 88 (4):2472–2493. https://doi.org/10.1529/biophysj.104.051938
66. Zuckerman DM, Chong LT (2017) Weighted ensemble simulation: review of methodology, applications, and software. Annu Rev Biophys 46:43–57. https://doi.org/10.1146/annurev-biophys-070816-033834
67. Lomize MA, Lomize AL, Pogozheva ID, Mosberg HI (2006) OPM: orientations of proteins in membranes database. Bioinformatics 22 (5):623–625. https://doi.org/10.1093/bioinformatics/btk023
68. Woolf TB, Roux B (1996) Structure, energetics, and dynamics of lipid-protein interactions: a molecular dynamics study of the gramicidin A channel in a DMPC bilayer. Proteins 24 (1):92–114. https://doi.org/10.1002/(SICI)1097-0134(199601)24:1<92::AID-PROT7>3.0.CO;2-Q
69. Bu L, Im W, Brooks IIICL (2007) Membrane assembly of simple helix homo-oligomers studied via molecular dynamics simulations. Biophys J 92(3):854–863. https://doi.org/10.1529/biophysj.106.095216
70. Hunt JF, Earnest TN, Bousché O et al (1997) A biophysical study of integral membrane protein folding. Biochemistry 36 (49):15156–15176. https://doi.org/10.1021/bi970146j
71. Raunest M, Kandt C (2011) dxTuber: detecting protein cavities, tunnels and clefts based on protein and solvent dynamics. J Mol Graph Model 29(7):895–905. https://doi.org/10.1016/j.jmgm.2011.02.003
72. Comer J, Gumbart JC, Hénin J et al (2015) The adaptive biasing force method: everything you always wanted to know but were afraid to ask. J Phys Chem B 119(3):1129–1151. https://doi.org/10.1021/jp506633n
73. Romo TD, Grossfield A (2014) Unknown unknowns: the challenge of systematic and statistical error in molecular dynamics simulations. Biophys J 106(8):1553–1554. https://doi.org/10.1016/j.bpj.2014.03.007
74. MacCallum JL, Bennett WFD, Tieleman DP (2008) Distribution of amino acids in a lipid bilayer from computer simulations. Biophys J 94(9):3393–3404. https://doi.org/10.1529/biophysj.107.112805
75. Lin D, Grossfield A (2015) Thermodynamics of micelle formation and membrane fusion modulate antimicrobial lipopeptide activity. Biophys J 109(4):750–759. https://doi.org/10.1016/j.bpj.2015.07.011
76. Hazel A, Chipot C, Gumbart JC (2014) Thermodynamics of deca-alanine folding in water. J Chem Theory Comput 10(7):2836–2844. https://doi.org/10.1021/ct5002076

77. Ngo VA, Kim I, Allen TW, Noskov SY (2016) Estimation of potentials of mean force from nonequilibrium pulling simulations using both minh-Adib estimator and weighted histogram analysis method. J Chem Theory Comput 12(3):1000–1010. https://doi.org/10.1021/acs.jctc.5b01050

78. Dickson CJ, Hornak V, Pearlstein RA, Duca JS (2017) Structure–kinetic relationships of passive membrane permeation from multiscale modeling. J Am Chem Soc 139(1):442–452. https://doi.org/10.1021/jacs.6b11215

Part V

pH and Polarization Effects

Chapter 11

pK_a Calculations in Membrane Proteins from Molecular Dynamics Simulations

Nuno F. B. Oliveira, Tomás F. D. Silva, Pedro B. P. S. Reis, and Miguel Machuqueiro

Abstract

The conformational changes of membrane proteins are crucial to their function and usually lead to fluctuations in the electrostatic environment of the protein surface. A very effective way to quantify these changes is by calculating the pK_a values of the protein's titratable residues, which can be regarded as electrostatic probes. To achieve this, we need to take advantage of the fast and reliable pK_a calculators developed for globular proteins and adapt them to include the explicit effects of membranes. Here, we provide a detailed linear response approximation protocol that uses our own software (PypKa) to calculate reliable pK_a values from short MD simulations of membrane proteins.

Key words Protonation, Conformation, Lipid bilayer, Poisson–Boltzmann, Monte Carlo, Linear response approximation

1 Introduction

Membrane proteins (MPs) are key players in many biological processes and have been estimated to be 20 to 30% of all proteins encoded in the human genome [1]. These proteins can insert partially or present a full insertion across the lipid bilayer. It is remarkable how MPs live in such an anisotropic environment, which changes from very high dielectric water to polar head group region and very low dielectric region in the lipid tails, all over a few angstroms distance. These are the two-extreme media in terms of polarity, and MPs were evolved to take advantage of this scenario. The complexity of the embedding media provides ample possibilities for electrostatic, hydrogen bond, and Van der Waals-type interactions. It is no surprise that physicochemical properties like pH or ionic strength can also influence the MPs conformational

Irina S. Moreira et al. (eds.), *Computational Design of Membrane Proteins*, Methods in Molecular Biology, vol. 2315, https://doi.org/10.1007/978-1-0716-1468-6_11,

space and overall stability. However, these effects will probably be harder to predict since the electrostatic interactions and protonation events will differ significantly from the bulk solution [2, 3].

In recent years, simulations of MPs have been sprouting in the literature due to steady advances in computer hardware, new methodologies, and force field development. These simulations are currently being used to support experimental findings, to unravel protein functions, and to help validate protein crystal structures [4]. The correct description of the electrostatic interactions is essential for the reliability of these simulations, and the inclusion of pH effects, by setting the most representative protonation states, is a challenging task. Therefore, it is very important that we adapt our protocols, developed to calculate pK_a values in well-solvated globular proteins, to allow the inclusion of membrane patches [5] and obtain a better description of the electrostatic environment in MPs.

Within the spectrum of pK_a predictor methods, those that use the Poisson-Boltzmann (PB) equation are usually regarded as the most accurate. By treating the solvent as a high dielectric medium while modeling the protein atoms as charges with a low dielectric, the PB framework is able to reliably derive the electrostatic-dependent quantities such as the solvation energy. Our newly developed pK_a predictor (PypKa) [5] implements this framework in a high-throughput computing tool that can be easily used from both a command-line interface (CLI) and an application programming interface (API). PypKa streamlines pK_a calculations in proteins obtained from both the PDB repository and MD simulations (supports GROMOS, AMBER, and CHARMM naming schemes) that can be used with solvated proteins as well as MPs. Its robust pipeline includes a preprocessing module, which interfaces with PDB2PQR [6], to handle different naming schemes, missing atoms, and the removal of solvent and cofactors, the PB solver DelPhi [7], including validated charges, radii, and PB parameter defaults [2, 8–10] to obtain the solvation and titratable sites interaction contributions and a parallelizable Monte Carlo routine to sample protonation states at different pH values. Using a single structure to estimate pK_a values is a rough approximation that only holds when the input structure is close to the ensemble average of both protonation states, which assumes that the site's environment is not strongly altered by changes in the protonation. Therefore, pK_a estimations in titratable sites located on protein domains undergoing conformational transitions or in regions that are not completely solvent-exposed are severely affected and will be harder to estimate. MPs usually feature a complex conformational space, and many of their key titratable sites are located at the membrane interface. In these systems, the single conformation pK_a calculations will most likely fail, and the inclusion of conformational sampling is of the utmost importance to improve the quality of the predictions. PypKa can be easily incorporated into a linear response approximation (LRA) method [11–14] that averages the pK_a values of a site over conformations sampled with MD simulations of the protonated and deprotonated states. With this approach, it is possible to overcome the biasing effect of sampling a single protonation state, and by bridging together the two ensembles, it can capture some of the protonation-coupled conformation space around the selected site.

The mitochondrial ATP/ADP carrier (AAC) is a large MP inserted on the inner mitochondrial membrane which has its ADP/ATP transport function modulated by pH [15]. This MP has two distinct forms: C-state, where its channel opens to the intermembrane space and is responsible for the transport of ADP to the matrix; and M-state, where the channel opens to the mitochondrial matrix and is ready to shuttle ATP from the matrix to the intermembrane space. The AAC internal channel used for ATP and ADP transport is rich in positively charged residues, thus taking advantage of the negative charges in these substrates (−4 and −3 charges, respectively). The electrostatic interactions involved in this process are pivotal for the high affinity and selectivity of the AAC. Many titratable amino acid residues are located in the channel and at the entrance on both sides of this MP, which are expected to change their protonation states upon binding to the substrates. These protonation-coupled conformational transitions and/or substrate binding events are very common among MPs, which emphasizes the need to develop new methods to calculate pK_a values, hence describing the electrostatic environment around these proteins.

Here, we provide a best practices protocol to calculate pK_a values in MPs in a fast and rather simple setup using our newly developed tool PypKa [5]. To that end, we devised a method based on the LRA approach to circumvent the biasing effect introduced in MD simulations [11–14]. Two residues located at the channel entrance of the ADP/ATP mitochondrial transporter were selected, one on each membrane side. Asp246 and Glu291 are located on the mitochondrial matrix and cytoplasmic sides, respectively (Fig. 1). Since the protein channel used (C-state) is opened to the intermembrane space in the cytoplasmic side, this results in a better solvent-exposed Glu291 and a more altered environment for Asp246. With these two distinct examples, we aim at helping not only to set up the protocol needed for the calculations but also to provide a detailed interpretation of the data obtained.

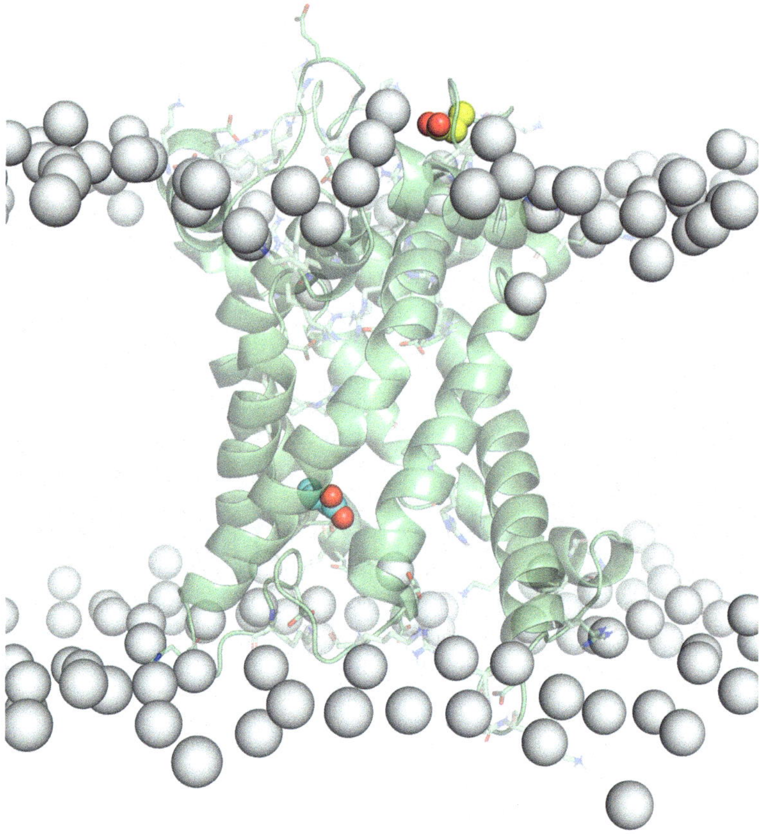

Fig. 1 Representative conformation of the mitochondrial ADP/ATP transporter in the C-state (PDBid:1OKC) [15] embedded in a POPC bilayer viewed from the side. The protein is shown as a green cartoon and the lipids are shown as gray spheres with the phosphorus atoms shown without transparency. Residues Asp246 and Glu291 are shown as yellow and cyan spheres, respectively

2 Software

1. All MD simulations were performed with GROMACS 2018.6 software [16, 17].
2. All pK_a calculations were performed with PypKa (https://github.com/mms-fcul/PypKa) [5].
3. Pymol [18] and gnuplot [19] were used to generate the figures.
4. A regular text editor such as emacs or notepad is also required.
5. A Unix shell and git to access the protocol files that are located at GitHub (https://github.com/mms-fcul/membrane-lra). Example:

   git clone https://github.com/mms-fcul/membrane-lra.git.

3 Methods

3.1 MD System Setup and Parameters

1. The transporter ATP/ADP carrier (PDBid:1OKC) was inserted into a pre-equilibrated 256 POPC membrane hydrated with SPC water molecules [20]. The clashing lipid and water molecules were removed or moved, resulting in a final 200 lipid bilayer membrane (100 in each leaflet).
2. Three different setups were built by varying the protonation state of Asp246 and Glu291: both ionized and either one of them in the neutral form (*see* **Note 1**).
3. GROMACS 2018.6 software [16, 17] and the GROMOS 54A7 force field [21] were employed in all MD simulations.
4. Particle Mesh Ewald (PME) [22, 23] with a cutoff of 1.0 nm for direct contributions [24] was used to treat long-range electrostatic interactions.
5. A Verlet atomistic scheme was used with a cutoff of 1.0 nm for the Van der Waals interactions.
6. Lipids and MP were constrained using the parallel linear constraint solver p-LINCS [25], while water molecules were constrained using the SETTLE algorithm [26].
7. The energy of each system was minimized using the steepest descent algorithm in two stages. A first stage consisting of 20k steps with a maximum step size of 0.001 nm, where no constraints were applied; and a second with 10k steps and maximum step size of 0.001 nm with all bonds constrained.
8. The systems were initialized in two steps. On the first step (0.1 ns NVT), velocities were generated by the v-rescale thermostat [27] with a coupling constant of 0.01 ps and according to a Maxwell distribution at 310 K. On the second step (1 ns NPT; *see* **Note 2**), pressure was kept constant semi-isotropically at 1 bar using the Parrinello-Rahman barostat [28, 29] with a coupling constant of 5.0 ps and compressibility of 4.5×10^{-5} bar^{-1} (*see* **Note 3**).

3.2 MD Simulations

1. A long MD simulation (300 ns) of the AAC system, with Asp246 and Glu291 in their most common ionized forms, was performed to allow for an effective equilibration of the lipid–protein interface (*see* **Note 4**).
2. We followed the RMSD, radius of gyration, DSSP, total system area (*see* **Note 5**), and the ADP/ATP channel volume (*see* **Note 6**) to ensure the MP stability in the membrane and the overall system equilibration.
3. After ensuring system equilibration, short MD simulations (10 ns; *see* **Note 7**) of the three systems (with different protonation states) were performed using the parameters described in the previous section.

4. For each system, we extracted snapshots (MP + lipids) at 100 ps intervals, which results in 100 configurations for subsequent pK_a calculations.

3.3 pK_a Calculations

1. pK_a calculations were performed on every extracted snapshot of the three systems. Our python-based program PypKa [5] performed all calculations and obtained pK_a values and protonation curves for all sites, including Asp246 and Glu291 residues. The obtained pK_a values are biased towards the protonation state of each site and should not be directly used as a macroscopic pK_a estimation. To mitigate this biasing effect, in the LRA formalism, these calculations are performed on the conformational ensembles of both protonation states. Furthermore, since all sites are titrating, their effects on the site of interest are only included in the pH-dependent average protonation data, which should be used to estimate the macroscopic pK_a values.
2. To run a PypKa calculation, it is required a gro/pdb file of the system and a settings file. The default settings file should look like the following:

```
## General parameters
structure = snapshot_001.gro
ncpus = -1 # to use all cores or specify a number of cpu cores
available to your machine.
clean_pdb = False #True if your pdb file requires a
pre-processing procedure to be compatible with PypKa.
sites_A = all # either specify the titrating residues (e.g.,:
1N, 246, 291, 292C) or calculate for "all" residues.
output = pka_output.out.
titration_output = titration_001.txt # optional for PypKa,
but required for LRA calculations.
## Poisson-Boltzmann parameters
epsin = 4.
ionicstr = 0.1.
pbc_dimensions = 2 # "0" for globular proteins, "2" for
membrane systems.
## Monte Carlo parameters
pH = -5,15 # pH range used for titration.
```

The list above only displays the required default settings to make a PypKa calculation. To run the program, type on your terminal: *"pypka < settings_filename > ."* Additional Monte Carlo and Poisson-Boltzmann parameters can be tweaked—for more information: https://pypka.readthedocs.io/en/latest/parameters.html.

3. Among the default settings, the last four parameters require more attention:
   (a) The *"epsin"* parameter refers to the dielectric constant value to be used in the Poisson-Boltzmann calculations and this value is dependent on the system/environment. In LRA, the dielectric constant also reflects the protonation-induced conformational rearrangements around the titrating site (*see* **Note 8**).
   (b) The *"ionicstr"* parameter asks for the desired ionic strength for the medium. This value reflects the electrostatic contribution of ions in solution as an implicit continuum effect.
   (c) The *"pbc_dimensions"* parameter selects the number of dimensions where periodic boundary conditions are applied. For solvated proteins, the parameter is set to zero, while for lipidic systems, it is set to two dimensions, since periodic boundary conditions are necessary on the *xy* plane.
   (d) The "*pH*" setting enables the user to delimit a pH range where the residues' titration curves and pK_a estimations are run. If, in the output, the result for a residue is "Not in range," one should decrease/increase the lower/higher pH value. Additionally, the step between the pH values can be defined;
   (e) When performing LRA calculations, the user will require the average protonation (instead of the instant pK_a values), hence, an additional entry is required: *"titration_output."* In this file, the average protonation is written, for each residue, at each step of the selected pH range.
4. Following the PypKa calculations on all systems snapshots, LRA can be used to estimate the macroscopic pK_a value of each residue for which we sampled the conformations of both protonation states. In a typical LRA calculation, at each pH value, we average the calculated protonation from all snapshots (100 in our case) and use it in the Henderson-Hasselbalch equation to obtain the pH-dependent pK values (Fig. 2). Once we obtain the pH vs pK curves for both protonation states, we can average them out to obtain the final LRA curve. The residue macroscopic pK_a value is obtained at the intersection between the $x = y$ line and the LRA curve (Fig. 2), which corresponds to the point where pH = pK.
5. From the LRA formalism, we can also calculate the reorganization energy (λ)—the free energy required to bring a protonated residue to the conformational environment of the deprotonated counterpart, and vice-versa. If the λ value is small (usually within 2 pK units), it means that there is no significant conformational reorganization induced by the protonation. However,

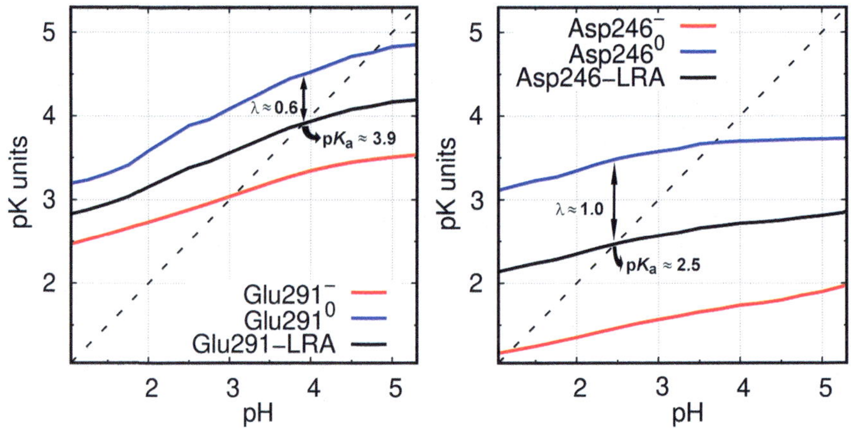

Fig. 2 LRA pK_a curves obtained for Glu291 (left) and Asp246 (right). The red and blue lines refer to the two MD conformational ensembles where the acidic residues are ionized or in their neutral forms, respectively. The pK values are obtained from the average protonation at each pH values applying the Henderson-Hasselbalch formalism. The black line is the average between the two-colored lines and the final pK_a value can be obtained at the point where pH = pK. The reorganization energies (λ) are also shown for comparison. In our example, Glu291 is better water-solvated which results in a pK_a value (3.9) similar to the Glu pentapeptide Null model (4.25) [30]. The Asp 246, on the other hand, is located on the closed channel entrance bringing it closer to several cationic residues. This positive electrostatic effect is well-captured with the lowering of the pK_a value (2.5) compared with the Null model (3.94) [31]

if λ is too high, it jeopardizes the applied linearization procedure and it will require an increase in the dielectric constant to help capture this protonation-induced conformational transition. If a dielectric constant of 8 (*see* **Note 8**) cannot decrease the λ value, then this approach will probably result in valueless pK_a estimations and should not be used.

4 Notes

1. By neutralizing each of these residues and comparing with their ionized counterpart, we can sample the conformational ensembles of both protonation states and use them in the LRA calculations of the pK_a values.
2. A low time step (1 fs) was used in this step to help the pressure equilibration and avoid errors in the LINCS constraints algorithm.
3. For MD production, the temperature and pressure coupling constants were updated to 0.1 and 1.0 ps, respectively.
4. Due to the large system size and the AAC being in the Apo state, longer equilibration times (hundreds of nanoseconds) are required to obtain an equilibrated MP. This is usually required when embedding a protein in a lipid bilayer and only needs to be performed once.

5. The total area of the system over time was computed by multiplying the xx and yy components of the simulation box vectors at each time step.
6. The ADP/ATP channel volume was estimated by defining the entrance and exiting reference atoms and counting the number of water molecules located within their *z* Cartesian coordinates.
7. The length of these production MD segments depends mainly on the system studied. Short 10 ns segments are usually enough to sample local side chain reorganizations due to protonation. Larger protonation-triggered conformational transitions would require longer production MD segments, but would also strain the limits of the LRA methodology.
8. The dielectric constant (ε) takes a slightly different role in LRA calculations. Usually, in rigid body calculations, an increase in ε is used to compensate for a lack of polarization effects and conformational freedom. Therefore, in LRA, by introducing conformational variability, one would expect that the use of $\varepsilon = 2$ should suffice to deal with the polarization effects. However, since there are usually conformational reorganizations introduced upon protonation changes, the dielectric constant assumes a role of bridging the two ensembles together. It is common to use $\varepsilon = 4$ in LRA calculation of well-solvated residues or increase this value up to 8, to deal with systems that change significantly their electrostatic environment around the (de)protonated site.

Acknowledgments

This work was supported by FCT grants CEECIND/02300/2017, SFRH/BD/136226/2018, SFRH/BD/140886/2018, PTDC/BIA-BFS/28419/2017, PTDC/FIS-OUT/28210/2017, UIDB/04046/2020, and UIDP/04046/2020. This work also used the EGI infrastructure with the dedicated support of IN2P3-IRES, and NCG-INGRID-PT.

References

1. Krogh A, Larsson B, von Heijne G et al (2001) Predicting transmembrane protein topology with a hidden Markov model: application to complete genomes. J Mol Biol 305:567–580. https://doi.org/10.1006/jmbi.2000.4315
2. Teixeira VH, Vila-Viçosa D, Pedro BP et al (2016) pKa values of titrable amino acids at the water/membrane interface. J Chem Theory Comput 12:930–934. https://doi.org/10.1021/acs.jctc.5b01114
3. Vila-Viçosa D, Silva TFD, Slaybaugh G et al (2018) Membrane-induced pKa shifts in wt-pHLIP and its L16H variant. J Chem Theory Comput 14:3289–3297. https://doi.org/10.1021/acs.jctc.8b00102
4. Goossens K, De Winter H (2018) Molecular dynamics simulations of membrane proteins: an overview. J Chem Inf Model 58:2193–2202. https://doi.org/10.1021/acs.jcim.8b00639

5. Reis, P. B. P. S., Vila-Viçosa, D., Rocchia, W., Machuqueiro, M. (2020) PypKa: a flexible Python module for Poisson–Boltzmann based pKa calculations, J Chem Inf Model, 60:4442-4448. https://pubs.acs.org/doi/10.1021/acs.jcim.0c00718
6. Dolinsky TJ, Nielsen JE, McCammon JA et al (2004) PDB2PQR: an automated pipeline for the setup of Poisson-Boltzmann electrostatics calculations. Nucleic Acids Res 32:W665–W667. https://doi.org/10.1093/nar/gkh381
7. Rocchia W, Alexov E, Honig B (2001) Extending the applicability of the nonlinear Poisson–Boltzmann equation: multiple dielectric constants and multivalent ions. J Phys Chem B 105:6507–6514. https://doi.org/10.1021/jp010454y
8. Teixeira VH, Cunha CA, Machuqueiro M et al (2005) On the use of different dielectric constants for computing individual and pairwise terms in poisson-boltzmann studies of protein ionization equilibrium. J Phys Chem B 109:14691–14706. https://doi.org/10.1021/jp052259f
9. Vila-Viçosa D, Teixeira VH, Baptista AM et al (2015) Constant-pH MD simulations of an oleic acid bilayer. J Chem Theory Comput 11:2367–2376. https://doi.org/10.1021/acs.jctc.5b00095
10. Santos HAF, Vila-Viçosa D, Teixeira VH et al (2015) Constant-pH MD simulations of DMPA/DMPC lipid bilayers. J Chem Theory Comput 11:5973–5979. https://doi.org/10.1021/acs.jctc.5b00956
11. Lee FS, Chu ZT, Warshel A (1993) Microscopic and semimicroscopic calculations of electrostatic energies in proteins by the POLARIS and ENZYMIX programs. J Comput Chem 14:161–185. https://doi.org/10.1002/jcc.540140205
12. Eberini I, Baptista AM, Gianazza E et al (2004) Reorganization in apo- and holo-β-lactoglobulin upon protonation of Glu89: molecular dynamics and pKa calculations. Proteins 54:744–758. https://doi.org/10.1002/prot.10643
13. Machuqueiro M, Campos SRR, Soares CM et al (2010) Membrane-induced conformational changes of kyotorphin revealed by molecular dynamics simulations. J Phys Chem B 114:11659–11667. https://doi.org/10.1021/jp104418g
14. Teixeira VH, Capacho ASC, Machuqueiro M (2016) The role of electrostatics in TrxR electron transfer mechanism: a computational approach. Proteins 84:1836–1843. https://doi.org/10.1002/prot.25166
15. Bidon-Chanal A, Krammer E-M, Blot D et al (2013) How do membrane transporters sense pH? The case of the mitochondrial ADP–ATP carrier. J Phys Chem Lett 21:3787–3791. https://doi.org/10.1021/jz401847d
16. Spoel DVD, Van Der Spoel D, Lindahl E et al (2005) GROMACS: fast, flexible, and free. J Comput Chem 26:1701–1718. https://doi.org/10.1002/jcc.20291
17. Abraham MJ, Murtola T, Schulz R et al (2015) GROMACS: high performance molecular simulations through multi-level parallelism from laptops to supercomputers. SoftwareX 1–2:19–25. https://doi.org/10.1016/j.softx.2015.06.001
18. The PyMOL Molecular Graphics System, Version 2.0 Schrödinger, LLC. https://pymol.org/2/
19. Williams T, Kelley C et al (2020) Gnuplot 5.4 - an interactive plotting program. http://www.gnuplot.info/
20. Hermans J, Berendsen HJC, Van Gunsteren WF et al (1984) A consistent empirical potential for water-protein interactions. Biopolymers 23:1513–1518. https://doi.org/10.1002/bip.360230807
21. Schmid N, Eichenberger AP, Choutko A et al (2011) Definition and testing of the GROMOS force-field versions 54A7 and 54B7. Eur Biophys J 40:843–856. https://doi.org/10.1007/s00249-011-0700-9
22. Darden T, York D, Pedersen L (1993) Particle mesh Ewald: AnN·log(N) method for Ewald sums in large systems. J Chem Phys 98:10089. https://doi.org/10.1063/1.464397
23. Essmann U, Perera L, Berkowitz ML et al (1995) A smooth particle mesh Ewald method. J Chem Phys 103:8577. https://doi.org/10.1063/1.470117
24. Poger D, Mark AE (2012) Lipid bilayers: the effect of force field on ordering and dynamics. J Chem Theory Comput 8:4807–4817. https://doi.org/10.1021/ct300675z
25. Hess B (2008) P-LINCS: a parallel linear constraint solver for molecular simulation. J Chem Theory Comput 4:116–122. https://doi.org/10.1021/ct700200b
26. Miyamoto S, Kollman PA (1992) Settle: an analytical version of the SHAKE and RATTLE algorithm for rigid water models. J Comput

Chem 13:952–962. https://doi.org/10.1002/jcc.540130805

27. Bussi G, Donadio D, Parrinello M (2007) Canonical sampling through velocity rescaling. J Chem Phys 126:014101. https://doi.org/10.1063/1.2408420
28. Nosé S, Klein ML (1983) Constant pressure molecular dynamics for molecular systems. Mol Phys 50:1055–1076. https://doi.org/10.1080/00268978300102851
29. Parrinello M, Rahman A (1981) Polymorphic transitions in single crystals: a new molecular dynamics method. J Appl Phys 52:7182. https://doi.org/10.1063/1.328693
30. Thurlkill RL, Grimsley GR, Martin Scholtz J et al (2006) pK values of the ionizable groups of proteins. Protein Sci 15:1214–1218. https://doi.org/10.1110/ps.051840806
31. Grimsley GR, Scholtz JM, Pace CN (2009) A summary of the measured pK values of the ionizable groups in folded proteins. Protein Sci 18:247–251. https://doi.org/10.1002/pro.19

Chapter 12

Poor Person's pH Simulation of Membrane Proteins

Chitrak Gupta, Umesh Khaniya, John W. Vant, Mrinal Shekhar, Junjun Mao, M. R. Gunner, and Abhishek Singharoy

Abstract

pH conditions are central to the functioning of all biomolecules. However, implications of pH changes are nontrivial on a molecular scale. Though a rigorous microscopic definition of pH exists, its implementation in classical molecular dynamics (MD) simulations is cumbersome, and more so in large integral membrane systems. In this chapter, an integrative pipeline is described that combines Multi-Conformation Continuum Electrostatics (MCCE) computations with MD simulations to capture the effect of transient protonation states on the coupled conformational changes in transmembrane proteins. The core methodologies are explained, and all the software required to set up this pipeline are outlined with their key parameters. All associated analyses of structure and function are provided using two case studies, namely those of bioenergetic complexes: NADH dehydrogenase (complex I) and V_o domain of V-type ATPase. The hybrid MCCE-MD pipeline has allowed the discovery of hydrogen bond networks, ligand binding pathways, and disease-causing mutations.

Key words Molecular dynamics, Multi-Conformation Continuum Electrostatics, Hydrogen bond network, Proton transfer, Potential of mean force, Complex I, V-ATPase

1 Introduction

The widely accepted *sequence-structure-function* paradigm manifests distinct mechanistic pathways to mitigate changes in the native conditions of a protein. For example, protein unfolding and refolding can, in addition to temperature and ions, also be induced by switching the pH of the solution [1]. This observation stresses the role of proton concentration in mediating secondary and tertiary structures [2, 3], and therefore controlling overall protein stability [4–6]. pH also determines the charge of acidic and basic surface residues, that is critical for protein-protein association in multimeric complexes [7]. Long-range molecular recognition in crowded cellular environments is therefore highly pH-dependent [8, 9]. Charge transport across cell membranes is driven by membrane potential and pH gradients [10] coupled with

Irina S. Moreira et al. (eds.), *Computational Design of Membrane Proteins*, Methods in Molecular Biology, vol. 2315, https://doi.org/10.1007/978-1-0716-1468-6_12, © Springer Science+Business Media, LLC, part of Springer Nature 2021

conformational changes. The so-called proton motive force generated by the integral membrane electron transfer chains of photosynthesis and of aerobic and anaerobic respiration is responsible for ATP synthesis in all living cells. Taken together, the proton buffering of biomolecules is indispensable for in silico, in vitro, or in vivo studies of their functions.

The formal definition of buffer pH $= -\log(a_{H^+})$ comes from macroscopic or bulk quantities. However, the single-molecule implications of pH are nontrivial. Given the finite size of the molecular problems, the microscopic definition of pH is tied to the pK_a of the amino acids in a probabilistic framework using the relationship [11].

$$p(\lambda) = \frac{10^{pK_a^{(i),\mu}}(\lambda', \lambda) - (n_\lambda - n_{\lambda'})pH}{\sum_{\lambda''} 10^{pK_a^{(i),\mu}}(\lambda', \lambda'') - (n_{\lambda''} - n_{\lambda'})pH}, \quad (1)$$

where control parameter λ implies the deprotonated state of an amino acid when set to 0 and the protonated state when set to 1. n_λ is the number of discrete steps required to reversibly introduce a proton on the protonation site, and $p(\lambda)$ is the probability of finding a site protonated, deprotonated, or partially protonated. The macroscopic proton motive force that drives cellular functions will have to be translated into a microscopic pH gradients to observe how the latter impacts single-protein function.

In this chapter, by using two exemplary bioenergetic proteins, namely NADH dehydrogenase (or complex I) and V/A-ATPases (analogous to F-ATPases or complex V), we describe molecular dynamics (MD) simulation to showcase how a pH gradient is created across the mitochondrial membrane by complex I and utilized by complex V or its analogues. The MD simulations are complemented by methodologies that monitor the microscopic pH and pK_a shifts of the amino acids in response to the evolving protein conformations. Normally such pK_a changes, seen in proton transfer reactions, are monitored either using the quantum mechanics/molecular mechanics (QM/MM) computations of protonation state changes [12, 13] or nonequilibrium MD based on free-energy perturbation schemes, that is, the constant pH simulations [14]. However, the system size and membrane environment make such simulations intractable for the examples at hand. For example, a protein with N interacting acidic and basic residues can exist in N charged states with 2^N possible protonation microstates, differing in proton number, position, and protonation energy. Empirical electrostatic methods such as Propka, though straightforward to use, fail to capture the membrane environment of the proteins [15, 16]. To address this issue, we monitor the cross-talk between single-molecule dynamics and microscopic changes in pH by using MCCE (multiconformation continuum electrostatics), a Monte

Carlo simulation-based scheme for finding the likely protonation states of all amino acids and ligands in the protein [17]. We show how an iterative cycle switching between the converged MD and MC simulations finds the molecular sites and resolves the molecular mechanism for the generation and utilization of the proton motive force (Fig. 1).

We have reported pH simulations of the respiratory complex I (Fig. 2a), which revealed the reduction-driven conformational transitions in the enzyme [18, 19]. Complex I is the first and the largest protein complex of the mitochondrial respiratory chain that harvests electrons from NADH to reduce quinone, while pumping protons across the inner mitochondrial membrane. Details of the mechanism of such electron transfer, specifically, the interplay between reduction of iron-sulfur clusters, large-scale conformational transitions, and hydrated proton transfer, remain fragmentary. We identified the redox switches that allosterically couple the sites of NADH reduction and quinone binding [18, 19]. pH simulations revealed how the coupling between quinone binding and long-range hydrogen-bond networks is responsible for proton pumping and generation of a proton gradient across the mitochondrial membrane.

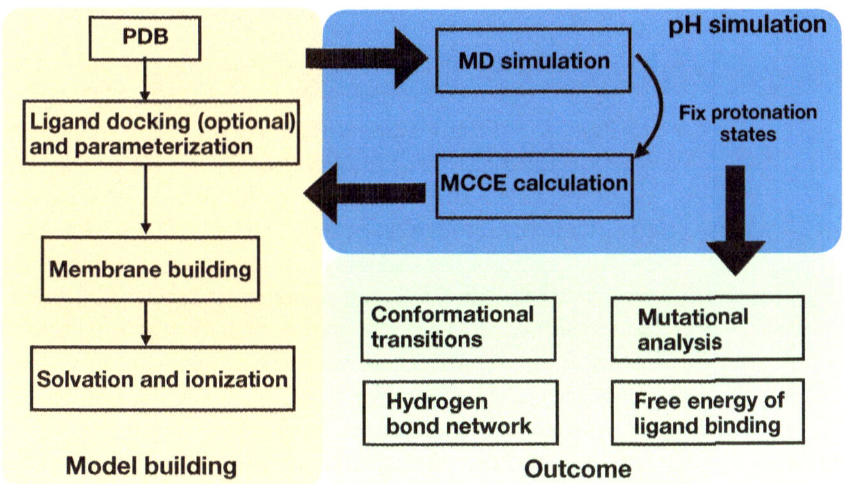

Fig. 1 Protocol for combined MD/MCCE simulation Model building: The initial structure is docked with the ligand (if needed), and all non-protein species are parameterized for MD simulation. A lipid membrane is generated and the protein-ligand system is embedded into it. This system is then solvated and ionized to generate the complete simulation. *pH simulation:* Unbiased MD simulation is performed with this system, where protonation states of all protonatable residues are fixed. Selected snapshots from this trajectory are then used for MCCE calculations in order to obtain favorable protonation states. These MCCE results can then be used to initiate new MD simulations after re-solvating and re-ionizing. *Outcome:* The simulation trajectories thus obtained can be used to study conformational transitions, mutation analysis, hydrogen bond networks, and free energy of ligand binding. Methods for each of these are described here, and examples are provided from two case studies

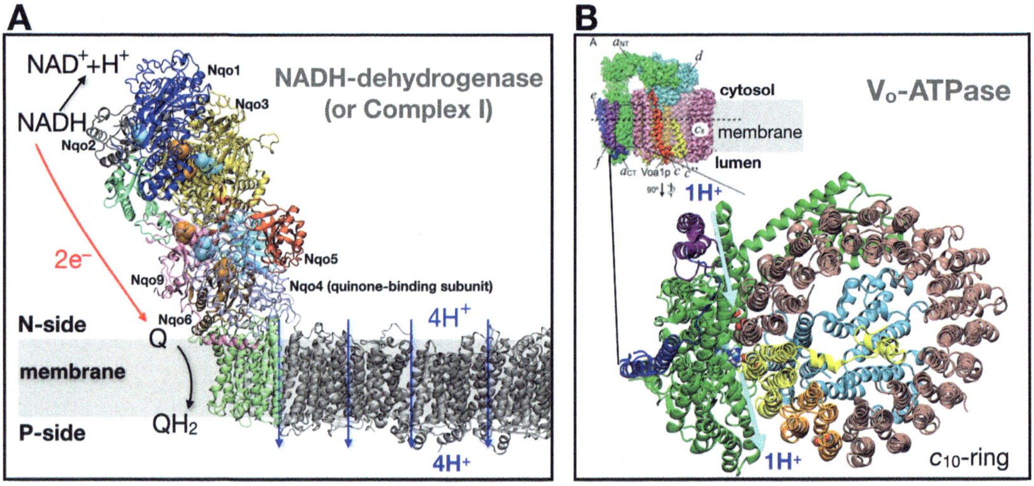

Fig. 2 The systems used for case study. (**a**) Crystal structure of complex I. Soluble domain is colored by subunits. NADH-binding subunit is shown in blue. Iron-sulfur clusters are shown in orange (reduced) or cyan (oxidized). Quinone headgroup-binding subunit is shown in ribbon representation for clarity. The transmembrane part is shown in silver, except for the quinone tail-coordinating subunit Nqo8, which is shown in green. (**b**) The integral membrane *c*-ring of the V_o-ATPase showing 10 two-helix subunits, and protein entry and exit pathways at the subunit *a*–subunit *c* interface. (Inset) 2.7 Å map of V_oND showing subunits *a* (green), *d* (cyan), *c*8 (pink), c' (orange), c" (yellow), *e* (blue), *f* (purple), and V_oa1p (red). The densities for two of the *c* subunits are removed to allow a view inside the c-ring

Seeking the mechanisms of consuming the proton motive force for ATP turnover, we have also reported the MD simulation of V_o ATPase, identifying water-mediated proton transport and autoinhibition mechanisms (Fig. 2b). Rotary V-ATPases drive transmembrane proton transport through a V_o proton channel subcomplex, the mechanism of which remains elusive. Our results demonstrated thermal equilibrium between the two rotary states of the subcomplex, and identified the role of protonation state of specific glutamic acid residues in coupling water-mediated proton transfer with V_o rotation. In combination with results from cryo electron microscopy (cryoEM) experiments, we rationalized a mechanism for passive proton transport in the V_o ATPase.

In subsequent sections, we provide a guide to MCCE-MD simulations, software required, and description of input parameters. We then present a case study discussing how complex I and complex V converts pH changes to conformational transitions, a process that is essential to life [20]. The necessary scripts and files to build the system and perform simulations are freely available to users on GitHub [21]. Note, while preparing this chapter it is assumed that the user feels comfortable using the UNIX command line and has basic experience with molecular dynamics using NAMD and scripting in VMD.

2 Materials

This section describes the input files and the software tools required to perform pH simulation of mitochondrial proteins.

2.1 Input Files

Following files are needed as input data:

1. **PDB**: The starting structural coordinates of the given protein are provided by the Protein Data Bank (PDB) file. PDB files for the protein of interest can be accessed and downloaded from the PDB database https://www.rcsb.org. For example, the starting 3D coordinates for V_o ATPase, the biomolecule in one of the case studies in this book chapter, were obtained using RCSB entry 6C6L (https://www.rcsb.org/structure/6C6L). For the other case study, we used an unpublished structure of complex I, which was modified from the RCSB entry 4HEA (https://www.rcsb.org/structure/4hea).
2. **Force field parameters**: Force field files contain numerical parameters (masses, charges, and spring constants) needed to set up a potential energy function for evaluating how the atoms from different chemical constituents move due to bonded, angular, dihedral, and nonbonded (electrostatic or van der Waals) interactions. These files typically consist of a topology file and a parameter file which has been generated either by spectroscopic data, quantum mechanical calculations, or through comparison of atoms with novel connectivity to known parameters. Topology files contain information about atom names, bond angles, and charges and end in the `.inp` file suffix for the CHARMM series of force fields. Parameter files contain constants required for an energy function, namely bond angle force constants and equilibrium angles bond and angle constants, and typically ends in the `.par` file suffix for the CHARMM simulations. More information on these files can be found at https://www.ks.uiuc.edu/Training/Tutorials/science/forcefield-tutorial/forcefield-html/node6.html or https://www.ks.uiuc.edu/Training/Tutorials/namd/namd-tutorial-unix-html/node25.html. Generation of a parameter file for a novel chemical constituent can be done with the CGenFF server (https://cgenff.umaryland.edu).

2.2 Software

The user should have access to the following software tools to run a large system simulation. MCCE parameters are contained in .ftpl files. They include a listing of all possible protonation and redox states to be sampled for each residue type. These choices are called conformers. For each conformer type the atomic connectivity, atomic partial charges, atom radius, and van der Waals parameters are given. If there are multiple tautomer states, each is treated as a separate conformer. In addition, for residues with multiple protonation and redox states, the pK_a and E_m in solution as well as the solution solvation energy of each redox, protonation and tautomer state is given. Small molecules such as ions can equilibrate with the protein. These ligands have a topology file that includes a conformer that has moved into the water.

1. **MCCE**: MCCE is a biophysics simulation program that combines continuum electrostatics and molecular mechanics. In this program, the protein side chain motions are simulated explicitly while the dielectric effect of solvent and bulk protein material is modeled by continuum electrostatics. It can be downloaded from https://gunnerlab.github.io/Stable-MCCE.

   Changes in protonation and redox state represent changes in chemical species, which cannot occur in standard MD simulation. Thus, MD chooses a single protonation state for each group, while Monte Carlo analysis finds the equilibrium ensemble of protonation states. In MCCE the changes in side chain and ligand position sample many positions, but they are discrete and on a rigid backbone. Here MD, with its continuous motions can explore space more effectively. The coupling together of Monte Carlo sampling of protonation states with MD moves should allow us to find the equilibrium distribution of these states in a conformational equilibrated protein. The overall goal is to follow the distribution of protonation states through the protein while the protein changes conformation due to the reaction. Constant pH MD is an emerging technique to accomplish this goal [22–30]. However, charge changes needed to equilibrate the protonation number and distribution creates a major disturbance slowing convergence, especially for the proteins of interest here where many buried sites have closely coupled ionization states.

   A different approach uses separate cycles of MD to generate a diversity of conformations and MC to find the equilibrium destitution of states in a modest number of frames. It is well established that this method can lead to errors as MD will stabilize the initial, imposed protonation state [24, 31–33]. This is one advantage of carrying out MCCE on crystal structures and seeding MD trajectories with proton distributions favored in MCCE. Another is to cluster the trajectory to find different positions for the protonatable residues of interest, using snapshots from each trajectory cluster for MCCE analysis. This method can find minority conformations in the trajectory that take different protonation states. However, it should be recognized that an MD trajectory will always tend to generate conformations that favor the input charge state.
2. **VMD**: Visual Molecular Dynamics (VMD) is a computer program, designed for molecular modeling and visualization of biological systems [34]. It can be downloaded from http://www.ks.uiuc.edu/Research/vmd/.

3. **NAMD**: NAMD can be used to simulate large systems consisting of millions of atoms and is noted for its parallel efficiency, scaling to thousands of nodes (CPUs or GPUs). The software is widely used with CHARMM force fields but is also compatible with other popular force fields like AMBER and OPLS. It can be downloaded from http://www.ks.uiuc.edu/Research/namd/.
4. **Charm++**: Charm+ + is a C+ + based system for efficient parallelization of computational tasks. It was designed with efficient portability and latency tolerance in mind, being widely used by scientific and engineering applications for efficient computation across the entire range of hardware, from local computer clusters to large supercomputers. It can be downloaded as a binary and installed from http://charm.cs.illinois.edu/software.

All software tools used for performing pH simulations are free for academic use and binary executable are available for common operating systems and machine types.

2.3 Hardware

1. **Recommended hardware**: MCCE runs well on any modern 64-bit Linux and Mac computers or clusters. 4GB computer memory and 10 GB spare disk space are recommended. NAMD runs on parallel computing platforms, either on a CPU- cluster or a GPU-cluster. Time and computational resources required to run NAMD depends on system size. As a first step, the user should benchmark using a short simulation, which gives an estimate of the time required for the full simulation to complete.
2. **Hardware used for case study**: For these case studies, we ran MCCE on a Linux 24-CPU core server with 64 GB memory. NAMD jobs were benchmarked on a GPU cluster. The simulations with 0.5–1 M system-size ran optimally on approximately 1000 GPU nodes, giving a performance of 50–100 ns/day on OLCF's Summit supercomputer.

3 Methods: Description of the Components of This Protocol

This section provides an overview of the steps involved in the performing MCCE-Molecular Dynamics simulations for integral membrane systems.

3.1 Model Building

1. Parameterization and docking. As explained in Subheading 12.4, the first step is to parameterize non-protein components, for example Fe–S clusters and quinone in case of respiratory complex I. For respiratory complex I, we docked the quinone into the binding site, as described in Subheading 12.4. The

next step is to create a structure" file (PSF), corresponding to the PDB file. The PSF file contains the following information about a molecule: atoms, partial charges of each atom, masses, bonds, angles, dihedrals, improper angles, and van der Waals terms. This step can be accomplished with the VMD's AutoPSF plugin (Extensions → Modeling → Automatic PSF Builder). By default the AutoPSF plugin in VMD 1.9.3 or higher uses the CHARMM36m force field parameters [35] for protein, nucleic acids, carbohydrates, lipid, ions, and water. However, other popular force-fields such as AMBER can be implemented as well.

2. Modeling membrane. The ligand-bound protein model is embedded into the lipid membrane using standard protocols on CHARMM-GUI [36–39]. Complex I is embedded in a simple POPC membrane, while the V_o-ATPase membrane is composed of PC:PE:PI:PS lipids in 48%:21%:27%:4% stoichiometry. For most problems the standard CHARMM-GUI protocols are adequate. However, for the hollow or ring-shaped proteins such as the V_o, one needs to model lipid patches inside the protein-excluded regions. We construct "lipid plugs" to address such issues.

   Starting from a membrane-embedded model of V_o, a grid density is created from the c-ring using the `mdff sim` command producing a .dx file (grid density resolutions of ~5 Å work the best). In VMD, candidate models for the lipid plug are created by manually placing a variable number of lipid molecules inside the pore formed by the c-ring. Inside the c-ring, lipid molecules are orientated perpendicular to the plane of the membrane for both the inner and outer leaflets. The candidate models are then refined with MDFF using the simulated density map to bias the lipids toward regions not occupied by the c-ring atoms. This procedure is atypical in MDFF, where simulated densities are usually converted to grid potentials so that the additional potential energy bias forces the molecule into the density, refining the search model to fit the map. Once the candidate models have been refined, then the lipid plugs are incorporated into the membrane-embedded model of the c-ring and minimization, and a short equilibrium simulation are performed. The model that completely plugs the c-ring pore and has the lowest potential energy is then chosen as the best model.

3. Molecular Dynamics Flexible Fitting (MDFF). The protocol for MDFF has been described in detail [40–42]. In a nutshell, starting with an experimentally determined or synthetic density map, a 3D grid potential is generated and is described as follows:

$$U_{\text{grid}}(\mathbf{r}) = \begin{cases} \zeta\left(1 - \dfrac{\Phi(\mathbf{r}) - \Phi_{\text{thr}}}{\Phi_{\max} - \Phi_{\text{thr}}}\right) & \text{if}\,\Phi(\mathbf{r}) \geq \Phi_{\text{thr}}, \\ \zeta & \text{if}\,\Phi(\mathbf{r}) < \Phi_{\text{thr}}\,. \end{cases} \tag{2}$$

where $\Phi(\mathbf{r})$ is the density map at a point $\mathbf{r}$, ζ referred to as gscale is the scaling factor that regulates the coupling strength between the map and model, Φ_{thr} is a threshold for disregarding background density due to bulk water, and $\Phi_{\max} = \max(\Phi(\mathbf{r}))$.

In MDFF, an initial protein or membrane model is refined employing MD, where the traditional potential energy surface is modified by U_{grid}. The grid potential directs the model to the density map, while simultaneously following constraints from the traditional force fields. This real space refinement performed by MDFF allows for fitting the protein or lipid densities with atomically detailed models.

4. Solvation and ionization. To generate a system with explicit solvent, we use the solvate plugin in VMD (Extensions → Modeling → Add Solvation Box). Additional care needs to be taken when performing ReMDFF in explicit solvent to neutralize the system by adding counter-ions to the simulation system. The neutralizing ions can be added using VMD's autoionize plugin (Extensions → Modeling → Add Ions). After each step, that is, the solvation of the protein and neutralization of the system, a PDB and PSF file will be automatically generated by VMD in the working directory.

3.2 pH Simulations

1. MCCE calculation. MCCE, developed in the Gunner lab, uses Monte Carlo (MC) sampling to bring the protonation and redox states, side chain and ligand positions, buried waters, and ion occupancy, into equilibrium at a given pH, hydration energy and ligand chemical potential [43, 44]. MCCE is initiated with coordinates from an experimental structure or MD snapshot, following which side chain rotamers can be sampled, or conformers can be restrained for "isosteric sampling" of polar protons and tautomer positions [44]. A microstate has one conformer for each residue and ligand and is analogous to an MD time step (although, as MC moves follow the Metropolis-Hastings algorithm in the λ-space of Eq. 1, there is no information about temporal order). Electrostatic energies are found with the Continuum Electrostatics Poisson Boltzmann equation [45, 46]. Models derived from MCCE are fed back in MD to recover kinetic information.

2. MD simulations of biomolecular systems rely on the accurate integration of the equation of motion $m_i \frac{d^2 R_i}{dt^2} = -\nabla_{R_i} \mathcal{U}(R_1, R_2, \ldots)$, where m_i is the mass of atom i, R_i is its

position, and $\mathcal{U}$ is the total potential energy that is a function of atomic positions and, thereby, couples the motion of atoms. This results in time-dependent trajectories for all atoms of the biomolecule [47]. To avoid surface effects at the simulated system's boundary, periodic boundary conditions are utilized. This permits the efficient computation of non-truncated electrostatic interactions by the FFT-based particle-mesh Ewald method. NAMD [47] generates trajectories via the symplectic time-reversible velocity-Verlet integrator with multiple time-stepping. The (stochastic) Langevin equation is used to generate a Boltzmann distribution. Isobaric (NPT) ensemble simulations employ a computationally efficient "Langevin-Hoover" method [47].

3.3 Analysis

1. Mutation analysis. Alanine scanning was performed on the residues lining the interfaces of the soluble subunits of complex I using the ROSETTA software [48, 49]. In this, residues are computationally mutated to an alanine, and the accompanying change in free energy ($\Delta\, \Delta G$) is calculated by a Monte Carlo-based algorithm. For comparison, we used the list of mutations of complex I from the work of Sazanov et al. [50]
2. Potential of Mean Force (PMF). Unbinding of the menaquinone to complex I was accelerated by using steered molecular dynamics (SMD) [51]. Here, a moving harmonic constraint is added to the potential energy function to apply an external force to the quinone molecule. In our simulations, a constant-velocity scheme was used with a pulling speed of 10 Å/ns and a spring constant for the harmonic constraint of 500 pN/Å. The carbon atom of the last isoprenoid unit of the menaquinone tail was chosen for attaching the moving constraint, which leaves the headgroup free to move and therefore finds a low-energy pathway through complex I's quinone-access channel. Pulling in both directions was done to gain information on the potential pathway from independent simulations.

   The system was prepared in an orientation such that the x axis was parallel to the direction connecting the center of the binding pocket (residues 38 and 87 of subunit Nqo4) with the channel opening and the z axis was parallel to the membrane normal. The SMD direction was parallel to the membrane plane along the positive x-direction. The pulling distance was ~12 Å, which resulted in a total simulation time of *ca.* 250 ns. Four such pulling simulations were performed for systems (oxidized or reduced complex I, bound to quinone or quinol), cumulatively taking 1 μs.

   To reconstruct the PMF along the proposed pathway, bias-exchange umbrella sampling (BEUS) simulations were performed along the path derived from the SMD trajectories of oxidized or reduced menaquinone unbinding. Details of these steering simulations have been published separately [7]. The BEUS simulations for the four pathways are performed employing the distance of quinone headgroup from the binding pocket as the collective variable, as also used by others [52]. Thirty-four windows were employed along the dissociation path, and an exchange was attempted every 1 ps between an image and one of its two nearest neighboring images in an alternating fashion. Ten replicas per image are employed for 5 ns-long BEUS simulations. A force constant of 5 kcal/mol/Å^2 is employed to restrain geometrically the position of the umbrellas along the inter-heme distance axis. This protocol resulted in a roughly similar rate of exchange between neighboring windows (ranging from 22% to 30%).

Thereafter our generalized weighted histogram methodology [53] was employed to re-weight the ensemble and determine the potential of mean force (PMF) profiles. Convergence of the free-energy calculation is tested for each of the three cases using a resampling protocol. Fifty percent of the data is randomly chosen and two one-dimensional free-energy profiles are constructed employing the aforementioned weighted histogram method. The process is repeated 200 times to obtain a collection of 200 PMFs. After these 200 trials a converged mean profile is obtained across the 200 PMFs, and the standard deviation is considered as the error estimate.

3. Hydrogen bond analysis. The network for each snapshot is formed from the hydrogen bonds found in at least 0.1% of microstates in the Boltzmann distribution. In the default network analysis, two residues or ligands are viewed as being connected if they can be bridged by zero to four water molecules. A permissive network is built up here as it contains connections from different microstates in a single snapshot. In addition, in the default presentation, networks from different snapshots from the same trajectory are merged. Networks are visualized using Cytoscape [54]. Active proton transfer networks end at the surface of the protein. Surface residues are identified based on their solvent accessible surface in the first frame of a given snapshot. A residue is considered surface exposed if 20% of the whole residue is surface exposed. Conservation of residues is determined by HSSP multiple alignment of approximately 2500 sequences of complex I [55].

4 Case Study

We present two examples of pH simulation of mitochondrial membrane proteins, the first with respiratory complex I, and the second with V_O ATPase. In these case studies, the combined MD/MCCE approach is showcased as a general approach for performing pH simulation of both these proteins.

4.1 Respiratory Complex I

The first case study focuses on complex I. The initial molecular assay was built using the crystal structure of the entire respiratory complex I of *T. thermophilus* (pdb: 4HEA, with loop Nqo6 55–70 updated as in the structure with decyl-ubiquinone bound [PDB 6Y11], unpublished). The protonation state of titratable residues was assigned based on pK_a values predicted by MCCE (states labeled in Fig. 3a—left). For histidines, the protonation state of imidazole ring nitrogens (proton on either ϵ or δ nitrogens, or on both) was chosen based on the local chemical environment and the capacity to form hydrogen bonds. In addition, we based our choice of protonation states based on the recent QM work by Gamiz-Hernandez et al., who predict histidine to be deprotonated due to reduction of N2 [56]. The Dowser program was used to fill empty cavities by water within the complex transmembrane region, resulting in the addition of 121 water molecules. The molecular system was subsequently embedded in a hydrated patch of 1000 1-palmitoyl-2-oleoyl-sn-glycero-3-phosphocholines (POPC) in the presence of 150 mM of NaCl. The complete molecular assay consists of ~1,000,000 atoms embedded in a rectangular box of $28 \times 14 \times 24$ nm^3. The iron-sulfur (Fe–S) cluster CHARMM-like parameters were those derived by Chang and Kim from Density Function Theory calculations [57]. The flavin mononucleotide was described with the parameter set designed by Freddolino et al. [58]. The polar ring of menaquinone was aligned with that of decyl-ubiquinone in the X-ray structure (PDB ID: 6Y11), so the phenyl oxygens are interacting with residues His 38 and Tyr 87 of subunit Nqo4. The menaquinone tail was then refined using 10 ns of MDFF [42] simulations into the hollow quinone-binding cavity. The structure with the highest cross-correlation and lowest energy was chosen for MD simulation, which extended up to a cumulative time of 1.5 μs.

First, we discuss the transmembrane arm of the complex. Physically speaking, complex I is a result of evolution repurposing different proteins for new functions: the peripheral arm of complex I is derived from the soluble [NiFe] hydrogenases [59, 60], whereas the antiporter subunits of the membrane arm are derived from Mrp cation/H+ antiporters [61]. The intramembrane subunit, Nqo8, connects to the peripheral arm. It is the most conserved subunit in complex I [20]; however, it appears not to be related to other proteins. While there are three antiporter derived subunits, four protons are pumped by complex I [62]. Thus, the sequence and structure of the fourth channel is unique [20]. The crystal structures [63, 64] led to the suggestion that the fourth channel may use subunits Nqo10 and Nqo11 (NuoJ/K in *E. coli*), an assignment which was also supported by MD simulations carried out on the *E. coli* protein [13]. However, after the structure of the *T. thermophilus* complex I was solved [65], it was suggested that

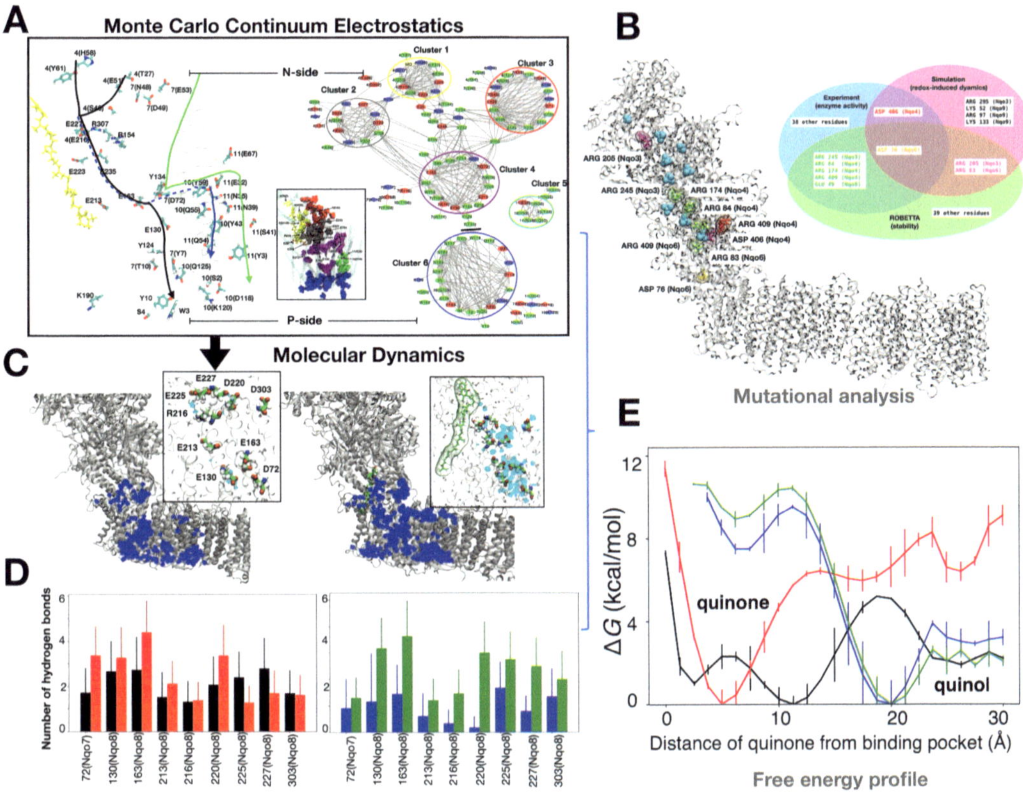

Fig. 3 Complex I application. (**a**) Left: Black line is the pathway predicted by us. Blue line (including blue dotted line) was predicted from inspection of the *T. thermophilus* structure by Baradaran et al. [65] Blue dotted line is path obtained by QM/MM-MD simulation on *T. thermophilus* Complex I by Di Luca et al. [75] Green line is consensus path predicted from inspection of the Complex I crystal structures from *E. coli* by Efremov et al. [64], *Y. lipolytica* complex I by Zickermann et al. [63] and from MD simulation of *T. thermophilus* complex I by Haapanen et al. [76] Right: MCCE hydrogen bond network for the menaquinone docked structure is shown on the right. Inset: Subunits Nqo4, Nqo7, Nqo8, Nqo10, and Nqo11 are shown as semi-transparent cylinders. Slide chains identified with each cluster (1: yellow, 2: gray, 3: red, 4: purple, 5: green, 6: blue). Several key residues are labeled. (**b**) Three categories of complex I mutations are shown as a Venn diagram (inset), and their locations on complex I are shown in the adjacent figure. Residues that affect function by stabilizing or destabilizing subunit-subunit interfaces are highlighted in green. Residues that affect function through redox-induced dynamics are highlighted in red. Asp 76 (Nqo6), the unique residue that affects function through stability as well as redox-sensitivity, is highlighted in yellow. Highlighted in pink are the residues that affect stability and redox motion but are heretofore not known to be functionally active. Iron-sulfur clusters are shown in blue for reference. (**c**) Solvation of the transmembrane region within apo (left) and menaquinone-bound (right) complex I. Water-mediated hydrogen-bond networks are shown in insets. A strong hydrogen-bonded network is seen upon quinone-binding which is missing in apo complex I. average and standard deviation of the number of hydrogen-bonds between these residues and water is shown for both the oxidized. (**d**) Average and standard deviation of the number of hydrogen bonds between these residues and water are shown for both the oxidized (left) and reduced (right) complex I in apo (black and blue) and quinone-bound (red and green) states. (**e**) Free-energy profile of quinone/quinol-binding (black: quinone in oxidized complex I, red: quinone in reduced complex I, blue: quinol in oxidized complex I, green: quinol in reduced complex I). Quinone shows minima near the binding pocket (His38 and Tyr87 of Nqo4), while quinol minima is shifted. Location of quinone minima changes when complex I is reduced

Nqo8 forms the fourth channel, now called the "E-channel" due to the abundance of glutamic acids in this proposed path.

Hydrogen bond networks were obtained with MC sampling of multiple MD snapshots [66] using MCCE [44], including only subunits Nqo4, Nqo7, Nqo8, Nqo10, and Nqo11 (Fig. 3a—right). The protein was inserted into a 33 Å thick rectangular slab using IPECE to emulate the membrane [67], a method which minimizes the burial of ionizable residues. Surface water molecules with >5% solvent accessible surface were replaced with continuum solvent. In MCCE a conformer represents a choice (degree of freedom) for a side chain (or ligand or water) that is included in the Monte Carlo (MC) sampling [44]. Here, the protein backbone and side chain carbon atoms were fixed during MCCE. Multiple "isosteric" conformers that change the position of polar hydrogens of side chain, protonation state of charged and polar residues, water protons, and occupancy of water binding sites were sampled in our calculation. MCCE uses GCMC (Grand Canonical Monte Carlo) sampling, which allows the number of water molecules within the complex to change by keeping constant chemical potential, volume, and temperature of a reservoir. The behavior of TIP3 water in GCMC sampling in MCCE has been recently benchmarked against the behavior of water in MD [68]. Eleven snapshots from the 1.5 μs MD simulation (representing the most likely configuration of water molecules) were subjected to MC sampling, collecting 17–35 million microstates for each snapshot. The network for each snapshot was formed from the hydrogen bonds found in at least 0.1% of microstates in the Boltzmann distribution. Two residues or ligands were viewed as being connected if they could be bridged by <=4 or <5 water molecules.

Our analysis showed a complex cluster of interconnected residues lining the E-channel, as shown in Fig. 3a. The left panel shows the amino acid side chains identified to be making up the network of residues that connect the N- to P-side of the protein. They are colored to represent the clusters identified by the network analysis, shown in the right panel, where the residues are divided into "clusters" and "connecting residues." We identified six internally connected clusters of polar and protonatable residues and waters. The network shown in Fig. 3a identifies an almost complete path between the N and P-side when menaquinone is bound. Overall, we see two clusters (2 and 3) which have residues that connect to the N-side of the E-channel. They surround the quinone containing cluster (cluster 1). Clusters 1, 2, and 3 connect to the deeply buried cluster 4. A small isolated cluster (cluster 5) is also found. Cluster 6 contains P-side surface exposed residues. With menaquinone bound there is a gap between clusters 4 and 6, which can stop proton flow across the membrane. This finding is in line with the observation that proton transport through the E-channel follows only after reduction of the menaquinone. Nonetheless, quinone binding sets up the network between clusters 1 through 4 which was missing in the apo complex (Fig. 3c, d). Taken together, in the apo form the E-channel within the transmembrane arm of the complex remains less hydrated. Menaquinone binding hydrates the N and P sides of the channel in preparation of proton transport, post reduction of the quinone species.

Presented in Fig. 3b, the MD simulation of complex I further elucidates three categories of residues in its peripheral arm based on whether or not they affect (1) interface stability, (2) redox induced dynamics, and (3) enzyme activity. ROSETTA alanine scanning [48, 49] of the residues at the interface of the soluble subunits revealed 47 residues with $|\Delta\,\Delta G| > 0.6$ kcal/mol, implying significant changes in stability. Three out of these 47 residues, namely, Asp76 (Nqo6), Arg205 (Nqo3), and Arg83 (Nqo6), were also found to be redox-sensitive in our simulations (Fig. 3b). Therefore, we predict that these residues have dual functions-maintaining complex I stability as well as mediating mechanical motion in response to iron-sulfur cluster reduction. Residue Asp 406 (Nqo4) was also found to be redox-sensitive. An alanine mutation of this residue has minimal impact on the stability of the complex; yet, such replacement is known to reduce complex I activity to 20% [69]. Thus, the redox-sensitive residues identified in our simulations, indeed control enzyme activity notwithstanding their contribution to interface stability.

Next, we look into the sites of complex I mutations that are essential for enzyme activity. A list of such mutations, along with their corresponding sites in *T. thermophilus*, has already been published [50]. Comparison of these mutation sites with our ROSETTA predictions confirmed six residues, namely, Arg245 (Nqo3), Arg84 (Nqo3), Arg174 (Nqo6), Glu49 (Nqo6), Arg409 (Nqo6), and Asp76 (Nqo6), that contribute to complex I stability. Remarkably, we find that mutations of two out eight of these residues (Arg245 and Arg174) are implicated in human diseases [33, 36]. Therefore, our combined MD-ROSETTA analysis detects specific residues that correlate reduced complex I stability with potential enzyme malfunctions. Finally, we seek residues that control all the three biochemical effects. Upon iron-sulfur cluster reduction, Asp76 (Nqo6) undergoes significant conformational transition to move away from the reduced cluster (N2) by ~24 Å. An alanine mutation of this residue over-stabilizes the interface by 0.75 kcal/mol and reduces the activity to only 6% [70]. Therefore, we discover that Asp76 is unique in that it co-modulates stability with redox-induced dynamics and enzyme activity. We note that the aforementioned Arg205 (Nqo3) and Arg83 (Nqo6) residues affect both stability and redox-driven motions, strongly suggesting a functional role. On the basis of these findings, we suggest these two locations for future mutation studies.

Free-energy calculations were performed to study the binding pathway of charge-carrier menaquinone and menaquinol into the protein complex. Notably, the quinone binding site of complex I is located at the hinge between the peripheral and transmembrane arms of the complex (Fig. 2a). As can be seen in Fig. 3e, the PMF reveals the preferred location of the substrate with respect to the binding site in an oxidized and reduced environment. When four of the seven iron-sulfur clusters (N3, N4, N6a, N2) are reduced, the PMF exhibits a single, deep minimum at 4.5 Å, which corresponds to a binding mode whereby both His38 and Tyr87 are hydrogen-bonded to the headgroup. It is worth noting that beyond 6 Å, the free-energy increases sharply in the reduced complex I, but decreases in the oxidized form, owing to favorable hydrophobic interactions promoted by tighter contacts of the quinone tail with the protein.

In stark contrast with the oxidized reactant, menaquinone form of the charge carrier, the reduced product, menaquinol form interacts with the protein following a markedly different binding mode. We found that the two hydroxyl groups are concomitantly hydrogen-bonded to the carboxylate moiety of residues Nqo8 Glu225 and Nqo3 Glu248. This alternate interaction pattern is reflected in a sharp minimum of the PMF, shifted from that of menaquinone by ca. 6 Å, irrespective of the redox state of the iron-sulfur clusters. In addition, the reduced complex has a more open cavity than the oxidized one; hence, the steric hindrance experienced by the quinol is expected to diminish, which, in turn, leads to a shallow free-energy profile between 6 and 10 Å from the binding pocket. The discrepant free-energy profiles obtained for menaquinone and menaquinol suggest a dynamic adaptation of the binding pocket in response to the redox state of the substrate for facilitating quinol release.

Our PMF calculations show that the global minimum of bound menaquinone is at 5 Å (Fig. 3e), much closer to the locus of the binding site, compared to the 35 Å minimum of ubiquinone [52, 71]. Thus, our model indicates that the transfer of menaquinone into the site of reduction is more spontaneous than that of ubiquinone—a possible design advantage making menaquinone the physiological substrate in *T. thermophilus.* Disparities between menaquinone and ubiquinone binding have already been observed in complex II [72]. Altogether, our simulations offer unforeseen insight at the cross-section of redox-induced protein conformational changes and their varied impact on binding charge carriers of different geometry.

4.2 V_o ATPase

V-ATPase is organized in cytosolic V_1-ATPase and membrane bound V_o proton channel subcomplexes that are coupled by a rotating central stalk and three peripheral stators. The V_o proton channel operates by a conserved mechanism that involves clockwise rotation (as seen from the cytosol) of a ring of ten *c*-ring subunits past the C-terminal transmembrane domain of subunit *a* (Fig. 2b).

Each *c*-ring subunit has one lipid-exposed glutamic acid residue that carries protons from a cytosolic to a luminal aqueous half-channel located within subunit *a* [73]. Direct proton-flow between the two half-channels is blocked by two arginine residues in the middle of the bilayer that transiently engage the *c* subunits' glutamates in a salt bridge during c-ring rotation.

Following local pK_a assessments using MCCE, we performed two separate MD simulations to explore the role of water dynamics in the transfer of the proton from essential c-ring glutamates to the luminal half-channel in *a* [74]. In eukaryotic V-ATPase, a critical residue involved in this process is a strictly conserved glutamic acid (aE789 in yeast) that is located at the entry of the luminal half-channel across from *c*(1)E137 (highlighted in Fig. 4a). In the first simulation, the donating *c*(1)E137 was protonated, and the receiving aE789 was deprotonated (Fig. 4b, left panel). In the second simulation, the protonation states of the two residues were reversed (Fig. 4b, right panel). The simulated dynamics of the water

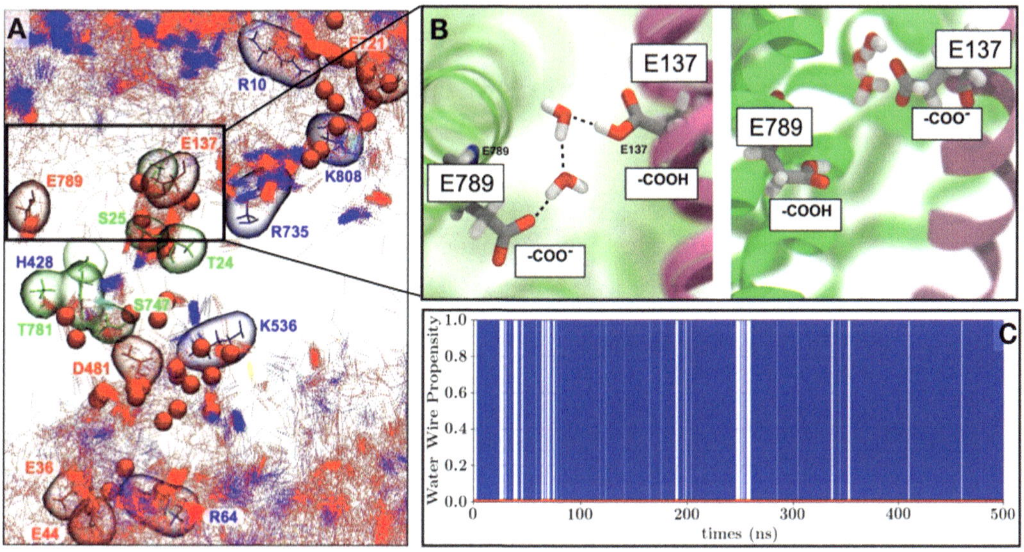

Fig. 4 V_o-ATPase application. (**a**) A scatter plot of the water-water and water-amino acid hydrogen bonds overlaid of 500 ns of MD of the pre-proton transfer state, highlighting the charged and polar amino-acids that directly interact with the water molecules. Hydrogen bonds between water and basic residues are in blue, and hydrogen bonds water to water and water to acidic residues are in red. Only a subset of basic residues is shown. Inset: (Left) Two views of a representative coordinate frame with water-wire connecting $c_{(1)}$E137 and aE789 in the pre-H+ -transfer state. The left image is seen parallel to the membrane towards the a_{CT}:c-ring interface. The right image is seen perpendicular to the membrane along the a_{CT}:c-ring interface. Hydrogen and oxygen atoms are shown in white and red, respectively. (Right) Representative coordinate frame with $c_{(1)}$E137 and aE789 seen in the post-H+ -transfer state. The protonated aE789 blocks access for water molecules from the luminal channel. A few water molecules are retained near the ionized $c_{(1)}$E137. (**b**) Occurrence of water wires with 5000 frames accumulated over 0.5 μs MD simulations indicated by blue vertical lines for the pre-H + -transfer state. No water wires were observed for the post-H+ -transfer state as indicated by the red line

molecules suggests that proton transfer from c-ring glutamic acid residues (*c*E137, *c*'E145, or *c*"E108) to *a*E789 occurs via a cluster of water molecules, whose lifetime depends on the protonation states of the two residues. At the onset of the transfer, when a proton resides on, for example, *c*(1)E137, a network of hydrogen-bonded water molecules (hereafter referred to as "water wire") connects the carboxyl groups of c(1)E137 and aE789 for >5% of the structures derived from three independent 500 ns simulations (Fig. 4c, blue vertical lines). These water wires enable transfer of the proton from c-ring essential glutamates to *a*E789. After the transfer, when the c-ring glutamate is ionized, and *a*E789 is protonated, no water wires are observed in the simulation (Fig. 4c, red line). This disruption of the water wire post proton transfer ensures that protons cannot leak back from the luminal half-channel into the *a*:*c*-ring interface once proton transfer has occurred. Taken together, dynamics of the water wires is coupled to the charge and conformational states of key acidic residues for controlling proton transport during *c*-ring rotation. The simulations show that proton transfer from the *c*-ring glutamates to the luminal half-channel in *a* involves transient water wires, reminiscent of the classic alternating-access mechanism of membrane transport.

Acknowledgements

The authors acknowledge start-up funds from the School of Molecular Sciences and Center for Applied Structure Discovery at Arizona State University, and the resources of the OLCF at the Oak Ridge National Laboratory, which is supported by the Office of Science at DOE under Contract No. DE-AC05-00OR22725, made available via the INCITE program. We also acknowledge NAMD and VMD developments supported by NIH (P41GM104601) and R01GM098243-02 for supporting our study of membrane proteins. AS acknowledges NSF (MCB-1942763) and NIH (R01GM095583). MG acknowledges NSF (MCB-1519640).

References

1. Tanford C, Roxby R (1972) Interpretation of protein titration curves. Application to lysozyme. Biochemistry 11:2192–2198
2. Dolphin GT, Baltzer L (1997) The pH-dependent tertiary structure of a designed helix–loop–helix dimer. Folding Design 2:319–330
3. Russo NVD, Estrin DA, Martí MA, et al. (2012) pH-dependent conformational changes in proteins and their effect on experimental pKas: the case of nitrophorin 4. PLOS Comput Biol 8:e1002761
4. Whitten ST, Wooll JO, Razeghifard R, et al. (2001) The origin of pH-dependent changes in m-values for the denaturant-induced unfolding of proteins1 1Edited by C. R. Matthews. J Mol Biol 309:1165–1175
5. Yang AS, Honig B (1993) On the pH dependence of protein stability. J Mol Biol 231:459–474

6. O'Brien EP, Brooks BR, Thirumalai D (2012) Effects of pH on proteins: predictions for ensemble and single-molecule pulling experiments. J Am Chem Soc 134:979–987
7. Singharoy A, Barragan AM, Thangapandian S, et al. (2016) Binding site recognition and docking dynamics of a single electron transport protein: cytochrome c2. J Am Chem Soc 138:12077–12089
8. Singharoy A, Maffeo C, Delgado-Magnero KH, et al. (2019) Atoms to phenotypes: molecular design principles of cellular energy metabolism. Cell 179:1098–1111.e23
9. Yu I, Mori T, Ando T, et al. (2016) Biomolecular interactions modulate macromolecular structure and dynamics in atomistic model of a bacterial cytoplasm. eLife 5:e19274
10. Anandakrishnan R, Zhang Z, Donovan-Maiye-R, et al. (2016) Biophysical comparison of ATP synthesis mechanisms shows a kinetic advantage for the rotary process. Proc Nat Acad Sci 113:11220–11225
11. Radak BK, Chipot C, Suh D, et al. (2018) Correction to Constant-pH molecular dynamics simulations for large biomolecular systems. J Chem Theory Comput 14:6748–6749
12. Hammes-Schiffer S (2015) Proton-coupled electron transfer: moving together and charging forward. J Am Chem Soc 137:8860–8871
13. Kaila VRI, Wikström M, Hummer G (2014) Electrostatics, hydration, and proton transfer dynamics in the membrane domain of respiratory complex I. Proc Nat Acad Sci 111:6988–6993
14. Radak BK, Chipot C, Suh D, et al. (2017) Constant-pH molecular dynamics simulations for large biomolecular systems. J Chem Theory Comput 13:5933–5944
15. Søndergaard CR, Olsson MHM, Rostkowski M, et al. (2011) Improved treatment of ligands and coupling effects in empirical calculation and rationalization of pKa values. J Chem Theory Comput 7:2284–2295
16. Olsson MHM, Søndergaard CR, Rostkowski M, et al. (2011) PROPKA3: consistent treatment of internal and surface residues in empirical pKa predictions. J Che Theory Comput 7:525–537
17. Georgescu RE, Alexov EG, Gunner MR (2002) Combining conformational flexibility and continuum electrostatics for calculating pKas in proteins. Biophys J 83:1731–1748
18. Gupta C, Khaniya U, Chan CK, et al. (2020) Charge transfer and chemo-mechanical coupling in respiratory complex I. J Am Chem Soc 142:9220–9230
19. Khaniya U, Gupta C, Cai X, et al. (2020) Hydrogen bond network analysis reveals the pathway for the proton transfer in the E-channel of T. thermophilus Complex I. Biochim Biophys Acta Bioenergetics 1861:148240
20. Sazanov LA (2015) A giant molecular proton pump: structure and mechanism of respiratory complex I. Nat Rev Mol Cell Biol 16:375–388
21. Gupta C, Shekhar M (2020) pH simulation of membrane systems. https://github.com/SingharoyLab/pH_simulation_membrane.git
22. Machuqueiro M, Baptista AM (2006) Constant-ph molecular dynamics with ionic strength effects: protonation-conformation coupling in decalysine. J Phys Chem B 110:2927–2933
23. Williams SL, Blachly PG, McCammon JA (2011) Measuring the successes and deficiencies of constant pH molecular dynamics: a blind prediction study. Proteins: Struct Funct Bioinf 79:3381–3388
24. Di Russo NV, Martí MA, Roitberg AE (2014) Underlying thermodynamics of pH-dependent allostery. J Phys Chem B 118:12818–12826
25. Swails JM, York DM, Roitberg AE (2014) Constant pH replica exchange molecular dynamics in explicit solvent using discrete protonation states: implementation, testing, and validation. J Chem Theory Comput 10:1341–1352
26. Cruzeiro VWD, Feliciano GT, Roitberg AE (2020) Exploring coupled redox and pH processes with a force-field-based approach: applications to five different systems. J Am Chem Soc 142:3823–3835
27. Itoh SG, Damjanović A, Brooks BR (2011) pH replica-exchange method based on discrete protonation states. Proteins: Struct Funct Bioinf 79:3420–3436
28. Goh GB, Hulbert BS, Zhou H, et al. (2014) Constant pH molecular dynamics of proteins in explicit solvent with proton tautomerism. Proteins: Struct Funct Bioinf 82:1319–1331
29. Damjanovic A, Miller BT, Okur A, et al. (2018) Reservoir pH replica exchange. J Chem Phys 149:072321
30. Vila-Viçosa D, Reis PBPS, Baptista AM, et al. (2019) A pH replica exchange scheme in the stochastic titration constant-pH MD method. J Chem Theory Comput 15:3108–3116
31. Warshel A, Dryga A (2011) Simulating electrostatic energies in proteins: perspectives and some recent studies of pKas, redox, and other crucial functional properties. Proteins: Struct Funct Bioinf 79:3469–3484

32. Zheng Y, Cui Q (2017) Microscopic mechanisms that govern the titration response and pKa values of buried residues in staphylococcal nuclease mutants. Proteins: Struct Funct Bioinf 85:268–281
33. Oliveira ASF, Campos SRR, Baptista AM, et al. (2016) Coupling between protonation and conformation in cytochrome c oxidase: insights from constant-pH MD simulations. Biochim Biophys Acta: Bioenergetics 1857:759–771
34. Humphrey W, Dalke A, Schulten K (1996) VMD: Visual molecular dynamics. J Mol Graph 14:33–38
35. Best RB, Zhu X, Shim J, et al. (2012) Optimization of the additive charmm all-atom protein force field targeting improved sampling of the backbone ϕ, ψ and side-chain $\chi 1$ and $\chi 2$ dihedral angles. J Chem Theory Comput 8:3257–3273
36. Jo S, Kim T, Iyer VG, et al. (2008) CHARMM-GUI: a web-based graphical user interface for CHARMM. J Comput Chem 29:1859–1865
37. Jo S, Lim JB, Klauda JB, et al. (2009) CHARMM-GUI Membrane builder for mixed bilayers and its application to yeast membranes. Biophys J 97:50–58
38. Lee J, Cheng X, Swails JM, et al. (2016) CHARMM-GUI input generator for NAMD, GROMACS, AMBER, OpenMM, and CHARMM/OpenMM simulations using the CHARMM36 additive force field. J Chem Theory Comput 12:405–413
39. Wu EL, Cheng X, Jo S, et al. (2014) CHARMM-GUI Membrane Builder toward realistic biological membrane simulations. J Comput Chem 35:1997–2004
40. Trabuco LG, Villa E, Mitra K, et al. (2008) Flexible fitting of atomic structures into electron microscopy maps using molecular dynamics. Structure 16:673–683
41. Trabuco LG, Villa E, Schreiner E, et al. (2009) Molecular dynamics flexible fitting: a practical guide to combine cryo-electron microscopy and X-ray crystallography. Methods 49:174–180
42. Singharoy A, Teo I, McGreevy R, et al. (2016) Molecular dynamics-based refinement and validation for sub-5 Å cryo-electron microscopy maps. eLife 5:e16105
43. Alexov EG, Gunner MR (1997) Incorporating protein conformational flexibility into the calculation of pH-dependent protein properties. Biophys J 72:2075–2093
44. Song Y, Mao J, Gunner MR (2009) MCCE2: Improving protein pKa calculations with extensive side chain rotamer sampling. J Comput Chem 30:2231–2247
45. Nicholls A, Honig B (1991) A rapid finite difference algorithm, utilizing successive over-relaxation to solve the Poisson–Boltzmann equation. J Comput Chem 12:435–445
46. Rocchia W, Alexov E, Honig B (2001) Extending the applicability of the nonlinear poisson boltzmann equation: multiple dielectric constants and multivalent ions. J Phys Chem B 105:6754–6754
47. Phillips JC, Braun R, Wang W, et al. (2005) Scalable molecular dynamics with NAMD. J Comput Chem 26:1781–1802
48. Kortemme T, Baker D (2002) A simple physical model for binding energy hot spots in protein–protein complexes. Proc Nat Acad Sci 99:14116–14121
49. Kortemme T, Kim DE, Baker D (2004) Computational alanine scanning of protein-protein interfaces. Sci STKE 2004:pl2
50. Sazanov LA (2007) Respiratory complex I: Mechanistic and structural insights provided by the crystal structure of the hydrophilic domain. Biochemistry 46:2275–2288
51. Aird A, Wrachtrup J, Schulten K, et al. (2007) Possible pathway for ubiquinone shuttling in rhodospirillum rubrum revealed by molecular dynamics simulation. Biophys J 92:23–33
52. Warnau J, Sharma V, Gamiz-Hernandez AP, et al. (2018) Redox-coupled quinone dynamics in the respiratory complex I. Proc Nat Acad Sci 115:E8413–E8420
53. Singharoy A, Chipot C, Moradi M, et al. (2017) Chemomechanical coupling in hexameric protein–protein interfaces harnesses energy within V-type ATPases. J Am Chem Soc 139:293–310
54. Shannon P, Markiel A, Ozier O, et al. (2003) Cytoscape: a software environment for integrated models of biomolecular interaction networks. Genome Res 13:2498–2504
55. Joosten RP, te Beek TAH, Krieger E, et al. (2011) A series of PDB related databases for everyday needs. Nucl Acids Res 39: D411–D419
56. Gamiz-Hernandez AP, Jussupow A, Johansson MP, et al. (2017) Terminal electron–proton transfer dynamics in the quinone reduction of respiratory complex I. J Am Chem Soc 139:16282–16288
57. Chang CH, Kim K (2009) Density functional theory calculation of bonding and charge parameters for molecular dynamics studies on [FeFe] hydrogenases. J Chem Theory Comput 5:1137–1145
58. Freddolino PL, Gardner KH, Schulten K (2013) Signaling mechanisms of LOV domains: new insights from molecular

dynamics studies. Photochem Photobiol Sci 12:1158–1170

59. Fontecilla-Camps JC, Volbeda A, Cavazza C, et al. (2007) Structure/function relationships of [NiFe]- and [FeFe]-hydrogenases. Chem Rev 107:4273–4303
60. Efremov RG, Sazanov LA (2012) The coupling mechanism of respiratory complex I—A tructural and evolutionary perspective. Biochim Biophys Acta: Bioenergetics 1817:1785–1795
61. Mathiesen C, Hägerhäll C (2002) Transmembrane topology of the NuoL, M and N subunits of NADH:quinone oxidoreductase and their homologues among membrane-bound hydrogenases and bona fide antiporters. Biochim Biophys Acta: Bioenergetics 1556:121–132
62. Ripple MO, Kim N, Springett R (2013) Mammalian complex I pumps 4 protons per 2 electrons at high and physiological proton motive force in living cells. J Biol Chem 288:5374–5380
63. Zickermann V, Wirth C, Nasiri H, et al. (2015) Mechanistic insight from the crystal structure of mitochondrial complex I. Science 347:44–49
64. Efremov RG, Sazanov LA (2011) Structure of the membrane domain of respiratory complex I. Nature 476:414–420
65. Baradaran R, Berrisford JM, Minhas GS, et al. (2013) Crystal structure of the entire respiratory complex I. Nature 494:443–448
66. Cai X, Haider K, Lu J, et al. (2018) Network analysis of a proposed exit pathway for protons to the P-side of cytochrome c oxidase. Biochim Biophys Acta: Bioenergetics 1859:997–1005
67. Song Y, Mao J, Gunner MR (2003) Calculation of proton transfers in bacteriorhodopsin bR and M intermediates. Biochemistry 42:9875–9888
68. Zhang Y, Haider K, Kaur D, et al. (2020) Characterizing the water wire in the Gramicidin channel found by Monte Carlo sampling using continuum electrostatics and in molecular dynamics trajectories with conventional or polarizable force fields. J Theor Comput Chem 2020:2042001
69. Kashani-Poor N, Zwicker K, Kerscher S, et al. (2001) A central functional role for the 49-kDa subunit within the catalytic core of mitochondrial complex I. J Biol Chem 276:24082–24087
70. Garofano A, Zwicker K, Kerscher S, et al. (2003) Two aspartic acid residues in the PSST-homologous NUKM subunit of complex I from Yarrowia lipolytica are essential for catalytic activity. J Biol Chem 278:42435–42440
71. Hoias Teixeira M, Menegon Arantes G (2019) Balanced internal hydration discriminates substrate binding to respiratory complex I. Biochim Biophys Acta: Bioenergetics 1860:541–548
72. Maklashina E, Cecchini G (2010) The quinone-binding and catalytic site of complex II. Biochim Biophys Acta: Bioenergetics 1797:1877–1882
73. Roh SH, Stam NJ, Hryc CF, et al. (2018) The 3.5-Å CryoEM structure of nanodisc-reconstituted yeast vacuolar ATPase Vo proton channel. Mol Cell 69:993–1004.e3
74. Roh SH, Shekhar M, Pintilie G, et al. (2020) Cryo-EM and MD infer water-mediated proton transport and autoinhibition mechanisms of v_o complex. Sci Adv 6:eabb9605
75. Luca AD, Gamiz-Hernandez AP, Kaila VRI (2017) Symmetry-related proton transfer pathways in respiratory complex I. Proc Nat Acad Sci 114:E6314–E6321
76. Haapanen O, Sharma V (2017) Role of water and protein dynamics in proton pumping by respiratory complex I. Sci Rep 7:7747

Chapter 13

Preparing and Analyzing Polarizable Molecular Dynamics Simulations with the Classical Drude Oscillator Model

Justin A. Lemkul

Abstract

Molecular dynamics (MD) simulations performed with force fields that include explicit electronic polarization are becoming more prevalent in the field. The increasing emergence of these simulations is a result of continual refinement against a range of theoretical and empirical target data, optimization of software algorithms for higher performance, and availability of graphical processing unit hardware to further accelerate the simulations. Polarizable MD simulations are likely to be most impactful in biomolecular systems in which heterogeneous environments or unique microenvironments exist that would lead to inaccuracies in simulations performed with fixed-charge, nonpolarizable force fields. The further adoption of polarizable MD simulations will benefit from tutorial material that specifically addresses preparing and analyzing their unique features. In this chapter, we introduce common protocols for preparing routine biomolecular systems containing proteins, including both a globular protein in aqueous solvent and a transmembrane model peptide in a phospholipid bilayer. Details and example input files are provided for preparation of the simulation system using CHARMM, performing the simulations with OpenMM, and analyzing interesting dipole moment properties in CHARMM.

Key words Molecular dynamics simulations, Induced dipoles, Polarizable force field, Drude oscillator, Protein modeling, Membrane modeling

1 Introduction

1.1 Molecular Dynamics Simulations

The application of molecular dynamics (MD) simulations in the investigations of biomolecules is widespread. MD simulations provide atomistic insight into mechanisms of conformational change, interaction, and transport; can be used to calculate free energies of binding; and can generate new hypotheses on biological phenomena. Simulation complexity can range from single proteins or nucleic acids in water to complexes including drugs or natural substrates, heterogeneous membrane systems, or entire virus capsids [1]. In MD simulations, the molecules being simulated may be represented atomistically (all atoms represented explicitly or with nonpolar hydrogen atoms subsumed into their parent carbon atom) or in a coarse-grain manner (chemical moieties are

Irina S. Moreira et al. (eds.), *Computational Design of Membrane Proteins*, Methods in Molecular Biology, vol. 2315, https://doi.org/10.1007/978-1-0716-1468-6_13,

encompassed by large particles with general features). Atomistic MD simulations are typically applied to detailed investigations of biomolecular conformational change, whereas coarse-grain simulations are often employed in simulations of complexation or large, heterogeneous membranes [2, 3]. Here, we focus on atomistic force fields and their applications to proteins in water and phospholipid membranes.

The most commonly used atomistic biomolecular force fields—AMBER [4–15], CHARMM [16–30], GROMOS [31–40], and OPLS [41–47]—are atomistic force fields that represent electrostatic interactions in a pairwise-additive manner. Multibody effects arising due to electronic polarization are not explicitly represented but are instead included in the model via charge assignment, assuming a mean-field response. That is, charges are assigned such that the resulting molecular properties reflect the average polarization response to a high-dielectric medium (i.e., water in the case of biomolecular force fields). While this approach is sufficiently accurate for most folded proteins and nucleic acids, mean-field charge assignment will introduce some amount of error in low-dielectric media (including the interior of globular proteins [48]) and heterogeneous systems like lipid membranes.

Recently, several force fields have been developed that explicitly represent electronic polarization. The two most comprehensive force fields that are currently in use are AMOEBA [49], an induced dipole–multipole model, and the Drude force field from the MacKerell and Roux groups based on the classical Drude oscillator model [50]. AMOEBA encompasses proteins [51], nucleic acids [52], water [53–55], and ions [56, 57]. The Drude force field includes parameters for proteins [58], nucleic acids [59–61], several zwitterionic phospholipids [62, 63], carbohydrates [64, 65], two models of water [66, 67], and monovalent and divalent ions [68]. The Drude model has been employed in simulations of phospholipid bilayers, demonstrating that the membrane dipole potential depends strongly on the explicit representation of electronic polarization [69]. Here, we will focus on methods for preparing and analyzing MD simulations with the Drude force field.

1.2 The Classical Drude Oscillator Model

The Drude force field is based on the classical Drude oscillator model [70], in which electronic degrees of freedom are represented via auxiliary particles ("Drude oscillators") that carry negative charge. The Drude oscillators are connected to their parent atoms via harmonic springs, the same convention used in representing covalent bonds in most atomistic force fields. The partial charge on each Drude-atom pair is partitioned based on the atomic polarizability, α, of the parent atom, according to:

$$\alpha = \frac{q_{\mathrm{D}}^2}{k_{\mathrm{D}}} \tag{1}$$

where q_D is the charge on the Drude oscillator and k_D is the force constant of the Drude-atom bond. In the Drude force field, α values are initially taken from Miller's model of atomic hybrid polarizability [71] and are empirically tuned by targeting quantum mechanical dipole moments, molecular polarizabilities, and water interactions [50]. Only non-hydrogen atoms are treated as polarizable; the polarization response of hydrogen atoms, which is expected to be small, is implicitly represented by tuning the α values of heavy atoms to which the hydrogen atoms are bonded. In principle, the force constant, k_D, can also be tuned but by convention is fixed at 1000 kcal mol^{-1} Å^{-2}. Thus, tuning α only affects q_D for simplicity in parameter fitting.

The functional form of the Drude force field is similar to that of the CHARMM nonpolarizable force field, including terms for bond stretching, angle bending, periodic dihedral rotation, van der Waals interactions represented with a 12-6 Lennard-Jones potential, and electrostatic interactions expressed via the Coulomb equation. The inclusion of Drude oscillators leads to only small modifications of the functional form. The bonds between Drude oscillators and their parent atoms are treated like any other, with a harmonic potential that gives the self-polarization energy:

$$U_{\text{self}}(\boldsymbol{d}) = \frac{1}{2} k_D \boldsymbol{d}^2 \tag{2}$$

where $\boldsymbol{d}$ is the displacement of the Drude oscillator from its parent atom; in other words, the bond length itself. Interested readers will note that the value of k_D is listed as 500 kcal mol^{-1} Å^{-2} in CHARMM-formatted parameter files due to the fact that CHARMM applies the half prefactor at the parameter level to all bonded force constants rather than performing an additional multiplication operation during the calculation of forces during an MD simulation.

While most atoms have their atomic polarizability treated isotropically (equal in all spatial dimensions), some interactions require the polarizability to be described anisotropically. In the Drude force field, hydrogen bond acceptors are treated in this fashion, which improves energetic and geometric properties that reflect the intrinsic directionality of hydrogen bonds [72]. These functional groups are also augmented by off-site "lone pairs," which are virtual interaction sites constructed from their parent atoms that carry negative charge. Forces on the virtual sites are interpolated back onto their constructing atoms. Thus, virtual sites are not dynamic particles but have their positions updated relative to the real atoms in the system. Lone pairs in the Drude force field improve the description of electron distribution around anisotropically polarizable atoms and contribute to the directionality of hydrogen bonding and ion interactions [72]. Anisotropic polarization is introduced by expanding the Drude-atom force constant k_D into a tensor $\mathbf{K}^{(D)}$ such that $\mathrm{tr}[\mathbf{K}^{(D)}]/3 = k_D$. The off-diagonal terms are set to zero, thus the expression of U_{self} becomes:

$$U_{\mathrm{self}} = \frac{1}{2}\left(\left[\boldsymbol{K}_{11}^{(D)}\right]d_1^2 + \left[\boldsymbol{K}_{22}^{(D)}\right]d_2^2 + \left[\boldsymbol{K}_{33}^{(D)}\right]d_3^2\right) \quad (3)$$

The values of d_1, d_2, and d_3 are the components of $\boldsymbol{d}$ in the local molecular frame rather than in absolute Cartesian coordinates (Fig. 1).

Unlike nonpolarizable force fields, in which nonbonded interactions within three covalent bonds are typically excluded, dipole–dipole interactions between 1-2 and 1-3 atom pairs (Fig. 1) are

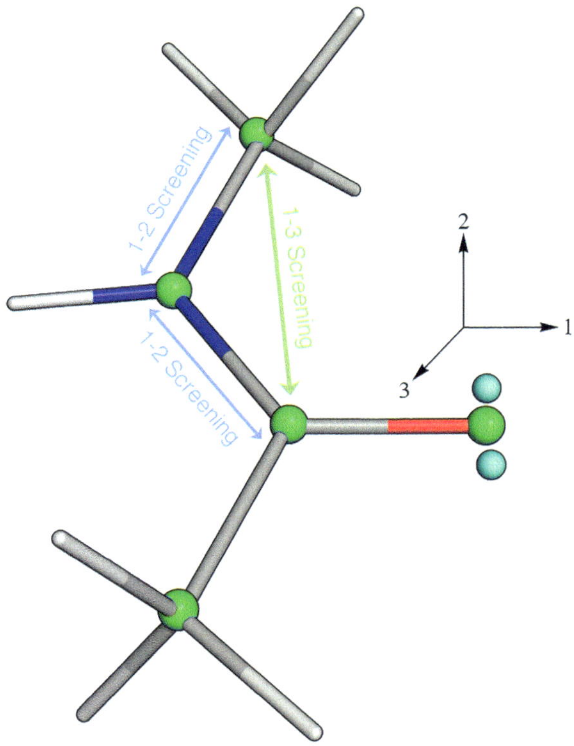

Fig. 1 Schematic of Thole screening of 1-2 and 1-3 dipole–dipole interactions and the frame of reference used for anisotropic polarization. The molecule shown is N-methylacetamide (NMA), colored by atom (H = white, C = gray, N = blue, O = red). Drude oscillators are shown as small green spheres and lone pairs are shown as small cyan spheres. Examples of 1-2 and 1-3 interactions that are subject to dipole screening are indicated (not all 1-2 and 1-3 pairs are illustrated). The principal axes of the local molecular frame around the carbonyl oxygen are shown in black and labeled. The carbon-oxygen bond defines axis 1 for the purpose of defining the axes. All other atomic polarizabilities in NMA are treated as isotropic

included in the Drude force field functional form, screened according to the function proposed by Thole [73]:

$$S_{ij}(r_{ij}) = 1 - \left[1 + \frac{a r_{ij}}{2(\alpha_i \alpha_j)^{1/6}}\right] \exp\left[\frac{-a r_{ij}}{(\alpha_i \alpha_j)^{1/6}}\right] \tag{4}$$

where α_i and α_j are the atomic polarizabilities of the two atoms participating in the dipole–dipole interaction. The value of a is the constant that defines the strength of the screening. Thole initially proposed a general value of 2.6, but this value is tunable and in the Drude force field is the sum of two individual atomic Thole values ($a = a_i + a_j$) [74]. Screening of dipole–dipole interactions applies to polarizable atoms that are treated isotropically and anisotropically.

Thole screening can also be applied in a "through-space" or nonbonded manner. The Drude force field applies Thole screening to ion–ion and ion–dipole interactions with monoatomic ions [75]. Doing so improves interaction energies and geometries in highly charged systems.

1.3 Essential Algorithms and Integration Scheme

Propagating dynamics in polarizable simulations relies on the Born-Oppenheimer approximation, which states that electronic degrees of freedom relax instantaneously to their ground state upon a change in nuclear configuration. This requirement can be satisfied in one of two ways. The first method achieves a self-consistent field (SCF) via relaxation of the induced dipoles, $\boldsymbol{d}_i$, after an update of the nuclear coordinates to satisfy:

$$\frac{\partial U}{\partial \boldsymbol{d}_i} = 0 \tag{5}$$

The SCF approach thus requires several cycles of energy minimization on the Drude oscillators per force evaluation on the nuclear coordinates and is computationally very expensive. Between 5 and 50 SCF cycles may be needed per time step, though in practice no more than 10 are typically needed for a simple system of a protein in water. To overcome this computational cost, Lamoureux and Roux introduced an extended Lagrangian integration method in which Drude oscillators are ascribed a small mass (typically 0.4 amu), allowing their positions to be integrated on equal footing with the real atoms of the system. The Born-Oppenheimer surface is approximated by coupling the Drude oscillators, and thus the induced dipoles, to a low-temperature relative thermostat [76]. The integration scheme was initially formulated using a modified velocity Verlet integrator implemented in CHARMM [77] but can also be carried out via Langevin dynamics [78].

A challenge in performing polarizable MD simulations is the possibility of "polarization catastrophe," a situation in which forces build up on the Drude oscillators and lead to unphysically large distances between the Drude oscillators and their parent atoms. Most modern simulations with the Drude force field employ a so-called "hard wall" constraint [62] that acts to reflect the Drude oscillator back toward its parent atom along the bond vector should the distance between them exceed a user-defined tolerance (typically 0.2 Å). The force field must be parametrized such that over-polarization is avoided, thus minimizing the likelihood of needing the hard wall constraint during an MD simulation. Careful tuning of α and Thole screening factors can achieve this requirement.

2 Software

2.1 Software and Hardware Used

The protocols described in this chapter have been developed and validated for correctness using CHARMM version 43b1 [77], NAMD version 2.13 [78, 79], and OpenMM version 7.3 [80], which include support for the Drude oscillator model [81]. Production simulations are generally performed using NVIDIA P100 or V100 GPU cards with CUDA version 8.0.61 or 9.1.85.

2.2 Other Software Implementations

The simulations described here can also be performed entirely within CHARMM, which has supported extended Lagrangian integration for Drude polarizable systems since 2003 using a modified velocity Verlet integrator [76]. Production simulations can also be carried out in NAMD using Langevin dynamics [78]. At the time of the writing of this chapter, GROMACS supports the core features of the Drude oscillator model and extended Lagrangian integration using the same velocity Verlet integrator as in CHARMM [82], but currently does not support through-space NBTHOLE screening [75] or NPT ensembles. These features should be implemented in the near future.

3 Simulation Methods

3.1 Summary of Example Systems

In this chapter, we will discuss a general protocol for preparing polarizable simulations of systems including proteins. One system is lysozyme in water with counterions. The other is a model transmembrane peptide, $WALP_{23}$, that is simulated in two environments, in water and in a POPC membrane. The contents of these systems are listed in Table 1, along with benchmark performance of each. Note that while simulations with the Drude force field are 3–4 times slower than nonpolarizable systems with the same composition, the use of GPU hardware makes them computationally tractable.

Table 1
System composition and simulation benchmark performance. Simulations performed on CPU used NAMD 2.13 as the MD engine. Simulations on P100 and V100 GPU hardware used OpenMM 7.3

System	Solution composition	Force field	Total no of particles	Hardware	Performance (ns/day)
Lysozyme	9633 water, 8 Cl^-	C36	30,867	24 CPU[a]	11.8
				P100	74.3
				V100	126
		Drude	51,546	24 CPU[a]	2.8
				P100	27.2
				V100	50.3
$WALP_{23}$	6068 water	C36	18,577	24 CPU[a]	15.3
				P100	166
				V100	250
		Drude	30,942	24 CPU[a]	4.8
				P100	39.3
				V100	80
$WALP_{23}$	116 POPC, 5041 water	C36	31,040	24 CPU[a]	10.3
				P100	98.7
				V100	165
		Drude	48,311	24 CPU[a]	2.6
				P100	24.5
				V100	54.6

[a]Intel E5-2680v3 Haswell processors (2.5 GHz)

All input files needed to perform the steps outlined in this chapter are available via GitHub at https://github.com/jalemkul/mimb2020. The procedures described here will include a description of important methodological details and commands that can be used to prepare the example systems.

3.2 System Construction

A diagram of the typical workflow in preparing a polarizable simulation with the Drude oscillator model is shown in Fig. 2. The first steps required in preparing a polarizable simulation system are no different than any other simulation; in fact, the coordinates are prepared and equilibrated using an additive force field to speed up the process of reaching a stable state. Simulated systems are initially constructed using the CHARMM-GUI online server [83] with the CHARMM36m force field [30] applied to the proteins, CHARMM-modified TIP3P for water [84–86], standard CHARMM ion parameters [87], and CHARMM36 for phospholipids [27, 28]. The topology for the protein is assigned after determining protonation states of all constituent amino acid side chains and termini, any disulfide linkages are assigned, and missing atoms such as hydrogen are constructed using the internal coordinate builder feature of the CHARMM program [77]. After the protein coordinates are prepared, a solvent box is added around the protein, which is centered in the unit cell. Transmembrane

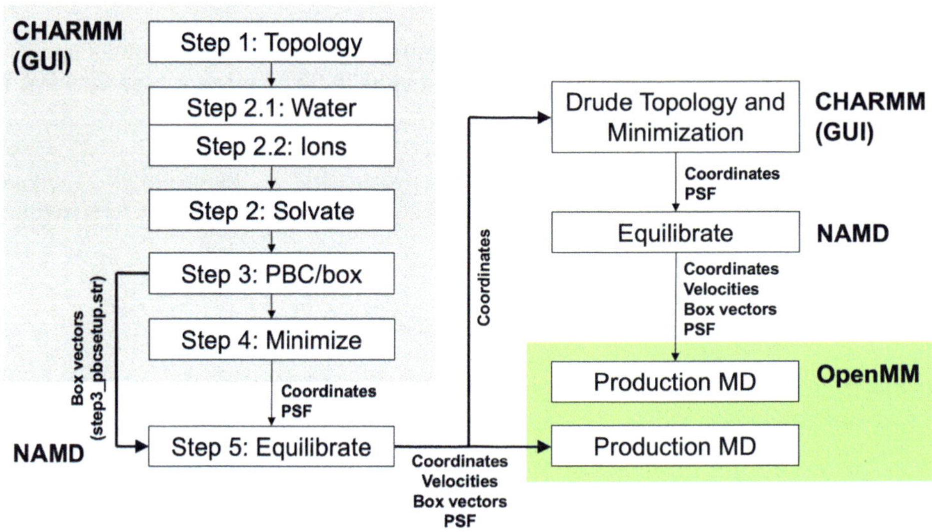

Fig. 2 Schematic of the general workflow outlined here. The system is initially constructed, minimized, and equilibrated using the CHARMM additive force field before being converted to the Drude polarizable model. Production simulations of nonpolarizable simulations can be performed in OpenMM as indicated but are not discussed here. Processes performed in CHARMM can also be carried out using the CHARMM-GUI web server

peptides are aligned with their α-helix axes coincident with the z-axis, which corresponds to the membrane normal, and lipids packed around the peptides using the CHARMM-GUI Membrane Builder protocol [88, 89]. Water molecules with any oxygen atom that overlap with the protein (or membrane, if presented) are deleted. Ions are added to neutralize the net charge of the protein and additional salt can be added if relevant (e.g., 100–150 mM for typical biological systems). The sequence of commands using the provided files in the `lysozyme_water/c36/build` directory on GitHub to build the lysozyme in water system is:

```
charmm -i step1_pdbreader.inp -o step1_pdbreader.out
charmm -i step2.1_waterbox.inp -o step2.1_waterbox.out
charmm -i step2.2_ions.inp -o step2.2_ions.out
charmm -i step2_solvator.inp -o step2_solvator.out
charmm -i step3_pbcsetup.inp -o step3_pbcsetup.out
```

3.3 Nonpolarizable Energy Minimization and Equilibration

We recommend performing energy minimization in two phases for maximum robustness. Energy minimization is performed in CHARMM. The first stage employs the steepest descent algorithm for 500 steps, followed by the second stage using the adopted-basis set Newton-Raphson (ABNR) method for 500 steps or until a gradient tolerance of 0.0001 kcal mol^{-1} $Å^{-1}$ is achieved. The following CHARMM command will perform energy minimization:

```
charmm -i step4_mini.inp -o step4_mini.out
```

The minimized coordinates and system topology (in X-PLOR PSF format) are then passed to NAMD for restrained equilibration, which is carried out for 1 ns under an NPT ensemble (see the provided `lysozyme_water/c36/equil/equil.in` file). Non-hydrogen atoms in the protein are restrained with a force constant of 5.0 kcal mol^{-1} $Å^{-2}$ by specifying this value in the B-factor field of a PDB supplied to NAMD. We create a restraint.pdb file using the provided `scripts/write_B-factors.pl` Perl script, which takes the minimized coordinates (in PDB format) and a force constant specified on the command line to generate the restraint reference file. Water and mobile ions are left unrestrained, as are lipids in the case of a membrane system. To generate the restraint file, issue this command:

```
perl write_B-factors.pl step4_mini.pdb 5.0
```

Temperature is regulated by a Langevin thermostat with a damping coefficient (γ) of 5.0 ps^{-1}, and pressure is regulated via the Langevin piston barostat [90] with a period of 200 fs and decay time of 100 fs. Target values for temperature and pressure can be set to any relevant value for the conditions being modeled; for the example provided here, temperature was set to 298 K and pressure to 1 atm (1.01325 bar). Per the convention of the CHARMM force field, covalent bonds to hydrogen atoms are constrained using SHAKE [91] and water molecules are held rigid using SETTLE [92]. The use of constraints allows an integration time step of 2 fs. Periodic boundary conditions are applied in all spatial dimensions.

The nonbonded treatment for the simulations follows the standard approach of the CHARMM force field. Van der Waals forces are smoothly switched to zero from 10–12 Å. Electrostatic forces are computed with the particle mesh Ewald (PME) method [93], using a real-space cutoff of 12 Å and Fourier grid points spaced at approximately 1 Å.

3.4 Polarizable Energy Minimization

The next step generates the topology of the system and performs energy minimization. Here, we generate the topology using the Drude-2019 protein force field [94], Drude-2017 phospholipid force field [63], SWM4-NDP water [66], and ion parameters by Yu et al. [68]. The equilibrated coordinates from nonpolarizable energy minimization serve as input for constructing Drude oscillators and lone pairs. The output coordinate file (in PDB format) from NAMD is read into CHARMM, which constructs these additional particles. Drude oscillators are constructed "on top" of their parent atoms, that is, they have identical initial coordinates. Lone pairs are constructed according to the geometric parameters listed in the residue topology definition.

The purpose of the polarizable energy minimization is to relax the induced dipoles in the field of the fixed-point charges of the real atoms. As such, all real atoms are restrained with a large force constant, typically on the order of 10^6–10^7 kcal mol^{-1} Å^{-2}. As with the nonpolarizable system preparation, energy minimization is carried out in two phases using steepest descents minimization (1000 steps) and the ABNR method (2000 steps or until a gradient tolerance of 0.0001 kcal mol^{-1} Å^{-1} is reached). A larger number of minimization steps is generally used in polarizable minimization to ensure sufficient self-consistent relaxation of the Drude displacements.

At this stage, the Drude displacements (dipole lengths) should be checked to see if any exceed 0.2 Å to be compatible with the "hard wall" constraint described below. Large Drude displacements may indicate over-polarization and potential instability at the outset of any subsequent MD simulation. The provided `scripts/check_drude_bond_lengths.pl` Perl script can be used for this purpose and can be used with PDB or CRD coordinate formats. To generate the Drude system and to check the Drude-atom bond lengths, issue these commands:

```
charmm -i write_drude_psf.inp -o drude.out
perl check_drude_bond_lengths.pl 1aki_drude_min.pdb 0.2
```

3.5 Polarizable Equilibration

Restrained equilibration is performed similarly to the nonpolarizable simulations described in Subheading 3.3. The same restraint scheme is applied, and the restraint reference file is generated in the same way as before. The same bond constraints are applied; however, in a Drude polarizable simulation, the fastest degrees of freedom are not bonds involving hydrogen atoms. Instead, the Drude-atom bonds are the highest-frequency motions in the system and therefore still limit the time step, which is set to 1 fs. Polarization catastrophe is prevented through the use of a "hard wall" constraint [62] that reflects the Drude oscillator back toward its parent atom along the Drude-atom bond vector if the bond length exceeds a user-defined tolerance, which is typically set to 0.2 Å.

The nonbonded treatment is similar to that of the CHARMM force field; however, in the case of the Drude polarizable force field, it is the van der Waals potential (not force) that is switched to zero from 10–12 Å. The treatment of electrostatics via PME is the same. Periodic boundary conditions are applied in all three spatial dimensions.

Equilibration is again performed for 1 ns under an NPT ensemble. Temperature is maintained via a dual Langevin thermostat (a modification of the original dual Nosé-Hoover algorithm [76]), under which the real atoms of the system are maintained at 298 K (or some other relevant temperature) with $\gamma = 5.0$ ps^{-1}. The Drude oscillators are coupled to a relative thermostat at 1 K and $\gamma = 20.0$ ps^{-1}. Pressure is maintained using the Langevin piston method as described in Subheading 3.3.

3.6 Polarizable Production Simulations

Unrestrained simulations are performed in OpenMM, which supports the algorithms necessary to perform MD simulations with the Drude oscillator model as of version 7.3 [81]. The output coordinate (equil.coor), velocity (equil.vel), and box vector (equil.xsc) files from NAMD are converted to an XML "restart" file that OpenMM takes as input. The script to perform this conversion is called `scripts/namd2omm.pl` and is written in Perl.

```
perl namd2omm.pl equil.coor equil.vel equil.xsc
```

The conversion script will generate a file generically named omm.rst that can be renamed as the user sees fit.

The constraints, nonbonded setup, thermostat, and integration scheme are identical to those applied during equilibration. A Monte Carlo barostat is applied to regulate pressure during production MD simulations, attempting box scaling every 25 integration steps. For lysozyme and $WALP_{23}$ in water, isotropic box scaling is employed, and for $WALP_{23}$ in POPC, the MembraneMonteCarloBarostat in OpenMM is used, which scales the box semi-isotropically. The `walp23_popc/drude/md/run.py` script provided in the example input files serves as the input for OpenMM.

4 Analysis

4.1 Overview of Analysis Concepts

Structural analysis of biomolecules in polarizable simulations is identical to methods applied in nonpolarizable systems. Here, we focus on several types of analysis that are particularly relevant to polarizable simulations. Of specific interest is the response of molecular dipole moments to changes in electric fields (such as polar and hydrophobic microenvironments), participation in hydrogen bonds, interactions with ions, etc. The analysis presented here focuses on dipole moment properties and reflects routine types of analysis that might be carried out on biomolecular systems.

4.2 Side-Chain Dipole Moments as a Function of Solvent Exposure in Lysozyme

The architecture of most globular proteins is such that polar and charged residues are oriented outward to face the aqueous solvent, while aliphatic amino acids are condensed in the interior of the tertiary structure. However, surface-exposed aliphatic residues and buried polar or charged residues do exist in many proteins. Moreover, the electrostatic properties of any amino acid are a function of its solvent exposure, backbone geometry, and interactions with the microenvironment formed by proximal amino acids. As such, we use the lysozyme simulation as a means to examine amino acid side-chain dipole response in a prototypical globular protein.

Here, we provide an example of computing the difference in side-chain dipole moment of amino acids in lysozyme (computed directly from the MD simulation trajectory) and the gas-phase dipole moment values in dipeptides of each residue in the same geometry. The difference between these two values for each residue reflects the average magnitude of dipole moment change as a function of solvent exposure and interactions with surrounding amino acids. It is possible to deconvolute these contributions but for the purposes of this chapter, we focus only on the total change in dipole moment as a function of solvent-accessible surface area (SASA). The steps necessary to perform these calculations are:

1. Compute the side-chain dipole moments of each amino acid. This task is achieved by invoking the COOR DIPO command in CHARMM for side-chain atoms.

```
charmm -i dipole.inp -o mu.out
```

2. Compute the fraction of available SASA for each amino acid. The relative SASA reflects the given residue's exposure; it is important to use the relative SASA rather than the total SASA since residues that are intrinsically larger would skew the results. That is, leucine might have a larger SASA than alanine due to its larger size, but both residues may have 0% solvent exposure if they are buried in the hydrophobic core of the structure. The total SASA is computed in CHARMM via CORREL tools by selecting just the amino acid of interest as the group for Connolly surface construction and output:

```
charmm -i total_sasa.inp -o total.out
```

   The actual SASA of the residue (in the context of the whole protein) is computed in the same manner but by choosing the whole protein as the group for surface construction and only the residue of interest for output:

```
charmm -i partial_sasa.inp -o partial.out
```

   The relative SASA is obtained by dividing the output of the second script by that of the first.

3. Calculate the geometry of each amino acid (ϕ, ψ, and all side-chain torsions) at every time step in the original MD trajectory. Dihedral angles can be computed via the QUICK command in CHARMM. For all residues, the backbone geometry needs to be recorded, as well as χ_n side-chain torsions, which depend on the residue.

```
charmm -i calc_dihedrals.inp -o geom.out
```

4. Construct a dipeptide of each amino acid in each geometry recorded in **step 3**, relax the Drude oscillators in vacuo while preserving these geometries, and compute the side-chain dipole moments of the energy-minimized structure. The dihedral angle values from **step 3** are read into a new script, which uses the CHARMM internal coordinate builder to construct the desired geometry. The dipeptide geometry is preserved by means of dihedral restraints.

```
charmm -i build_dipeptides.inp -o dipep.out
```

The resulting differences in dipole moment are expressed as a percentage, again to normalize for intrinsic differences in amino acid residues based on their size and chemical characteristics. For example, a change in side-chain μ of 0.4 D would be somewhat small for lysine (side-chain μ ~12 D) but quite large for leucine (side-chain μ ~0.6 D).

The results of these calculations are shown in Fig. 3, with each amino acid type (aliphatic, aromatic, polar neutral, basic/positive, and acidic/negative) colored separately. Aliphatic and aromatic residues generally have relative SASA <20%, indicating they are buried in the hydrophobic core of the protein according to the definition used by Huang et al. [48]. Some of these residues have relative SASA of nearly 40%, but there are very few above 20%, and the region of relative SASA from 20 to 50% is still considered the protein interior [48]. Interestingly, aliphatic residue side chains are almost uniformly depolarized relative to their gas-phase dipeptide values. This outcome can be rationalized by the fact that polarization response is believed to be greater in the gas phase due to repulsion between electron clouds in the condensed phase that opposes induction [95–100]. Such electronic repulsion is reflected in the results shown here. Aromatic residues, however, generally exhibit larger side-chain dipole moment values than in the gas phase, likely reflecting their ability to more favorably interact with water and polar/charged residues (e.g., cation–π interactions).

Polar neutral, acidic, and basic residues exhibit a wide range of solvent accessibility, reflecting tertiary structure features and packing between residues. Some polar residues are deeply buried and have negative values for dipole moment changes, reflecting depolarization as a function of burial and the damped polarization response mentioned above. Most of the polar neutral residues, however, have positive $\Delta|\mu|$, indicating polarization as a function of exposure to water and reflecting the less-dense environment at the surface of the protein relative to the interior. The corresponding values of $\Delta|\mu|$ range widely from 0 to 100%, emphasizing the considerable electronic plasticity of these species. In contrast, charged residues were almost uniformly polarized,

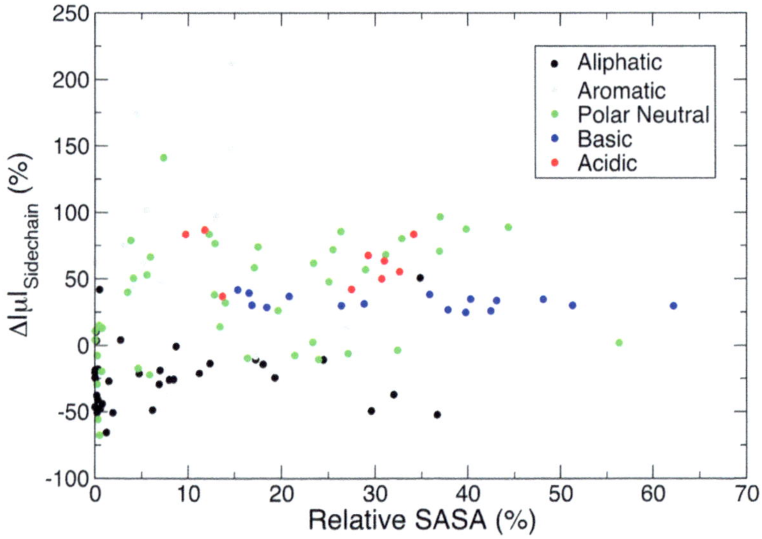

Fig. 3 Percent change in side-chain dipole moments relative to vacuum dipeptide values for all residues (except glycine) in lysozyme as a function of their relative solvent exposure

irrespective of their solvent exposure, though they tend to have high relative SASA, as expected for residues that generally appear on the surface of a globular protein (Fig. 3).

4.3 Differences in Peptide-Bond and Side-Chain Dipole Moments in the WALP$_{23}$ Peptide in Water and in a POPC Membrane

In contrast to globular proteins, membrane proteins often have aliphatic groups oriented towards their surfaces, to favorably interact with the hydrophobic interior of phospholipid membranes. As such, it is reasonable to expect that dipole properties in these species will vary markedly if simulated in water and in a membrane. What is less clear is if peptide-bond dipole moments are different in membrane proteins/peptides in water and in membranes. Here, we compute peptide-bond and side-chain dipole moments of WALP$_{23}$ in water and in a POPC membrane, and simply take the differences between the two. We do not consider the gas-phase values here.

1. Compute peptide-bond dipole moments, which include atoms in two consecutive residues (Fig. 4).

```
charmm -i dipole.inp -o backbone.out
```

2. Compute the side-chain dipole moments.

```
charmm -i sc_dipole.inp -o sc.out
```

The change in each dipole moment is expressed as a percent change relative to the values obtained in water, thus reflecting the induced polarization effect upon partitioning from water into the

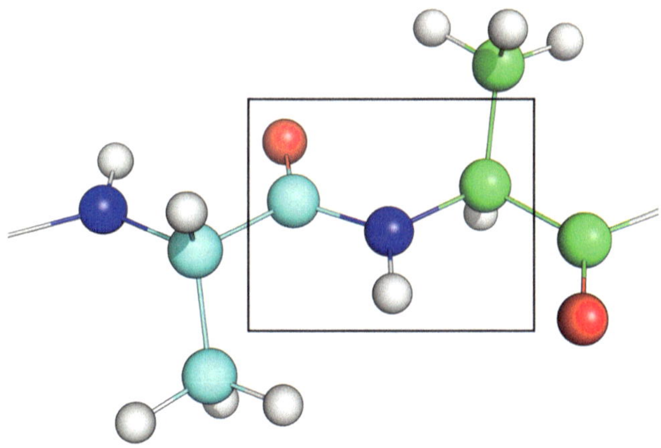

Fig. 4 Schematic of the atoms included in the calculation of peptide-bond dipole moments (enclosed in the black box). The carbonyl group of residue *i* (cyan carbon atoms) and the remaining atoms in residue *i* + 1 (green carbon atoms) comprise a net-neutral group. Drude oscillators and lone pairs associated with the indicated atoms are not shown but are included in the dipole moment calculation

POPC membrane, again normalizing for intrinsic differences in residue size and properties. The two CHARMM commands at the beginning of his Subheading produce data files containing the dipole moment of each residue as a function of time. Averages are computed and the differences are taken to generate the relative change. The results are plotted in Fig. 5. In general, both peptide-bond dipole moments (Fig. 5a) and side-chain dipole moments (Fig. 5b) are higher in water than in POPC, with a few exceptions. The peptide-bond dipole moment values shift by relatively smaller amounts (generally <5%) than the side-chain dipole moments (up to nearly 40%). In our 100-ns simulations, $WALP_{23}$ remained α-helical in both water and in POPC, so the peptide-bond hydrogen-bonding patterns were similar and shielded from water and/or POPC to similar extents. Nevertheless, the results suggest that the dipole moment magnitudes and thus the resulting strength of hydrogen bonding is slightly different in water than in POPC.

Side-chain dipole moments were more variable. Residues at the membrane–water interface (Trp2, Trp3, Trp21, Trp22, Ala23) were similarly polarized between the simulation in water and the simulation in POPC, though Trp2 was more polarized in POPC than in water. The 31.5% increase in the Trp2 side-chain dipole moment in POPC relative to the value in water suggests specific interactions with phospholipid headgroups that augment this value relative to solvation by pure water. Leucine residues in $WALP_{23}$

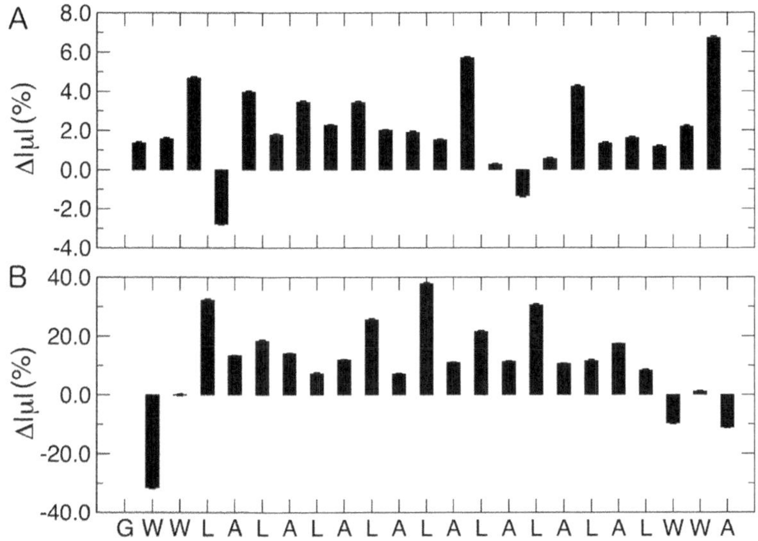

Fig. 5 Percent change in (**a**) peptide-bond and (**b**) side-chain dipole moments in the $WALP_{23}$ peptide in a POPC membrane. The reference values for the percent change were the values in water. Error bars were computed from the standard error of the dipole moment values

were also much more strongly polarized in water than alanine (Fig. 5b). These results emphasize that different aliphatic residues have different electronic plasticity. Moreover, nonpolarizable force fields will fail to capture these effects, leading to less-accurate representations of interactions with lipids that may be important for understanding dynamics in membrane-protein systems.

5 Conclusions

Here, we have provided a protocol for preparing and analyzing polarizable simulations using the classical Drude oscillator. We have focused on specific methods for equilibrating these systems using a nonpolarizable force field first, followed by additional minimization and equilibration with a polarizable topology. Analysis methods emphasized specific techniques relevant to polarizable systems, noting instances in which the microenvironment of an amino acid (e.g., protein interior and phospholipid membrane) modulates its dipole moment and thus its electrostatic interactions with its surroundings. We hope that adoption of these methods leads to more widespread use of polarizable force fields in heterogeneous systems. Despite some added computational cost, these simulations remain tractable and have the potential to provide novel insights into the balance of forces in biologically relevant systems.

Acknowledgments

This work was supported by the National Institutes of Health (grant R35GM133754). Computing time and resources were provided by Virginia Tech Advanced Research Computing.

References

1. Durrant JD, Kochanek SE, Casalino L et al (2020) Mesoscale all-atom influenza virus simulations suggest new substrate binding mechanism. ACS Cent Sci 6(2):189–196. https://doi.org/10.1021/acscentsci.9b01071
2. Marrink SJ, Corradi V, Souza PCT et al (2019) Computational modeling of realistic cell membranes. Chem Rev 119 (9):6184–6226. https://doi.org/10.1021/acs.chemrev.8b00460
3. Bond PJ, Holyoake J, Ivetac A et al (2007) Coarse-grained molecular dynamics simulations of membrane proteins and peptides. J Struct Biol 157(3):593–605. https://doi.org/10.1016/j.jsb.2006.10.004
4. Cornell WD, Cieplak P, Bayly CI et al (1995) A second generation force field for the simulation of proteins, nucleic acids, and organic molecules. J Am Chem Soc 117 (19):5179–5197. https://doi.org/10.1021/ja00124a002
5. Wang J, Cieplak P, Kollman PA (2000) How well does a restrained electrostatic potential (RESP) model perform in calculating conformational energies of organic and biological molecules? J Comput Chem 2 (12):1049–1074. https://doi.org/10.1002/1096-987X(200009)21:12<1049::AID-JCC3>3.0.CO;2-F
6. Duan Y, Wu C, Chowdhury S et al (2003) A point-charge force field for molecular mechanics simulations of proteins based on condensed-phase quantum mechanical calculations. J Comput Chem 24(16):1999–2012. https://doi.org/10.1002/jcc.10349
7. Lindorff-Larsen K, Piana S, Palmo K et al (2010) Improved side-chain torsion potentials for the Amber ff99SB protein force field. Proteins 78:1950–1958. https://doi.org/10.1002/prot.22711
8. Pérez A, Marchán I, Svozil D et al (2007) Refinement of the AMBER force field for nucleic acids: improving the description of α/γ conformers. Biophys J 92 (11):3817–3829. https://doi.org/10.1529/biophysj.106.097782
9. Banáš P, Hollas D, Zgarbová M et al (2010) Performance of molecular mechanics force fields for RNA simulations: stability of UUCG and GNRA hairpins. J Chem Theory Comput 6(12):3836–3849. https://doi.org/10.1021/ct100481h
10. Krepl M, Zgarbová M, Stadlbauer P et al (2012) Reference simulations of noncanonical nucleic acids with difference χ variants of the AMBER force field: quadruplex DNA, quadruplex RNA, and Z-DNA. J Chem Theory Comput 8(7):2506–2520. https://doi.org/10.1021/ct300275s
11. Woods RJ, Dwek RA, Edge CJ, Fraser-Reid B (1995) Molecular mechanical and molecular dynamical simulations of glycoproteins and oligosaccharides. 1. GLYCAM_93 parameter development. J Phys Chem 99 (11):3832–3846. https://doi.org/10.1021/j100011a061
12. Basma M, Sundara S, Çalgan D et al (2001) Solvated ensemble averaging in the calculation of partial atomic charges. J Comput Chem 22(11):1125–1137. https://doi.org/10.1002/jcc.1072
13. Maier JA, Martinez C, Kasavajhala K et al (2015) ff14SB: improving the accuracy of protein side chain and backbone parameters from ff99SB. J Chem Theory Comput 11 (8):3696–3713. https://doi.org/10.1021/acs.jctc.5b00255
14. Tian C, Ksavajhala K, Belfon KAA et al (2020) ff19SB: amino-acid-specific protein backbone parameters trained against quantum mechanics energy surfaces in solution. J Chem Theory Comput 16(1):528–552. https://doi.org/10.1021/acs.jctc.9b00591
15. Ivani I, Dans PD, Noy A et al (2015) Parmbsc1: a refined force field for DNA simulations. Nat Methods 13:55–58. https://doi.org/10.1038/nmeth.3658
16. Best RB, Zhu X, Shim J et al (2012) Optimization of the additive CHARMM all-atom protein force field targeting improved sampling of the backbone φ,ψ and side-chain χ_1 and χ_2 dihedral angles. J Chem Theory Comput 8(9):3257–3273. https://doi.org/10.1021/ct300400x

17. Foloppe N, MacKerell AD Jr (2000) All-atom empirical force field for nucleic acids: I. Parameter optimization based on small molecule and condensed phase macromolecular target data. J Comput Chem 21 (2):86–104. https://doi.org/10.1002/(SICI)1096-987X(20000130)21:2<86::AID-JCC2>3.0.CO;2-G
18. MacKerell AD Jr, Banavali NK (2000) All-atom empirical force field for nucleic acids: II. Application to solution MD simulations of DNA. J Comput Chem 21 (2):105–120. https://doi.org/10.1002/(SICI)1096-987X(20000130)21:2<105::AID-JCC3>3.0.CO;2-P
19. MacKerell AD Jr (2004) Empirical force fields for biological macromolecules: overview and issues. J Comput Chem 25(13):1584–1604. https://doi.org/10.1002/jcc.20082
20. MacKerell AD Jr, Bashford D, Bellott M et al (1998) All-atom empirical potential for molecular modeling and dynamics studies of proteins. J Phys Chem B 102 (18):3586–3616. https://doi.org/10.1021/jp973084f
21. MacKerell AD Jr, Feig M, Brooks IIICL (2004) Extending the treatment of backbone energetics in protein force fields: limitations of gas-phase quantum mechanics in reproducing protein conformational distributions in molecular dynamics simulations. J Comput Chem 25:1400–1415. https://doi.org/10.1002/jcc.20065
22. Guvench O, Greene SN, Kamath G et al (2008) Additive empirical force field for hexopyranose monosaccharides. J Comput Chem 29(15):2543–2564. https://doi.org/10.1002/jcc.21004
23. Guvench O, Hatcher E, Venable RM et al (2009) CHARMM additive all-atom force field for glycosidic linkages between hexopyranoses. J Chem Theory Comput 5 (9):2353–2370. https://doi.org/10.1021/ct900242e
24. Hart K, Foloppe N, Baker CM et al (2011) Optimization of the CHARMM additive force field for DNA: improved treatment of the BI/BII conformational equilibrium. J Chem Theory Comput 8(1):348–362. https://doi.org/10.1021/ct200723y
25. Hatcher E, Guvench O, MacKerell AD Jr (2009) CHARMM additive all-atom force field for aldopentofuranoses, methyl-aldopentofuranosides, and fructofuranose. J Phys Chem B 113(37):12466–12476. https://doi.org/10.1021/jp905496e
26. Hatcher ER, Guvench O, MacKerell AD Jr (2009) CHARMM additive all-atom force field for acyclic polyalcohols, acyclic carbohydrates, and inositol. J Chem Theory Comput 5(5):1315–1327. https://doi.org/10.1021/ct9000608
27. Klauda JB, Venable RM, Freites JA et al (2010) Update of the CHARMM all-atom additive force field for lipids: validation on six lipid types. J Phys Chem B 114 (23):7830–7843. https://doi.org/10.1021/jp101759q
28. Pastor RW, MacKerell AD Jr (2011) Development of the CHARMM force field for lipids. J Phys Chem Lett 2(13):1526–1532. https://doi.org/10.1021/jz200167q
29. Raman EP, Guvench O, MacKerell AD Jr (2010) CHARMM additive all-atom force field for Glycosidic linkages in carbohydrates involving Furanoses. J Phys Chem B 114 (40):12981–12994. https://doi.org/10.1021/jp105758h
30. Huang J, Rauscher S, Nawrocki G et al (2016) CHARMM36m: an improved force field for folded and intrinsically disordered proteins. Nat Methods 14(1):71–73. https://doi.org/10.1038/nmeth.4067
31. Chandrasekhar I, Kastenholz M, Lins RD et al (2003) A consistent potential energy parameter set for lipids: dipalmitoylphosphatidylcholine as a benchmark of the GROMOS96 45A3 force field. Eur Biophys J 32(1):67–77. https://doi.org/10.1007/s00249-002-0269-4
32. Daura X, Mark AE, van Gunsteren WF (1998) Parametrization of aliphatic CH_n united atoms of GROMOS96 force field. J Comput Chem 19(5):535–547. https://doi.org/10.1002/(SICI)1096-987X(19980415)19:5<535::AID-JCC6>3.0.CO;2-N
33. Lins RD, Hünenberger PH (2005) A new GROMOS force field for hexopyranose-based carbohydrates. J Comput Chem 26 (13):1400–1412. https://doi.org/10.1002/jcc.20275
34. Oostenbrink C, Soares TA, van der Vegt NFA, van Gunsteren WF (2005) Validation of the 53A6 GROMOS force field. Eur Biophys J 34 (4):273–284. https://doi.org/10.1007/s00249-004-0448-6
35. Oostenbrink C, Villa A, Mark AE, van Gunsteren WF (2004) A biomolecular force field based on the free enthalpy of hydration and solvation: the GROMOS force-field parameter sets 53A5 and 53A6. J Comput Chem 25 (13):1656–1676. https://doi.org/10.1002/jcc.20090
36. Reif MM, Hünenberger PH, Oostenbrink C (2012) New interaction parameters for

charged amino acid side chains in the GROMOS force field. J Chem Theory Comput 8 (10):3705–3723. https://doi.org/10.1021/ct300156h
37. Reif MM, Winger M, Oostenbrink C (2013) Testing of the GROMOS force-field parameter set 54A8: structural properties of electrolyte solutions, lipid bilayers, and proteins. J Chem Theory Comput 9(2):1247–1264. https://doi.org/10.1021/ct300874c
38. Schmid N, Eichenberger AP, Choutko A et al (2011) Definition and testing of the GROMOS force-field versions 54A7 and 54B7. Eur Biophys J 40:843–856. https://doi.org/10.1007/s00249-011-0700-9
39. Schuler LD, Daura X, van Gunsteren WF (2001) An improved GROMOS96 force field for aliphatic hydrocarbons in the condensed phase. J Comput Chem 22 (11):1205–1218. https://doi.org/10.1002/jcc.1078
40. Soares TA, Hünenberger PH, Kastenholz MA et al (2005) An improved nucleic acid parameter set for the GROMOS force field. J Comput Chem 26(7):725–737. https://doi.org/10.1002/jcc.20193
41. Harder E, Damm W, Maple J et al (2016) OPLS3: a force field providing broad coverage of drug-like small molecules and proteins. J Chem Theory Comput 12(1):281–296. https://doi.org/10.1021/acs.jctc.5b00864
42. Jorgensen WL, Maxwell DS, Tirado-Rives J (1996) Development and testing of the OPLS all-atom force field on conformational energetics and properties of organic liquids. J Am Chem Soc 118(45):11225–11236. https://doi.org/10.1021/ja9621760
43. Jorgensen WL, Tirado-Rives J (1988) The OPLS potential function for proteins. Energy minimization for crystals of cyclic peptides and Crambin. J Am Chem Soc 110 (6):1657–1666. https://doi.org/10.1021/ja00214a001
44. Kaminski GA, Duffy EM, Matsui T, Jorgensen WL (1994) Free energies of hydration and pure liquid properties of hydrocarbons from the OPLS all-atom model. J Phys Chem 98 (49):13077–13082. https://doi.org/10.1021/j100100a043
45. Kaminski GA, Friesner RA, Tirado-Rives J, Jorgensen WL (2001) Evaluation and reparametrization of the OPLS-AA force field for proteins via comparison with accurate quantum chemical calculations on peptides. J Phys Chem B 105(28):6474–6487. https://doi.org/10.1021/jp003919d
46. Robertson MJ, Tirado-Rives J, Jorgensen WL (2015) Improved peptide and protein torsional energetics with the OPLS-AA force field. J Chem Theory Comput 11 (7):3499–3509. https://doi.org/10.1021/acs.jctc.5b00356
47. Shivakumar D, Harder E, Damm W et al (2012) Improving the prediction of absolute solvation free energies using the next generation OPLS force field. J Chem Theory Comput 8(8):2553–2558. https://doi.org/10.1021/ct300203w
48. Huang J, Lopes PEM, Roux B, MacKerell AD Jr (2014) Recent advances in polarizable force fields for macromolecules: microsecond simulations of proteins using the classical Drude oscillator model. J Phys Chem Lett 5 (18):3144–3150. https://doi.org/10.1021/jz501315h
49. Ponder JW, Wu C, Ren P et al (2010) Current status of the AMOEBA polarizable force field. J Phys Chem B 114(8):2549–2564. https://doi.org/10.1021/jp910674d
50. Lemkul JA, Huang J, Roux B, MacKerell AD Jr (2016) An empirical polarizable force field based on the classical Drude oscillator model: development history and recent applications. Chem Rev 116(9):4983–5013. https://doi.org/10.1021/acs.chemrev.5b00505
51. Shi Y, Xia Z, Zhang J et al (2013) Polarizable atomic multipole-based AMOEBA force field for proteins. J Chem Theory Comput 9 (9):4046–4063. https://doi.org/10.1021/ct4003702
52. Zhang C, Lu C, Jing Z et al (2018) AMOEBA polarizable atomic multipole force field for nucleic acids. J Chem Theory Comput 14 (4):2084–2108. https://doi.org/10.1021/acs.jctc.7b01169
53. Wang L-P, Head-Gordon T, Ponder JW et al (2013) Systematic improvement of a classical molecular model of water. J Phys Chem B 117 (34):9956–9972. https://doi.org/10.1021/jp403802c
54. Laury ML, Wang L-P, Pande VS et al (2015) Revised parameters for the AMOEBA polarizable atomic multipole water model. J Phys Chem B 119(29):9423–9437. https://doi.org/10.1021/jp510896n
55. Ren P, Ponder JW (2003) Polarizable atomic multipole water model for molecular mechanics simulation. J Phys Chem B 107 (24):5933–5947. https://doi.org/10.1021/jp027815+
56. Grossfield A, Ren P, Ponder JW (2003) Ion solvation thermodynamics from simulation with a polarizable force field. J Am Chem

Soc 125(50):15671–15682. https://doi.org/10.1021/ja037005r

57. Jiao D, King C, Grossfield A et al (2006) Simulation of Ca^{2+} and Mg^{2+} solvation using polarizable atomic multipole potential. J Phys Chem B 110(37):18553–18559. https://doi.org/10.1021/jp062230r
58. Lopes PEM, Huang J, Shim J et al (2013) Polarizable force field for peptides and proteins based on the classical Drude oscillator. J Chem Theory Comput 9(12):5430–5449. https://doi.org/10.1021/ct400781b
59. Lemkul JA, MacKerell AD Jr (2017) Polarizable force field for DNA based on the classical Drude oscillator: I. Refinement using quantum mechanical base stacking and conformational energetics. J Chem Theory Comput 13(5):2053–2071. https://doi.org/10.1021/acs.jctc.7b00067
60. Lemkul JA, MacKerell AD Jr (2017) Polarizable force field for DNA based on the classical Drude oscillator: II. Microsecond molecular dynamics simulations of duplex DNA. J Chem Theory Comput 13(5):2072–2085. https://doi.org/10.1021/acs.jctc.7b00068
61. Lemkul JA, MacKerell AD Jr (2018) Polarizable force field for RNA based on the classical Drude oscillator. J Comput Chem 39(32):2624–2646. https://doi.org/10.1002/jcc.25709
62. Chowdhary J, Harder E, Lopes PEM et al (2013) A polarizable force field of dipalmitoylphosphatidylcholine based on the classical Drude model for molecular dynamics simulations of lipids. J Phys Chem B 117(31):9142–9160. https://doi.org/10.1021/jp402860e
63. Li H, Chowdhary J, Huang L et al (2017) Drude polarizable force field for molecular dynamics simulations of saturated and unsaturated Zwitterionic lipids. J Chem Theory Comput 13(9):4535–4552. https://doi.org/10.1021/acs.jctc.7b00262
64. Aytenfisu AH, Yang M, MacKerell AD Jr (2018) CHARMM Drude polarizable force field for glycosidic linkages involving pyranoses and furanoses. J Chem Theory Comput 14(6):3132–3143. https://doi.org/10.1021/acs.jctc.8b00175
65. Jana M, MacKerell AD Jr (2015) CHARMM Drude polarizable force field for aldopentofuranoses and methyl-aldopentofuranosides. J Phys Chem B 119:7846–7859. https://doi.org/10.1021/acs.jpcb.5b01767
66. Lamoureux G, Harder E, Vorobyov IV et al (2006) A polarizable model of water for molecular dynamics simulations of biomolecules. Chem Phys Lett 418:245–249. https://doi.org/10.1016/j.cplett.2005.10.135
67. Yu W, Lopes PEM, Roux B, MacKerell AD Jr (2013) Six-site polarizable model of water based on the classical Drude oscillator. J Chem Phys 138:034508. https://doi.org/10.1063/1.4774577
68. Yu H, Whitfield TW, Harder E et al (2010) Simulating monovalent and divalent ions in aqueous solution using a Drude polarizable force field. J Chem Theory Comput 6(3):774–786. https://doi.org/10.1021/ct900576a
69. Harder E, MacKerell AD Jr, Roux B (2009) Many-body polarization effects and the membrane dipole potential. J Am Chem Soc 131(8):2760–2761. https://doi.org/10.1021/ja806825g
70. Drude P, Millikan RA, Mann RC (1902) The theory of optics. Longmans, Green, and Co, New York
71. Miller KJ (1990) Additivity methods in molecular polarizability. J Am Chem Soc 112(23):8533–8542. https://doi.org/10.1021/ja00179a044
72. Harder E, Anisimov VM, Vorobyov IV et al (2006) Atomic level anisotropy in the electrostatic modeling of lone pairs for a polarizable force field based on the classical Drude oscillator. J Chem Theory Comput 2(6):1587–1597. https://doi.org/10.1021/ct600180x
73. Thole BT (1981) Molecular polarizabilities calculated with a modified dipole interaction. Chem Phys 59(3):341–350. https://doi.org/10.1016/0301-0104(81)85176-2
74. Harder E, Anisimov VM, Whitfield T et al (2008) Understanding the dielectric properties of liquid amides from a polarizable force field. J Phys Chem B 112(11):3509–3521. https://doi.org/10.1021/jp709729d
75. Luo Y, Jiang W, Yu H et al (2013) Simulation study of ion pairing in concentrated aqueous salt solutions with a polarizable force field. Faraday Discuss 160:135–149. https://doi.org/10.1039/c2fd20068f
76. Lamoureux G, Roux B (2003) Modeling induced polarization with classical Drude oscillators: theory and molecular dynamics simulation algorithm. J Chem Phys 119(6):3025–3039. https://doi.org/10.1063/1.1589749
77. Brooks BR, Brooks IIICL, MacKerell AD Jr et al (2009) CHARMM: the biomolecular simulation program. J Comput Chem

30:1545–1614. https://doi.org/10.1002/jcc.21287
78. Jiang W, Hardy DJ, Phillips JC et al (2011) High-performance scalable molecular dynamics simulations of a polarizable force field based on classical Drude oscillators in NAMD. J Phys Chem Lett 2:87–92. https://doi.org/10.1021/jz101461d
79. Phillips JC, Braun R, Wang W et al (2005) Scalable molecular dynamics with NAMD. J Comput Chem 26(16):1781–1802. https://doi.org/10.1002/jcc.20289
80. Eastman P, Swails J, Chodera JD et al (2017) OpenMM 7: rapid development of high performance algorithms for molecular dynamics. PLoS Comput Biol 13(7):e1005659. https://doi.org/10.1371/journal.pcbi.1005659
81. Huang J, Lemkul JA, Eastman PK, MacKerell AD Jr (2018) Molecular dynamics simulations using the Drude polarizable force field on GPUs with OpenMM: implementation, validation, and benchmarks. J Comput Chem 39 (21):1682–1689. https://doi.org/10.1002/jcc.25339
82. Lemkul JA, Roux B, van der Spoel D, MacKerell AD Jr (2015) Implementation of extended Lagrangian dynamics in GROMACS for polarizable simulations using the classical Drude oscillator model. J Comput Chem 36(19):1473–1479. https://doi.org/10.1002/jcc.23937
83. Lee J, Cheng X, Swails JM et al (2016) CHARMM-GUI input generator for NAMD, GROMACS, AMBER, OpenMM, and CHARMM/OpenMM simulations using the CHARMM36 additive force field. J Chem Theory Comput 12(1):405–413. https://doi.org/10.1021/acs.jctc.5b00935
84. Jorgensen WL, Chandrasekhar J, Madura JD et al (1983) Comparison of simple potential functions for simulating liquid water. J Chem Phys 79(2):926–935. https://doi.org/10.1063/1.445869
85. Neria E, Fischer S, Karplus M (1996) Simulation of activation free energies in molecular systems. J Chem Phys 105:1902–1921. https://doi.org/10.1063/1.472061
86. Durell SR, Brooks BR, Ben-Naim A (1994) Solvent-induced forces between two hydrophilic groups. J Phys Chem 98:2198–2202. https://doi.org/10.1021/j100059a038
87. Beglov D, Roux B (1994) Finite representation of an infinite bulk system: solvent boundary potential for computer simulations. J Chem Phys 100:9050–9063. https://doi.org/10.1063/1.466711
88. Wu EW, Cheng X, Jo S et al (2014) CHARMM-GUI membrane builder toward realistic biological membrane simulations. J Comput Chem 35(27):1997–2004. https://doi.org/10.1002/jcc.23702
89. Lee J, Patel DS, Ståhle J et al (2019) CHARMM-GUI membrane builder for complex biological membrane simulations with glycolipids and lipoglycans. J Chem Theory Comput 15(1):775–786. https://doi.org/10.1021/acs.jctc.8b01066
90. Feller SE, Zhang Y, Pastor RW, Brooks BR (1995) Constant pressure molecular dynamics simulation: the Langevin piston method. J Chem Phys 103(11):4613–4621. https://doi.org/10.1063/1.470648
91. Ryckaert J-P, Ciccotti G, Berendsen HJC (1977) Numerical integration of the Cartesian equations of motion of a system with constraints: molecular dynamics of *n*-alkanes. J Comput Phys 23:327–341. https://doi.org/10.1016/0021-9991(77)90098-5
92. Miyamoto S, Kollman PA (1992) SETTLE: an analytical version of the SHAKE and RATTLE algorithms for rigid water molecules. J Comput Chem 13:952–962. https://doi.org/10.1002/jcc.540130805
93. Darden T, York D, Pedersen L (1993) Particle mesh Ewald: an N·log(N) method for Ewald sums in large systems. J Chem Phys 98 (12):10089–10092. https://doi.org/10.1063/1.464397
94. Lin F-Y, Huang J, Pandey P et al (2020) Further optimization and validation of the classical Drude polarizable protein force field. J Chem Theory Comput 16 (5):3221–3239. https://doi.org/10.1021/acs.jctc.0c00057
95. Giese TJ, York DM (2004) Many-body force field models based solely on pairwise Coulomb screening do not simultaneously reproduce correct gas-phase and condensed-phase polarizability limits. J Chem Phys 120 (21):9903. https://doi.org/10.1063/1.1756583
96. Kaminski GA, Stern HA, Berne BJ, Friesner RA (2004) Development of an accurate and robust polarizable molecular mechanics force field from ab initio quantum chemistry. J Phys Chem A 108(4):621–627. https://doi.org/10.1021/jp0301103
97. Morita A (2002) Water polarizability in condensed phase: *Ab initio* evaluation by cluster approach. J Comput Chem 23 (15):1466–1471. https://doi.org/10.1002/jcc.10169
98. Morita A, Kato S (1999) An *ab initio* analysis of medium perturbation on molecular polarizabilities. J Chem Phys 110(24):11987. https://doi.org/10.1063/1.479167

99. in het Panhuis M, PLA P, Munn RW, Ángyán JG (2001) Distributed polarizability of the water dimer: field-induced charge transfer along the hydrogen bond. J Chem Phys 114 (18):7951. https://doi.org/10.1063/1.1361247

100. Tu Y, Laaksonen A (2000) The electronic properties of water molecules in water clusters and liquid water. Chem Phys Lett 329 (3-4):283–288. https://doi.org/10.1016/S0009-2614(00)01026-5

Part VI

Membrane Permeability and Transport

Chapter 14

In Silico Prediction of Permeability Coefficients

Ricardo J. Ferreira

Abstract

In silico simulations of biological systems are of the significant importance to obtain insights on specific processes that experimental protocols have difficulty to elucidate. More particularly, and to ensure that a given molecule is able to reach its cellular target, the development of computational methods able to quickly estimate the cellular permeabilities for small molecules can become an important tool in the early stages of drug development. Herein, a computational protocol for predicting permeability coefficients, concerning both membranes and proteins, is presented and discussed.

Key words Inhomogeneous solubility-diffusion model, Resistance to permeation, Permeability coefficients, Membranes, Proteins, Molecular dynamics, Umbrella sampling, Python

1 Introduction

In biological processes, understanding how small molecules are transported across lipid membranes is of the utmost importance. While small uncharged molecules are able to freely diffuse through the lipid bilayer, molecules with higher molecular weights or with charged groups at physiological pH must cross the lipid bilayers through channels or transport proteins [1]. Therefore, the calculation of permeability coefficients is a quantitative measurement of the ability of a given molecule to cross through the intended biological barrier. More important, when applied in an early stage of the drug development pipeline, predicting if a promising compound will be able to reach its intracellular target *a priori* becomes an obvious advantage, thus avoiding most of the drawbacks that are only detected at a much later stage of clinical testing [2].

While experimentally more challenging [3], estimating permeabilities using molecular dynamics (MD) simulations are relatively fast, and do not require too demanding computational resources while providing mechanistic insights about the process of permeation itself. Initially used to model water transport through membranes [4], later studies applied similar approaches to investigate

Irina S. Moreira et al. (eds.), *Computational Design of Membrane Proteins*, Methods in Molecular Biology, vol. 2315, https://doi.org/10.1007/978-1-0716-1468-6_14,

how gases or other small organic compounds permeate through the lipid bilayer [5–9]. A similar setup was used to map the free-energy landscape for permeation through aquaglyceroporins [10], ion channels [5, 11], or, more recently, Gram-negative outer membrane porins [12, 13].

2 Theory

The inhomogeneous solubility-diffusion (ISD) model, developed by Marrink and Berendsen [4], was derived from a previous homogeneous solubility-diffusion (HSD) model to overcome several of the limitations of the latter and to further enhance its predictive power. Modeled as an irreversible process from the diffusional theory of transport, the ISD approach allows the estimation of the permeability coefficient for crossing an inhomogeneous medium such as a lipid membrane or, more recently, protein channels. Herein, the permeability coefficient is calculated as the inverse of the resistance to permeation, according to the following relation:

$$\frac{1}{P} = R^p = \int_{z_1}^{z_2} \frac{e^{[\beta \Delta G(z)]}}{D(z)} dz, \tag{1}$$

in which $\beta = 1/k_B T$, with k_B as the Boltzmann constant, T the temperature, $\Delta G(z)$ the potential of mean force (PMF), $D(z)$ the local diffusivity coefficient, and z the relative position of the molecule along the coordinate of interest (*see* **Note 1**). While $\Delta G(z)$ and $D(z)$ can be obtained in several ways (*see* **Note 2**), by using an umbrella sampling (US) approach both parameters can be obtained from each umbrella window from a relatively simple setup (see Methods).

The local diffusivity coefficient, $D(z)$, can be calculated directly from the US windows by using the following equation [14, 15]:

$$D(z = z) = \frac{var(z)}{\tau_z}. \tag{2}$$

In the above equation, $\langle z \rangle$ is the average of the coordinate path z in each US window, $\text{var}(z) = \langle z^2 \rangle - \langle z \rangle^2$ is the variance, and τ_z is the autocorrelation time. The most frequent approach to obtain the integrated autocorrelation time (IACT, $\tau_{\text{i,int}}$) derives from calculating its autocorrelation function $R_i(\Delta t)$, but a faster approach [16, 17] also allows a rather accurate estimation of τz from the variance of the mean value z and the interval of the time series data points (Δt) as (*see* **Note 3**):

$$\tau_z \approx \left[\frac{n\text{var}(\overline{z})}{\text{var}(z)} - 1 \right] \frac{\Delta t}{2}. \tag{3}$$

3 Methods

3.1 Systems Preparation

Prior to permeability calculations, equilibrated systems with (a) a membrane patch containing one or more lipids of interest, or (b) a protein embedded in the membrane patch, are required. The following items provide the necessary information for assembling each of the systems in GROMACS software package. For unexperienced users, the online tutorials by Justin A. Lemkul [18] are advised.

3.1.1 Membrane-Only Systems

Pre-equilibrated membrane systems for the united-atom GROMOS96 forcefield can be readily obtained from the Automated Topology Builder (ATB) and Repository (https://atb.uq.edu.au/) [19] (*see* **Note 4**). Concerning all-atom forcefields, the best option is to download them from CHARMM-GUI Archive (http://www.charmm-gui.org/) [20] or, alternatively, to build the membrane patch using the "Membrane Builder" [21] module also available in CHARMM-GUI server. In both cases, the topologies for each class of lipid are also available for download. Membrane patches available in public repositories are already pre-equilibrated, but quite often these patches contain a nonideal number of lipids (128 to 256 lipids, half in each leaflet). The size of the membrane patch can be increased through the command (herein exemplified for *dppc128.gro*):

user@unix:~$ *gmx genconf -f dppc128.gro -nbox 2 2 1 -o dppc512.gro -renumber*

Herein, the *-nbox* command specifies the number of additional boxes to be created. While *-nbox 2 2 1* creates a symmetrical lipid patch, rectangular membranes can be obtained by specifying a different number of boxes in the *xy* plane (e.g., *-nbox 3 2 1* or *-nbox 2 3 1*) (*see* **Note 5**). Following, the number of lipids (and waters) must be updated in the *topol.top* file (*see* **Note 6**), and the whole system re-equilibrated through (a) an energy minimization, (b) temperature equilibration, (c) pressure equilibration, and (d) production runs (*see* **Note 7**).

3.1.2 Protein-Membrane Systems

If using an all-atom CHARMM36 forcefield, you may build your protein-membrane system directly in CHARMM-GUI [22] and then proceed in the computer after downloading the system. For other forcefields as AMBER or GROMOS, the assembly of protein-membrane systems must be done by the user, as exemplified below (using *gmx membed* command available in GROMACS) (*see* **Note 8**).

The first step is to setup the system by correcting the relative position of the protein regarding the lipid bilayer (*see* **Note 9**). For most proteins, the location of the lipid membrane and its orientation can be retrieved from the OPM database (https://opm.phar.umich.edu) or, if not in the database, can be inferred using the PPM server [23] in order to estimate the proteins' rotational and translational positions of its transmembrane regions within the membrane. Usually, as the membrane patch is larger than the protein to be embedded, the protein coordinates must be translated to the center of the membrane. For a membrane patch with box coordinates $x = 16.0$ nm, $y = 15.0$ nm, and $z = 10.0$ nm, this can be accomplished using *gmx editconf* (*see* **Note 10**):

user@unix:~$ *gmx editconf -f protein.gro -center 8.0 7.5 5.0 -box 16.0 15.0 10.0 -o protein_recentered.gro*

Following, g_membed requires as input a concatenated file containing the membrane patch (lipids + waters) and the protein. Such input file can be generated using bash scripting:

user@unix:~$ *head -n -1 protein_recentered.gro > protein_membrane.gro && tail -n +3 membrane.gro >> protein_membrane.gro*

As the atom number is incorrect in the newly generated file, the user must manually edit the *protein_membrane.gro* with a text editor, count the number of lines discarding the first two lines (file header and atom number) and the last line (box vectors), and update the atom number (*see* **Note 11**).

Following, after updating the topology *topol.top* to reflect the above changes (e.g., adding the protein and/or the lipid topologies and number of molecules), the protein of interest can be inserted in the target membrane with the commands (under GROMACS 4.6.7) (*see* **Note 12**):

user@unix:~$ *gmx grompp -f minim.mdp -c protein_membrane.gro -p topol.top -o into_mem.tpr.*

user@unix:~$ *g_membed -f into_mem.tpr -c protein_membrane.gro -xyinit real 0.5 -xyend real 1 -asymmetry.*

user@unix:~$ *mdrun -membed membed.dat -s into_mem.tpr -c protein_membrane.gro -o intomem.trr -e intomem.edr -nt 1 -cpt 1 -mn index.ndx -testverlet.*

While the first command generates a suitable file for input, the second creates the *membed.dat* file with additional parameters for embedding the protein. In the last command, it must be noted that (a) the index is only required if dealing with mixed membranes, i.e., if there is only one lipid type you can choose that group as the one where the membrane will be inserted, and (b) the *-testverlet* switch is only used if running GPU calculations (the most commonly used nowadays) (*see* **Note 13**).

3.1.3 Topologies and Small Molecule Parameterizations

CHARMM-GUI [24] or CGenFF (https://cgenff.umaryland.edu) [25] can provide all-atom CHARMM topologies for small molecules, and concerning AMBER and GROMOS, while for the former suitable topologies can be obtained using Antechamber (http://ambermd.org/AmberTools.php) [26] or acpype (https://github.com/alanwilter/acpype) [27], for the latter the Automated Topology Builder (ATB) and Repository (https://atb.uq.edu.au) [20] or the ProDRG Server (http://prodrg2.dyndns.org/) [28] can be used (*see* **Note 14**).

3.2 Steered Molecular Dynamics

The first step in estimating permeabilities through either a lipid membrane and/or a protein pore is to define a "probable path" by which the molecule of interest gain access to the intracellular medium by traversing the biological barrier. While in a membrane two main energy barriers are to be expected (crossing the lipid headgroups and drug "flip-flop" between membrane leaflets), within a protein or a water-filled pore the main energy barrier will mostly depend on the distribution of charged residues along the channel and on the pore radius. Therefore, the initial steered molecular dynamics should be slow enough to allow an adequate adjustment between the permeant and the components of the biological barrier (e.g., lipids or protein residues) but fast enough to be performed with reasonable computational resources. While more fluid membranes and/or protein channels with large pores may allow faster pulling rates, more tightly packed membranes (e.g., bacterial membranes) or narrow channels (ion channels, some bacterial porins) may require a more thorough fine-tuning of the parameters.

3.2.1 Initial Coordinates

Having already equilibrated systems concerning the biological barrier of interest and a test molecule (previously equilibrated and simulated in water), the first step is to define coordinates for the initial position of the ligand. This position should be greater than the cutoff radius defined in the simulation parameters. Thus, for the system with coordinates $x = 16.0$ nm, $y = 15.0$ nm, and $z = 10.0$ nm in which the lipid–water interface (or protein–water interface) is located at $z = 6.5$ nm, and assuming a 1.2 nm cutoff, the ligand should ideally be placed with its center of mass around $6.5 + (1.2 \times 2) = \sim 8.9$ nm, removing all overlapping waters. The translation of the ligand into the new coordinates can be done with the following commands:

user@unix:~$ *gmx editconf -f ligand.gro -o lig_centered.gro -box 16.0 15.0 10.0 -center 8.0 7.5 8.9*

Concatenate the two files using a similar command to that on Subheading 3.1.2, name it *system_assembled.gro* and remove overlapping waters with:

user@unix:~$ *gmx select -s system_assembled.gro -on n_index.ndx -select 'group "System" and not same residue as (group "SOL" and within 0.2 of resname LIG)'*

user@unix:~$ *gmx editconf -f system_assembled.gro -n n_index.ndx -o system_stripped.gro*

Herein, the first command generates a new index (*n_index.ndx*) file in which all overlapping solvent molecules are removed, later used to generate the *system_stripped.gro* as the input for the steered MD simulation run. As above, the topology *topol.top* must also be updated with the new number of water molecules, e.g., obtained, for instance, with:

user@unix:~$ *cat system_stripped.gro |grep OW | wc.*

3.2.2 Run Parameters (See **Note 15**)

By taking advantage of the pull code included in the GROMACS package (http://manual.gromacs.org/documentation/2020/reference-manual/special/pulling.html) [29], we can now proceed in adding to the molecular dynamics' parameter file (.mdp) several lines defining how the molecule will be "pulled" along the simulation. Below is an example of the pull code (test molecule referred as LIG):

```
# PULL CODE EXAMPLE
 pull = yes
 pull_ngroups = 1
 pull_ncoords = 1
 pull_group1_name = LIG
 pull_nstxout = 10
 pull_nstfout = 10
 pull_coord1_type = umbrella
 pull_coord1_geometry = direction-periodic
 pull_coord1_origin = 8.0 7.5 8.9
 pull_coord1_groups = 0 1
 pull_coord1_vec = 0 0 -1
 pull_coord1_rate = 0.000079
 pull_coord1_k = 1000
 pull_coord1_start = no
```

Herein, one pulling group is defined to pull the test molecule from its original coordinate (*pull_coord1_origin*) along its absolute *z* vector (*pull_coord1_groups*, *pull_coord1_vec*) (*see* **Note 16**). To ensure that final position of the test molecule is above the 1.2 nm cutoff previously specified, we can assume that in this example the molecules' center of mass reaches bulk water at $z = 1.0$ nm. Therefore, for a total path of 7.9 nm ($p = 8.9$ nm $-$ 1.0 nm), a pulling rate of 0.000079 nm.ps^{-1} (*pull_coord1_rate*) is needed to obtain an initial sampling path. By default, a force constant of 1000 kJ.mol^{-1}.nm^{-2} can be used, but sometimes this value must be reduced if any distortions occur on the biological barrier (e.g., membrane deformation).

3.2.3 Analysis

After a successful run, it is always good practice to visually evaluate the obtained path and to analyze the file outputs. Quite often, concerning proteins, experimental data is available and allows the user to better evaluate if the obtained path includes residues described to be important for drug binding. While visualization can be performed in VMD [30], the output files *pullx.xvg* and *pullf.xvg* can be rendered as two-dimensional plots with *xmgrace* (https://plasma-gate.weizmann.ac.il/Grace/). It is advisable to thoroughly check (a) the maximum force used in the pulling run (*pullf.xvg*), (b) the roughness of the path (*pullx.xvg*), and (c) if any distortions occurred that can be attributed to the pulling of the molecule through the biological barrier (in VMD). When encountering any problem, it is advisable to repeat the pulling run under more mild conditions (e.g., by doubling the simulation run to 200 ns, thus decreasing the pulling rate, or by reducing the force constant) (*see* **Note 17**).

3.3 Umbrella Sampling

Having obtained an initial "permeation path," the next step is to choose a reasonable number of points along the *z* coordinate that can be used as starting configurations for each of the umbrella sampling (US) windows. This is a critical step because (a) undersampling the path may not allow a thorough exploitation of the free-energy basin along the permeation path and (b) oversampling may consume additional computational resources without any practical benefits for the calculations. Therefore, defining the number of bins is one of the most critical steps in umbrella sampling.

3.3.1 Number of Bins

There are a number of ways to define an adequate number of bins for sampling any permeation path. For water permeation through a lipid membrane, 30 bins (or windows) were used in the original study by Marrink and Berendsen [4]. In later studies, such number greatly varied, ranging from 7 [10] or 28 [5] to 153 [11], implying window spacings varying between 0.25 and 1.0 Å [6]. A more practical approach relies on determining the maximum axis length of the test molecule, using that information to guide the user in choosing an adequate number of bins. Two examples are provided, for a total "permeation path" of ~7.9 nm (Fig. 1).

Herein, while linezolid has ~17 Å length in its maximal axis, glucose only has ~9 Å. Therefore, for a proper sampling of glucose permeation path, a ~ twofold larger number of bins would be required. If considering a path with ~8.0 nm length, and to assure a proper overlapping between bins (*see* **Note 18**), for glucose the minimum number of bins would be 20 (bin width of 0.5 nm). For

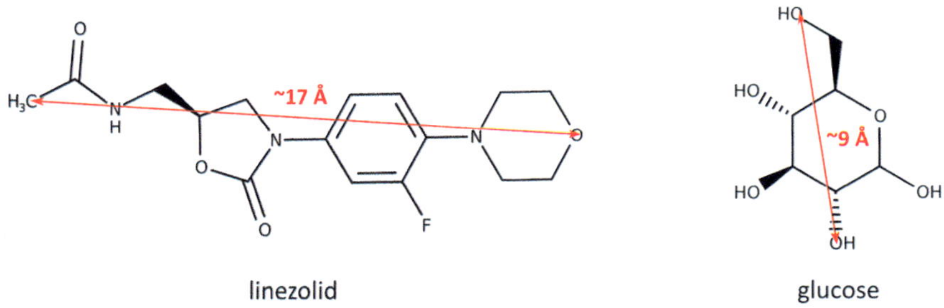

Fig. 1 Linezolid and glucose molecular size (longest axis)

linezolid, 8 to 10 bins (bin width of 1.0 nm) would theoretically be sufficient to properly sample the permeation path.

Nonetheless, the above approach may be difficult to accomplish while predicting a large number of molecules. In such cases, the best approach is to use (a) an average bin width considering all molecules or (b) the same number of bins for all molecules (if not computationally too expensive) (*see* **Note 19**).

3.3.2 Run Parameters

The pull code is similar to the one present in Subheading 3.2.2, only with minor changes: (a) the *pull_coord1_origin* is removed (only needed for the initial steered MD run), (b) a reference group is used instead of an absolute coordinate system, (c) *pull_coord1_geometry* is changed to *distance* (to the reference group), and (d) the pulling rate is changed to zero while keeping the force constant at 1000 kJ.mol^{-1}.nm^{-2}.

```
# PULL CODE EXAMPLE
 pull = yes
 pull_ngroups = 1
 pull_ncoords = 1
 pull_group1_name = LIG
 pull_group2_name = DPPC # or Protein
 pull_nstxout = 10
 pull_nstfout = 10
 pull_coord1_type = umbrella
 pull_coord1_geometry = distance
 pull_coord1_dim = N N Y
 pull_coord1_groups = 1 2
 pull_coord1_vec = 0 0 -1
 pull_coord1_rate = 0
 pull_coord1_k = 1000
 pull_coord1_start = no
 pull_coord1_init = 0.0
```

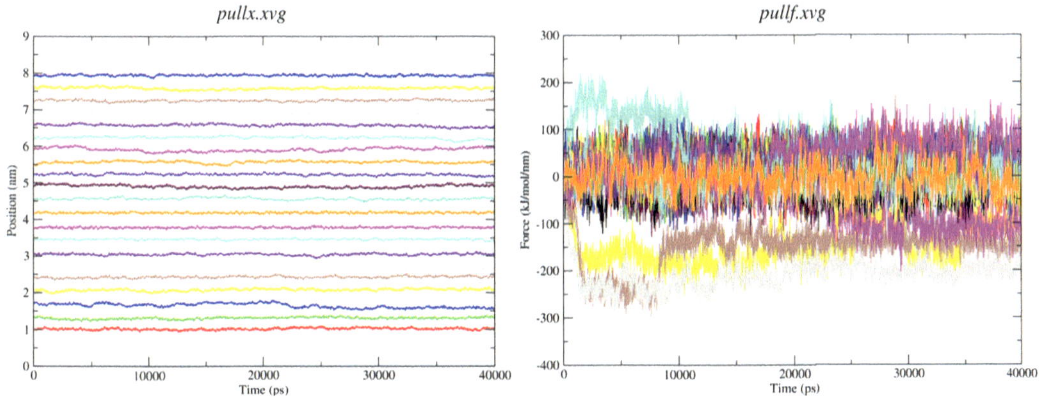

Fig. 2 Examples for the output contained in *pullx.xvg* and *pullf.xvg* files, obtained from 40 ns US simulation runs

By keeping the molecule spatially constrained around its center of mass (CoM) *z* coordinate while allowing it to freely diffuse in the *xy* plane (*pull_coord1_dim*), it is possible to sample the free-energy basin around the molecules' CoM by integrating the forces needed to keep the molecule at that particular distance from the reference group, in each of the previously defined US windows. It is advisable to use, at least, 50 ns MD simulation in each window, but for all-atom forcefields each run can be easily extended to no more than 200 ns. It is advisable to perform longer runs in each US window while in the initial assessment of the system, only reducing the length of the run when there is enough information that convergence is achieved within shorter simulation times.

3.3.3 Analysis

All the information required for calculating the permeability coefficients in the following steps are contained in two files, *pullf.xvg* (the force required to keep the molecule in that particular position, in $kJ.mol^{-1}.nm^{-2}$) and *pullx.xvg* (the average fluctuation of the molecule around its initial position, nm). Below are two examples of both *pullf.xvg* and *pullx.xvg* files generated from a 21 bin US simulations (*see* **Note 20**) (Fig. 2).

From Fig. 2, it can be seen that, after 20 ns, most of the US windows have reached equilibration. Therefore, using the last 10–20 ns would in theory be enough to obtain reliable data to estimate the permeability coefficient for this particular molecule (*see* **Note 21**).

3.4 Permeability Calculations

Although possible to calculate the permeabilities in a spreadsheet for a single molecule, for multiple molecules and/or biological barriers such approach is unfeasible. One of the best ways to process the data in an automated way is in Python. This way, the user can readily process the US data in order to obtain the variables to insert in the ISD model equation. While from the several *pullx* files (one for each US window), the variance, var(z), and the autocorrelation time (τ_z) are estimated and from the *pullf* files the potential of mean force (PMF) is obtained. It is also useful to perform an error estimation of the PMF data, which can be done by bootstrapping the umbrella sampling windows sampled at intervals equal to the autocorrelation times τ_z in each umbrella window [13, 31].

3.4.1 Variance and τ_z

Both values are obtained from the *pullx.xvg* file. However, for obtaining the autocorrelation times through the BlockAverage approach, a function in Python is required *a priori* (based on the blockAverage.py function by shane5ul, https://gist.github.com/shane5ul):

```
# block averaging function (needed for estimating autocorrela-
tion times)
def blockAverage(datastream, maxBlockSize=0):
 Nobs = len(datastream)
 minBlockSize = 1;
 if maxBlockSize == 0:
  maxBlockSize = int(Nobs/5);
 NumBlocks = maxBlockSize - minBlockSize
 blockMean = numpy.zeros(NumBlocks)
 blockVar = numpy.zeros(NumBlocks)
 blockCtr = 0
 for blockSize in range(minBlockSize, maxBlockSize):
  Nblock = int(Nobs/blockSize)
  obsProp = numpy.zeros(Nblock)
 for i in range(1,Nblock+1):
  ibeg = (i-1) * blockSize
  iend = ibeg + blockSize
  obsProp[i-1] = numpy.mean(datastream[ibeg:iend])
  blockVar[blockCtr] = numpy.var(obsProp)/(Nblock - 1)
  blockCtr += 1
 return blockVar[-1]
```

Following, to obtain variances and the autocorrelation times for all US windows, we can use the following code:

```
# open and sort files by name
pullx = sorted(glob.glob('*pullx*.xvg'))
pullx.sort(key=lambda f: int(filter(str.isdigit, f)))

# initialize variables
iact = []
variances = []

# obtain variances and autocorrelation times
for f in pullx:
 x = []
```

```
 pullfile = open(f, 'r')
 pulllines = pullfile.readlines()
 # 1 ps for each 50 lines, z in nm
 for l in pulllines[start:end:50]:
  time, z = l.split('\t')
  x.append(float(z))
 # calculate variances
 variances.append(numpy.var(x))
 # calculate autocorrelation times, tau in ps
 tau = 0.0
 dt = 1.0
 tau = (((len(x)* blockAverage(x, maxBlockSize=0)) / numpy.var
(x))-1) * (dt / 2)
 iact.append(tau)
 pullfile.close()
```

The above python code is an example on how variances and autocorrelation times can be obtained directly from the data file (*see* **Note 22**). Both variables are conveniently stored as lists or, alternatively, can be written to a text file for later analysis. The autocorrelation times τ_z will be later used to decorrelate the forces within the *pullf.xvg* files, thus allowing the user to estimate errors using a bootstrap function on the PMF data.

3.4.2 Diffusivity Coefficients

Having calculated the variances and autocorrelation times for each of the US windows, the diffusivity coefficients can now be readily calculated with Eq. 2:

```
# initialize
b = 0
e = len(variances)

# diffusivity coefficients
while b < e:
 # diffusion in cm^2/s
 diffusion = ((float(variances[b]) * (10 ** (-14))) / (float
(iact[b]) * (10 ** (-12))))
 diffusions.append(float(diffusion))
 b = b + 1
```

Herein, the diffusivity coefficients are calculated in $cm^2.s^{-1}$, a unit often found in data obtained from experiments [32, 33], but the user may opt to remove the conversion factors to obtain the data in $nm.ps^{-1}$. However, this must also be taken into consideration when calculating the permeability coefficient.

3.4.3 Potential of Mean Forces (PMF)

Although not essential, it is advisable to implement any error estimation function in the permeability calculations. Herein, we accomplished this by bootstrapping the potential of mean force curve [31]. Thus, and prior to the calculation of the energetic term, a simple code was implemented as following (based on the bootstrapping function implemented in *g_mmpbsa* (https://github.com/RashmiKumari/g_mmpbsa):

```
# bootstrapping function (used for error estimation)
def BootStrap(x,step=1000):
 if(numpy.mean(x)) == 0:
 return 0.000
 else:
  avg =[]
  x = numpy.array(x)
  n = len(x)
  idx = numpy.random.randint(0,n,(step,n))
  sample_x = x[idx]
  avg = numpy.sort(numpy.mean(sample_x,1))
 return avg
```

Following, by decorrelating forces using the previously estimated autocorrelation time, we are able to generate 100 bootstrapped PMF profiles, from which the average PMF and error estimation is later obtained:

```
# open and sort files by name
pullf = sorted(glob.glob('*pullf*.xvg'))
pullf.sort(key=lambda f: int(filter(str.isdigit, f)))

# initialize variables
forces_array = []
b = 0

# bootstrapping
for f in pullf:
 meanff = []
 pullfile = open(f, 'r')
 pulllines = ff.readlines()
 # decorrelate forces for generating PMFs
 # 1 ps each 50 lines
 i = int(round(iact[b]) * 50)
 # to prevent when ACT = 0 ps (unlikely)
 if i == 0:
  i = 1
 for lines in fflines[start:end:i]:
  time, energy = lines.split('\t')
```

```
 # forces in kJ/mol/nm
  meanff.append(float(energy))
 forces_array.append(BootStrap(meanff,step=100))
 b = b + 1
```

Finally, for all the bootstrapped PMFs, we obtain the corresponding energies for each US windows and integrate them using a trapezoidal rule as follows:

```
# initialize variables
pmf_array = []
b = 0
e = len(forces_array)

# potential of mean force
while b < e:
 point = 0.0
 pmf = []
 new_list = zip(*forces_array)[b]
 bb = 1
 ee = len(new_list)
 pmf.append(new_list[0])
 point = new_list[0]
 while bb < ee:
 # bin centers in nm, forces in kJ/mol/nm
  point = (positions[bb] - positions[bb - 1]) * new_list[bb]
  pmf.append(point)
  bb = bb + 1
 # add to corresponding lists
 pmf_array.append(pmf)
 b = b + 1
```

It is possible, of course, to fine-tune the number of bootstrapped PMF curves in order to reduce its computational cost, thus speeding up the calculations. For the particular system for which such code was tested, 100 bootstrapped PMFs were found to be adequate for proper error estimation.

3.4.4 Resistance to Permeation and Permeability Coefficient

Having calculated both the diffusivity coefficients, $D(z)$, and the energetic term, $\Delta G(z)$, for all the US windows, we can now calculate the resistance to permeation (R^P) and, at the end, the permeability coefficient P as its inverse (*see* **Note 23**):

```
# initialize variables
resistance_array = []
us_window = 0
b = 0
e = len(positions) - 1
```

```
# resistance to permeation
while b < e:
 resistance = []
 for item in pmf_array:
  # gas constant R (in kJ/mol.K), T (in Kelvin), nm to cm
conversion
   us_window = ((positions[b] - positions[b + 1]) * (10 **
(-7))) * \
  ((numpy.exp((item[b]) / (0.0083144598 * 310)) / diffusions
[b]))
 # in cm/s (average PMF)
  resistance.append(window)
 # append the resistance sum (to allow averages and stdev)
 resistance_array.append(numpy.sum(resistance))
 b = b + 1

# permeabilities
mean = (1 / numpy.mean(resistance_array))
stdev = (1 / numpy.std(resistance_array))
```

Herein, a conversion factor (from nm to cm) is additionally required in order to apply the trapezoidal rule needed for integrating the obtained values in each US window. It is also worth noticing that temperature ($T = 310$ K) must also match the one used in all US simulations.

4 Notes

1. Both $\Delta G(z)$ and $D(z)$ can be accurately estimated through MD simulations. To ensure that all z positions of the molecule are adequately sampled, some degree of overlapping between adjacent positions must occur. It is recommended to fine-tune the initial setup by changing the parameters controlling both the steered and US molecular dynamics simulations.
2. There are several methods by which both $\Delta G(z)$ and $D(z)$ can be obtained and, depending on the systems, some approaches may provide more reasonable results than others. For more comprehensive reviews, the users are referred to Refs. 7 and 8.
3. Based on the findings by Zhu and Hummer [11], while estimating $\langle z \rangle$ it is assumed that the autocorrelation function $C(t)$ disappears at a time scale much shorter than the overall trajectory length $n\Delta t$, therefore making the variance of z also dependent on τ. The user can always choose to obtain τ from its autocorrelation function, as in *gmx wham* (although more time-consuming). It is also worth noticing that, for *gmx wham*, IACTs are estimated by simple integration of the

autocorrelation functions (ACFs) while the ACFs are larger than 0.05, which is also prone to errors. The GROMACS developers now suggest a more sophisticated way to compute IACTs, by using *gmx analyze* to fit the ACTs to a double exponential and providing the results to *gmx wham* (although less robust) [29].

4. Since version 2.0, ATB server provides molecules with non-standard atom types (not included in the standard GROMOS 54A7 forcefield provided in GROMACS). The user must check if they are using modified topologies/forcefields when compared with the original forcefield and, if needed, these atom types must be added to the internal GROMACS parameter files.
5. The user must not replicate boxes in the *z* coordinate (always equal to 1), or a stacked membrane patch on top of the original one will be generated.
6. Be careful with the sequence of the residues and the one specified in the topol.top file. For instance, if the original arrangement in *topol.top* is:

```
[ molecules ]
DPPC 128
SOL 3234
```

After duplication, the number of DPPC and water molecules cannot be simply quadruplicated (example #1) because it mismatches the order in the *dppc512.gro* file. Instead, molecule ordering in *topol.top* should be as in example #2:

```
#1 #2
[ molecules ] [ molecules ]
DPPC 512 DPPC 128
SOL 12936 --> SOL 3234
 DPPC 128
 SOL 3234
 DPPC 128
 SOL 3234
 DPPC 128
 SOL 3234
```

If the user opts to change the number of DPPC (and SOL) molecules, it must manually update the *dppc512.gro* file in order to maintain a sequential order for the DPPC lipids.

7. Examples of molecular dynamics parameter files, such as *minim.mdp*, *nvt.mdp*, *npt.mdp*, and *md.mdp*, can be found at Justin Lemkul's Online Tutorial pages (http://www.mdtutorials.com/gmx/) [18].

8. There are other software that also allow the insertion of proteins in membranes, as InflateGRO (http://www.mdtutorials.com/gmx/membrane_protein/index.html, step 3) [18], InflateGRO2/LAMBADA [34] or even manually.
9. It is advisable to maintain the membrane patch at the *xy* plane (larger protein axis oriented towards the *z* axis).
10. It is important to retain the original *xy* coordinates of the membrane patch, when using an already equilibrated system (merely adjusting the box *z* size, if needed). When both the protein and the lipid bilayer are not equilibrated, it is advisable to create a slightly larger box, thus allowing the lipids to adequately pack around the protein during the equilibration MD runs.
11. It is advisable to open the file under a visualization software, e.g., VMD [30], to verify that the file was correctly generated. The size of the box must also be corrected if the protein is larger along its *z* coordinate than the initial box.
12. Users may encounter problems with energy exclusion, freeze groups, and removal of group cutoffs in newer versions of GROMACS, rendering this version (4.6.7) as the most reliable. GROMACS developers hope to have this corrected in later versions (number of ranks to run now corrected in GROMACS 2021.1.
13. User must pay attention to the command output because the number and type of molecules to be removed from the system (e.g., overlapping with protein, usually lipids, water molecules and ions) will be printed to *stdout*. These values may have to be manually updated in *topol.top*.
14. While ATB currently uses nonstandard atom types (check **Note 4**), in ProDRG partial charges may not be optimal. Users are advised to follow the paper by Lemkul et al. [35] and/or to use AM1-BCC charges, which can be calculated in AnteChamber [27].
15. Until GROMACS version 4.6.7, the *g_wham* code allowed *pull_geometry* = *position*, enabling the usage of positional coordinates along the proposed permeation path (usually *z* axis) to restrain the molecules center of mass. However, from GROMACS 5.0.x version forward, this keywork became deprecated. By using absolute coordinates, the user can still perform the US runs (*see* **Note 16**), but a set of absolute coordinates along the permeation path is needed to obtain the correct PMF profile.
16. Although similar to *pull-coord1-geometry* = *direction*, by using *direction-periodic* the distance is allowed to be larger than half the box size while avoiding the definition of a reference pulling group. However, with this geometry, the box must not be

dynamic (*compressibility* = *4.5e-5 0,* for *xy* and *z* dimensions, respectively) in the pull dimensions (*z*), and the pull force is not added to virial. For additional information, the user is referred to the GROMACS manual [29].

17. For very narrow pores or tightly packed membranes, including spatial constraints on key atoms (e.g., phosphorous) only in *z*, thus allowing *xy* motions, helps in keeping the biological structures in place.
18. Allowing bins to partially overlap, while keeping the molecule spatially constrained in its *z* position (around its center-of-mass), enables the exploitation of common section of the free-energy landscape by the test molecule, thus improving the reliability of the calculations.
19. It is advisable to initiate studies with the largest and shortest molecules, thus evaluating if the bin width is adequate or if it needs to be modified.
20. Extending the length of the US runs in each window may be needed, especially to initially evaluate the systems and to better understand the behavior of the system.
21. By using *distance* as a keyword for the US procedure, the user must keep tracking of the original *z* coordinates (from the initial center of mass) to obtain the correct PMFs in Subheading 3.4.3.
22. In this example, the time interval *dt* was chosen to closely match the integrated autocorrelation times obtained from an autocorrelation function (as in *gmx wham*).
23. For each US window, the diffusivity coefficient is identical for all 100 bootstrapped PMF curves. Although neglecting the last point (due to the trapezoidal rule), if the last US window is in bulk water (e.g., at a sufficient distance from the protein and/or membrane), such fact should not contribute significantly to the obtained resistance/permeability.

References

1. Yang NJ, Hinner MJ (2015) Getting across the cell membrane: an overview for small molecules, peptides, and proteins. In: Gautier A, Hinner MJ (eds) Site-specific protein labeling: methods and protocols. Springer, New York, NY, pp 29–53
2. Fortuna A, Alves G, Falcão A (2007) The importance of permeability screening in drug discovery process: PAMPA, Caco-2 and rat everted gut assays. Curr Top Pharmacol 11 (2):63–86
3. Kleinzeller A, Deamer D, Fambrough D (eds) (1999) Membrane permeability: 100 years since Ernest Overton, 1st edn. Academic Press, New York
4. Marrink S-J, Berendsen HJC (1994) Simulation of water transport through a lipid membrane. J Phys Chem 98(15):4155–4168. https://doi.org/10.1021/j100066a040
5. Hub JS, Winkler FK, Merrick M, de Groot BL (2010) Potentials of mean force and permeabilities for carbon dioxide, ammonia, and water flux across a rhesus protein channel and lipid membranes. J Am Chem Soc 132 (38):13251–13263. https://doi.org/10.1021/ja102133x

6. Awoonor-Williams E, Rowley CN (2016) Molecular simulation of nonfacilitated membrane permeation. Biochim Biophys Acta 1858(7 Pt B):1672–1687. https://doi.org/10.1016/j.bbamem.2015.12.014
7. Lee CT, Comer J, Herndon C et al (2016) Simulation-based approaches for determining membrane permeability of small compounds. J Chem Inf Model 56(4):721–733. https://doi.org/10.1021/acs.jcim.6b00022
8. Venable RM, Krämer A, Pastor RW (2019) Molecular dynamics simulations of membrane permeability. Chem Rev 119(9):5954–5997. https://doi.org/10.1021/acs.chemrev.8b00486
9. Vila-Viçosa D, Victor BL, Ramos J et al (2017) Insights on the mechanism of action of INH-C 10 as an antitubercular prodrug. Mol Pharm 14(12):4597–4605. https://doi.org/10.1021/acs.molpharmaceut.7b00719
10. Hénin J, Tajkhorshid E, Schulten K, Chipot C (2008) Diffusion of glycerol through Escherichia coli aquaglyceroporin GlpF. Biophys J 94 (3):832–839. https://doi.org/10.1529/biophysj.107.115105
11. Zhu F, Hummer G (2012) Theory and simulation of ion conduction in the pentameric GLIC channel. J Chem Theory Comput 8 (10):3759–3768. https://doi.org/10.1021/ct2009279
12. Acosta-Gutiérrez S, Ferrara L, Pathania M et al (2018) Getting drugs into Gram-negative bacteria: rational rules for permeation through general porins. ACS Infect Dis 4 (10):1487–1498. https://doi.org/10.1021/acsinfecdis.8b00108
13. Ferreira RJ, Kasson PM (2019) Antibiotic uptake across Gram-negative outer membranes: better predictions towards better antibiotics. ACS Infect Dis 5(12):2096–2104. https://doi.org/10.1021/acsinfecdis.9b00201
14. Hummer G (2005) Position-dependent diffusion coefficients and free energies from Bayesian analysis of equilibrium and replica molecular dynamics simulations. New J Phys 7:34. https://doi.org/10.1088/1367-2630/7/1/034
15. Woolf TB, Roux B (1994) Conformational flexibility of o-phosphorylcholine and o-phosphorylethanolamine: a molecular dynamics study of solvation effects. J Am Chem Soc 116(13):5916–5926. https://doi.org/10.1021/ja00092a048
16. Flyvbjerg H, Petersen HG (1989) Error estimates on averages of correlated data. J Chem Phys 91:461–466. https://doi.org/10.1063/1.457480
17. Zhu F, Hummer G (2012) Theory and simulation of ion conduction in the pentameric GLIC channel. J Chem Theory Comput 8 (10):3759–3768. https://doi.org/10.1021/ct2009279
18. Lemkul J (2018) From proteins to perturbed Hamiltonians: a suite of tutorials for the GROMACS-2018 molecular simulation package [article v1.0]. LiveCoMS 1(1):5068. https://doi.org/10.33011/livecoms.1.1.5068
19. Koziara KB, Stroet M, Malde AK, Mark AE (2014) Testing and validation of the automated topology builder (ATB) version 2.0: prediction of hydration free enthalpies. J Comput Aided Mol Des 28(3):221–233. https://doi.org/10.1007/s10822-014-9713-7
20. Jo S, Kim T, Iyer VG, Im W (2008) CHARMM-GUI: a web-based graphical user interface for CHARMM. J Comput Chem 29 (11):1859–1865. https://doi.org/10.1002/jcc.20945
21. Wu EL, Cheng X, Jo S et al (2014) CHARMM-GUI membrane builder toward realistic biological membrane simulations. J Comput Chem 35(27):1997–2004. https://doi.org/10.1002/jcc.23702
22. Jo S, Kim T, Im W (2007) Automated builder and database of protein/membrane complexes for molecular dynamics simulations. PLoS One 2(9):e880. https://doi.org/10.1371/journal.pone.0000880
23. Lomize MA, Pogozheva ID, Joo H et al (2012) OPM database and PPM web server: resources for positioning of proteins in membranes. Nucleic Acids Res 40(Database issue):D370–D376. https://doi.org/10.1093/nar/gkr703
24. Kim S, Lee J, Jo S et al (2017) CHARMM-GUI ligand reader and modeler for CHARMM force field generation of small molecules. J Comput Chem 38(21):1879–1886. https://doi.org/10.1002/jcc.24829
25. Vanommeslaeghe K, Hatcher E, Acharya C et al (2010) CHARMM general force field: a force field for drug-like molecules compatible with the CHARMM all-atom additive biological force fields. J Comput Chem 31 (4):671–690. https://doi.org/10.1002/jcc.21367
26. Wang J, Wang W, Kollman PA, Case DA (2006) Automatic atom type and bond type perception in molecular mechanical calculations. J Mol Graph Model 25(2):247–260.

https://doi.org/10.1016/j.jmgm.2005.12.005
27. Sousa da Silva AW, Vranken WF (2012) ACPYPE - AnteChamber PYthon parser interfacE. BMC Res Notes 5:367. https://doi.org/10.1186/1756-0500-5-367
28. Schüttelkopf AW, van Aalten DMF (2004) PRODRG: a tool for high-throughput crystallography of protein-ligand complexes. Acta Crystallogr D Biol Crystallogr 60 (Pt 8):1355–1363. https://doi.org/10.1107/S0907444904011679
29. Abraham MJ, van der Spoel D, Lindahl E, Hess B, The GROMACS development team (2019) GROMACS User Manual version 2019. http://www.gromacs.org. Accessed 18 May 2020
30. Humphrey W, Dalke A, Schulten K (1996) VMD: visual molecular dynamics. J Mol Graph 14:33–38
31. Hub JS, de Groot BL, van der Spoel D (2010) g_wham—a free weighted histogram analysis implementation including robust error and autocorrelation estimates. J Chem Theory Comput 6(12):3713–3720. https://doi.org/10.1021/ct100494z
32. Nikaido H, Varra M (1985) Molecular basis of bacterial outer membrane permeability. Microbiol Rev 49(1):1–32
33. Nikaido H (2003) Molecular basis of bacterial outer membrane permeability revisited. Microbiol Mol Biol Rev 67(4):593–656. https://doi.org/10.1128/MMBR.67.4.593
34. Schmidt TH, Kandt C (2012) LAMBADA and InflateGRO2: efficient membrane alignment and insertion of membrane proteins for molecular dynamics simulations. J Chem Inf Model 52(10):2657–2669. https://doi.org/10.1021/ci3000453
35. Lemkul JA, Allen WJ, Bevan DR (2010) Practical considerations for building GROMOS-compatible small-molecule topologies. J Chem Inf Model 50(12):2221–2235. https://doi.org/10.1021/ci100335w

Chapter 15

Identification of Pan-Assay INterference compoundS (PAINS) Using an MD-Based Protocol

Pedro R. Magalhães, Pedro B. P. S. Reis, Diogo Vila-Viçosa, Miguel Machuqueiro, and Bruno L. Victor

Abstract

Pan-assay interference compounds (PAINS) are promiscuous molecules with multiple behaviors that interfere with assay readouts. Membrane PAINS are a subset of these compounds that influence the function of membrane proteins by nonspecifically perturbing the lipid membranes that surround them. Here, we describe a computational protocol to identify potential membrane PAINS molecules by calculating the effect that a given compound has on the bilayer deformation propensity.

Key words Molecular dynamics, Umbrella sampling, Potential of mean force

1 Introduction

Pan-assay interference compounds (PAINS) is a class of molecules that show activity in many assays by interfering with the assay readout, instead of specifically interacting with a target [1, 2]. It has been found that this interference behavior can occur in several different ways, including metal chelation, redox reactivity, aggregation, or membrane modulation [1]. In this work, we are focused on the so-called membrane PAINS, due to their broad range of actions in altering protein's function by lipid perturbation [3].

Several compounds that were initially hailed as miracle cures due to their wide range of uses were discovered to aggressively perturb biological membranes [1, 3]. Due to the negative economic impact, the investment in such false positives would have in the development of new therapeutics, and the early identification of such compounds is paramount. However, there are few works that propose techniques or approaches to accurately perform such tasks. One such tool was first employed by Ingólfsson and coworkers [3] to evaluate the effect that several compounds had on the stability of membranes, and the consequent impact on the function

Irina S. Moreira et al. (eds.), *Computational Design of Membrane Proteins*, Methods in Molecular Biology, vol. 2315, https://doi.org/10.1007/978-1-0716-1468-6_15,

of mechanosensitive membrane proteins. The authors initially confirmed that such compounds would have a mild perturbation impact on different bulk membrane properties such as area-per-lipid and bilayer thickness, making the use of such measurements completely insufficient to perform such evaluation. Therefore, the authors decided to quantify the changes made by several compounds to the bilayer deformation using an umbrella sampling (US) method to calculate the potential of mean force (PMF) of dragging a large bead across a bilayer. With such an approach, they discovered that some compounds, such as curcumin (CUR), had a drastic perturbation effect, while others, such as resveratrol (RES), only exhibited mild activity. Later on, Jesus et al. [4] used an adapted atomistic version of this approach to compare the membrane perturbation activity of nothofagin (NOT) to RES. As reported, contrary to NOT, resveratrol seemed to alter the energetics of membrane perturbation in both the headgroup and center of the bilayer regions.

The goal of this work is to present an adapted computational protocol from refs. [3, 4] which is optimized to identify and characterize membrane PAINS. To that end, we use replicates of conventional molecular dynamics (MD) simulations coupled to the biasing harmonic potentials of the US scheme. We also provide details on how to build bilayer systems of 128 1-palmitoyl-2-oleoyl-*sn*-glycero-3-phosphocholine (POPC) lipids in the presence and absence of the three previously mentioned compounds known to have varying degrees of membrane PAINS activity: CUR, NOT, and RES.

2 Materials

1. GROMACS 2018.6 [5–7], the software for MD simulations and analyses, is freely available at http://manual.gromacs.org/documentation/2018/download.html, and it already includes the GROMOS 54A7 [8] force field. Simulations were performed on different multicore machines, some of which had a graphics processing unit (GPU) card. We strongly recommend that each individual MD simulation is performed on a GPU-enabled machine (GTX 1070 ti or better) with at least eight physical CPU cores, since this setup produces approximately 100 ns/day with this system.
2. The automated topology builder (ATB) [9–11] is also freely available online at https://atb.uq.edu.au/. Regarding file manipulation, a regular text editor such as emacs or notepad++ is required. A program to generate plots, such as Gnuplot [12], and a molecular visualization program, such as PyMOL [13], are also recommended.

3 Methods

3.1 Software and Parameters

1. All simulations were performed using GROMACS 2018.6 and the GROMOS 54A7 force field.
2. The simple point-charge (SPC) water model was used [14].
3. Force field parameters for all three evaluated molecules (CUR, NOT, and RES) were obtained using the automated topology builder (ATB) server. Force field parameters for POPC are already included in GROMOS 54A7 [15, 16]. A zip file containing the parameters and the equilibrated structures of POPC with each compound is provided as supporting information.
4. Long-range electrostatic interactions were treated using the particle mesh Ewald (PME) method [17, 18] with a Fourier grid spacing of 0.12 nm and a cutoff of 1.0 nm for direct contributions.
5. Lennard-Jones interactions were calculated using a nonbonded neighbor pair list with a cutoff of 1.0 nm, allowing the use of a Verlet scheme [19].
6. Lipid and evaluated compound bonds were constrained using the parallel linear constraint solver P-LINCS [20], while water molecules were constrained using the SETTLE algorithm [21] (*see* **Note 1**).
7. Temperature coupling was performed using the v-rescale [22] thermostat at 298.15 K with a coupling constant of 0.1 ps.
8. A semi-isotropic Parrinello-Rahman barostat [23, 24] was used to keep the pressure constant at 1 bar with a coupling constant of 2.0 ps and compressibility of 4.5×10^{-5} bar^{-1}.

3.2 System Building and Initialization

1. The base system consisted of a 128 POPC lipid bilayer solvated by 5940 water molecules (*see* **Note 2**).
2. To build the systems for the simulation of the different compounds, we replaced 12 POPC molecules (evenly distributed between the two monolayers) by 12 molecules of CUR, NOT, or RES (*see* **Note 3**).
3. This resulted in four different starting systems: pure POPC, POPC+CUR, POPC+NOT, POPC+RES.
4. The energy of each system was minimized using the steepest descent algorithm [25] in two stages, both with a maximum number of allowed steps of 10,000: first, with no constraints and with maximum step size (emstep) of 0.0001 nm, and second, with all bonds constrained and with a maximum step size of 0.001 nm (*see* **Note 4**). In both cases, the tolerance (emtol) was set to 0 meaning the algorithm will stop when machine precision was achieved.

5. Each system was then initialized, i.e., velocities were generated according to a Maxwell distribution at gen-temp = 298.15 K, with varying values for the initial seed (gen-seed).
6. This initialization was performed for each of the four systems using three different values for gen-seed, which resulted in three separate replicate simulations per system, totaling 12 systems.
7. Each initialization was performed for 100,000 steps with a timestep of 0.002 ps using the MD integrator corresponding to 200 ps (*see* **Note 5**).

3.3 Regular MD Simulations/ Equilibration

1. Each system was then simulated for 200 ns using the parameters described in Subheading 3.1.
2. The stability of these equilibration runs was assessed by following the total area of the system over time. In the case of pure POPC, we can also compare its area-per-lipid with experimental values [15, 16].
3. The total area of the system over time is computed by multiplying the x and y components of the simulation box at each timestep. The area-per-lipid in pure POPC was computed dividing the total area by the number of lipid molecules in each monolayer (64).
4. Other parameters such as the bilayer thickness or lipid order parameter may also be computed to evaluate the equilibration of the different systems.

3.4 Steered MD Simulations

1. The goal of this step is to prepare each system for a set of umbrella sampling simulations. To perform such simulations, we will require a probe, with similar characteristics to the one described in ref. [4]. Therefore, the probe was built analogously to a benzene molecule with Lennard-Jones parameters C6 = 1.01×10^{-1} kJ mol^{-1} nm^{6} and C12 = 2.14×10^{-2} kJ mol^{-1} nm^{12}, and a mass of 78 Da.
2. To create this probe, we edited the force field files and add a new atom type named "PRB" to the atomtypes.dat file, and its corresponding nonbonded interactions with all other atom types in the force field in the ffnonbonded.itp file (using the values described in the previous step). These force field modifications are already included in the supporting information zip file.
3. Next, we selected a water molecule that was located far from the membrane and changed its name from OW to PRB in the gro file, deleting its corresponding HW1 and HW2 lines in the process. The new PRB line was moved to the end of the file, and the total number of atoms in the second line of the gro file was also corrected.

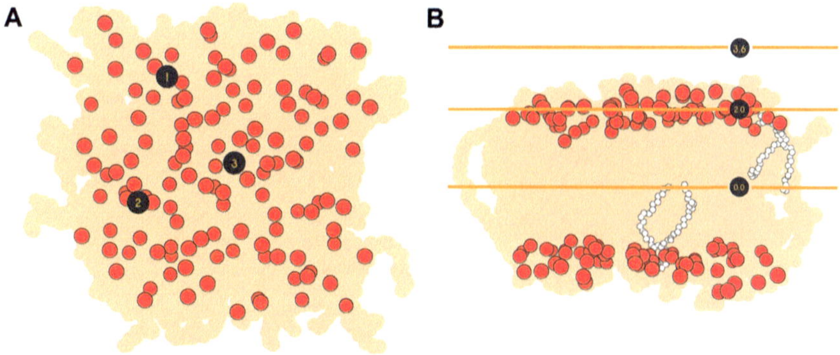

Fig. 1 POPC bilayer viewed from the top (**a**) depicted as a surface with the phosphorus atoms shown as red spheres and the probes, representing the different replicates, shown as numbered black spheres. A side view of the membrane is also shown (**b**), with the probe insertion positions relative to the center of the bilayer written inside the black spheres (in nm)

4. Next, we included the PRB in the topology of our system, while also correcting the total number of solvent molecules in the topology to account for the water molecule we converted into the probe.
5. To create replicates for this step, instead of using a process similar to the one described in Subheading 3.2, we simplified it by creating two new copies of each system (POPC, POPC +CUR, POPC+NOT, and POPC+RES), and then edited the *xy* coordinates of the probe, thereby ensuring that each probe will explore a different region of the membrane (Fig. 1a).
6. The potential energy of the resulting systems, containing the probe and the solvated POPC bilayer (with or without the 12 PAINS molecules) was then minimized following a protocol similar to the described in Subheading 3.2.
7. We added two different entries to the index file: one containing the probe and one containing the membrane P atoms. These are the two groups being used in the steered MD process, which we termed PROBE and PHOSPHOR.
8. We then ran a steered MD simulation for each system, in which the probe is gradually pulled in the *z* coordinate across the membrane normal while keeping the *xy* coordinates unrestrained. This was done for 2,000,000 steps (4 ns) with a rate of 0.001 nm/ps, which results in a full-steered MD trajectory with a range of probe's position of 4 nm (*see* **Note 6** for the remaining parameters).
9. Finally, from each system, we selected 37 initial structural configurations, with the probe placed ranging from the center of the membrane (0.0 nm) to the bulk water (3.6 nm), every 0.1 nm (**Note 7** and Fig. 1b).

3.5 Umbrella Sampling and PMF Calculations

1. For each of the 37 initial structures mentioned in Subheading 3.4, **step 9**, we performed umbrella sampling simulations for 3×100 ns.
2. The parameters used in this process are similar to the ones in **Note 6**, with a few small differences, which we have included in **Note 8**.
3. The PMF profiles were calculated using the weighted-histogram analysis method (WHAM) [26] implemented in GROMACS, and error calculations were performed using a Bayesian bootstrapping with 50 bootstrap iterations. An example of the resulting PMF is shown in Fig. 2 (*see also* **Note 9** and the Figure caption for extra information on the analysis and interpretation).

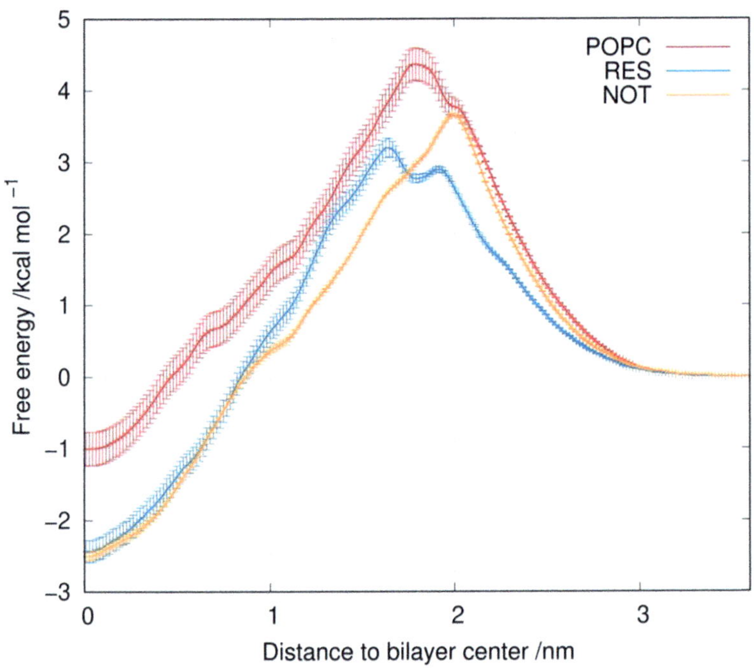

Fig. 2 PMF profile for translocating a probe across a lipid bilayer in a pure POPC bilayer (red), and in the presence of RES (blue) and NOT (yellow). Both have an effect on the bilayer deformation energy when compared to pure POPC, especially by lowering the energy barrier around the phosphorus atoms (~1.8 nm). The larger impact occurred for RES, which is considered a mild PAINS, while NOT has a smaller effect, which agrees with the observation that this is not a membrane PAINS [4]

4 Notes

1. Whenever we used more than eight CPU cores in our calculations, we manually restricted the value in the -rcon flag to 1.0 to allow for this high level of parallelization.
2. The membrane was built and solvated following a standard protocol like the one found in http://www.gromacs.org/Documentation/How-tos/Membrane_Simulations.
3. We randomly picked six POPC lipids to replace in each monolayer. This choice was curated by visual inspection of the bilayer to avoid the formation of large clusters. Once each lipid was identified, the starting PDB file of the solvated pure POPC membrane was simply modified using a text editor.
4. Energy minimization converged to machine precision at around 5000 and 1500 steps for stages one and two, respectively.
5. It is worth noting that the recipes for energy minimization and system initialization are not strict and can be easily adapted as long as the final result is a stable membrane system that runs smoothly and converges its total area.
6. We used the following options for the pull code:

```
pull = yes
pull_ngroups = 2
pull_ncoords = 1
pull_group1_name = PROBE
pull_group2_name = PHOSPHOR
pull_coord1_type = umbrella
pull_coord1_geometry = direction
pull_coord1_groups = 1 2
pull_coord1_dim = N N Y
pull_coord1_rate = 0.001
pull_coord1_k = 1000
pull_coord1_start = yes
pull-coord1-vec = 0.0 0.0 1.0
```

7. This was performed using a shell script and a python script (also provided in the zip file in SI), but could be done using any tool capable of reading the output from the steered MD simulations and sorting the results by the distance to the center of the bilayer.

8. These are the parameters used in the umbrella sampling simulations that differ from the ones mentioned in **Note 6**. The pull-coord1-init parameter varies from system to system, and it should reflect the distance to the center of the bilayer (in nm), i.e., it should be equal to 0.0 at that level and 3.6 when the probe is already at the water phase.

```
pull_coord1_rate = 0.0
pull_coord1_start = no
pull-coord1-init = ?.?
pull-nstxout = 1000
pull-nstfout = 1000
```

9. The PMF profile in Fig. 2 includes all three replicates per system. Each umbrella was 100 ns long, and the first 10 ns were discarded.

Acknowledgments

This work was supported by FCT projects PTDC/BIA-BFS/28419/2017 and UID/Multi/04046/2019, and individual grants CEECIND/02300/2017 and SFRH/BD/136226/2018.

References

1. Nelson KM, Dahlin JL, Bisson J et al (2017) The essential medicinal chemistry of curcumin: miniperspective. J Med Chem 60(5):1620–1637. https://doi.org/10.1021/acs.jmedchem.6b00975
2. Baell JB, Nissink JWM (2018) Seven year itch: pan-assay interference compounds (PAINS) in 2017 – utility and limitations. ACS Chem Biol 13(1):36–44. https://doi.org/10.1021/acschembio.7b00903
3. Ingólfsson HI, Thakur P, Herold KF et al (2014) Phytochemicals perturb membranes and promiscuously alter protein function. ACS Chem Biol 9(8):1788–1798. https://doi.org/10.1021/cb500086e
4. Jesus AR, Vila-Viçosa D, Machuqueiro M et al (2017) Targeting type 2 diabetes with C-glucosyl dihydrochalcones as selective sodium glucose co-transporter 2 (SGLT2) inhibitors: synthesis and biological evaluation. J Med Chem 60(2):568–579. https://doi.org/10.1021/acs.jmedchem.6b01134
5. Berendsen H, Van Der Spoel D, Van Drunen R et al (1995) GROMACS: a message-passing parallel molecular dynamics implementation. Comput Phys Commun 91(1–3):43–56. https://doi.org/10.1016/0010-4655(95)00042-E
6. Van Der Spoel D, Lindahl E, Hess B et al (2005) GROMACS: fast, flexible, and free. J Comput Chem 26(16):1701–1718. https://doi.org/10.1002/jcc.20291
7. Abraham MJ, Murtola T, Schulz R et al (2015) GROMACS: high performance molecular simulations through multi-level parallelism from laptops to supercomputers. SoftwareX 1:19–25. https://doi.org/10.1016/j.softx.2015.06.001
8. Schmid N, Eichenberger AP, Choutko A et al (2011) Definition and testing of the GROMOS force-field versions 54A7 and 54B7. Eur Biophys J 40(7):843–856. https://doi.org/10.1007/s00249-011-0700-9
9. Malde AK, Zuo L, Breeze M et al (2011) An automated force field topology builder (ATB) and repository: version 1.0. J Chem Theory Comput 7(12):4026–4037. https://doi.org/10.1021/ct200196m
10. Canzar S, El-Kebir M, Pool R et al (2013) Charge group partitioning in biomolecular

simulation. J Comput Biol 20:188–198. https://doi.org/10.1089/cmb.2012.0239
11. Koziara KB, Stroet M, Malde AK et al (2014) Testing and validation of the automated topology builder (ATB) version 2.0: prediction of hydration free enthalpies. J Comput Aided Mol Des 28:221–233. https://doi.org/10.1007/s10822-014-9713-7
12. Williams T, Kelley C, Bersch C et al (2019) Gnuplot 5.2 an interactive plotting program. http://gnuplot.info
13. The PyMOL Molecular Graphics System, Version 2.3.4, Schrödinger, LLC. https://pymol.org
14. Hermans J, Berendsen HJ, Van Gunsteren WF et al (1984) A consistent empirical potential for water--protein interactions. Biopolymers 23 (8):1513–1518. https://doi.org/10.1002/bip.360230807
15. Poger D, Van Gunsteren WF, Mark AE (2010) A new force field for simulating phosphatidylcholine bilayers. J Comput Chem 31 (6):1117–1125. https://doi.org/10.1002/jcc.21396
16. Poger D, Mark AE (2010) On the validation of molecular dynamics simulations of saturated and *cis*-monounsaturated phosphatidylcholine lipid bilayers: a comparison with experiment. J Chem Theory Comput 6(1):325–336. https://doi.org/10.1021/ct900487a
17. Darden T, York D, Pedersen L (1993) Particle mesh Ewald: an N· log (N) method for Ewald sums in large systems. J Chem Phys 98 (12):10089–10092. https://doi.org/10.1063/1.464397
18. Essmann U, Perera L, Berkowitz ML et al (1995) A smooth particle mesh Ewald method. J Chem Phys 103(19):8577–8593. https://doi.org/10.1063/1.470117
19. Páll S, Hess B (2013) A flexible algorithm for calculating pair interactions on SIMD architectures. Comput. Phys Commun 184 (12):2641–2650. https://doi.org/10.1016/j.cpc.2013.06.003
20. Hess B (2008) P-LINCS: a parallel linear constraint solver for molecular simulation. J Chem Theory Comput 4:116–122. https://doi.org/10.1021/ct700200b
21. Miyamoto S, Kollman PA (1992) Settle: an analytical version of the SHAKE and RATTLE algorithm for rigid water models. J Comput Chem 13(8):952–962. https://doi.org/10.1002/jcc.540130805
22. Bussi G, Donadio D, Parrinello M (2007) Canonical sampling through velocity rescaling. J Chem Phys 126(1):014101. https://doi.org/10.1063/1.2408420
23. Nosé S, Klein ML (1983) Constant pressure molecular dynamics for molecular systems. Mol Phys 50(5):1055–1076. https://doi.org/10.1080/00268978300102851
24. Parrinello M, Rahman A (1981) Polymorphic transitions in single crystals: a new molecular dynamics method. J Appl Phys 52 (12):7182–7190. https://doi.org/10.1063/1.328693
25. Luenberger DG, Ye Y (2008) Linear and nonlinear programming, vol 228, 4th edn. Springer, Boston, MA
26. Hub JS, De Groot BL, Van Der Spoel D (2010) g_wham a free weighted histogram analysis implementation including robust error and autocorrelation estimates. J Chem Theory Comput 6(12):3713–3720. https://doi.org/10.1021/ct100494z

Chapter 16

Transmembrane Anion Transport Mediated by Halogen Bonds: Using Off-Center Charges

Paulo J. Costa

Abstract

Synthetic anion transporters are promising therapeutic agents designed to emulate the specialized role of certain transmembrane proteins that maintain the ion concentration in cells. In the last few years, besides hydrogen bonds and ion pairs, halogen bonds have also been explored to promote the association between the synthetic molecule and the anion and their subsequent transport. This interaction is due to an anisotropic charge distribution on the halogen, and therefore, modeling halogen bonds is not a trivial task using classical force field methods that typically rely on point-charge models.

Herein, a computational protocol capable of dealing with halogen bonds is presented. This protocol takes advantage of the addition of an off-center particle during the charge fitting procedure, and the resulting set of charges can be used along with the classical force field parameters from GAFF or GROMOS 54A7.

Key words Halogen bonding, Biomolecular recognition, Electrostatic potential, Point-charges, Molecular dynamics simulations

1 Introduction

The regulation of ion concentration in cells is critical for human health and, in biological systems, ion transport across membranes is commonly performed by several specialized transmembrane proteins. The misregulation of these systems often leads to serious illness (channelopathies), e.g., Bartter's syndrome (sodium, potassium, and chloride) and cystic fibrosis (chloride) [1]. In this scope, the latter has received particular attention leading to the development of several synthetic molecules that eventually mimic the specialized membrane protein function by promoting anion transport. These promising therapeutic agents often rely on the formation of hydrogen bonds or ion pairs between the synthetic molecule and the anion [2]. Given their hydrophobic nature, directionality, and strength, halogen bonds were also added very recently to the *palette* of interactions that can be explored to build chloride anion transporters [3]. Indeed, halogen-bond mediated transport can be

Irina S. Moreira et al. (eds.), *Computational Design of Membrane Proteins*, Methods in Molecular Biology, vol. 2315,
https://doi.org/10.1007/978-1-0716-1468-6_16,

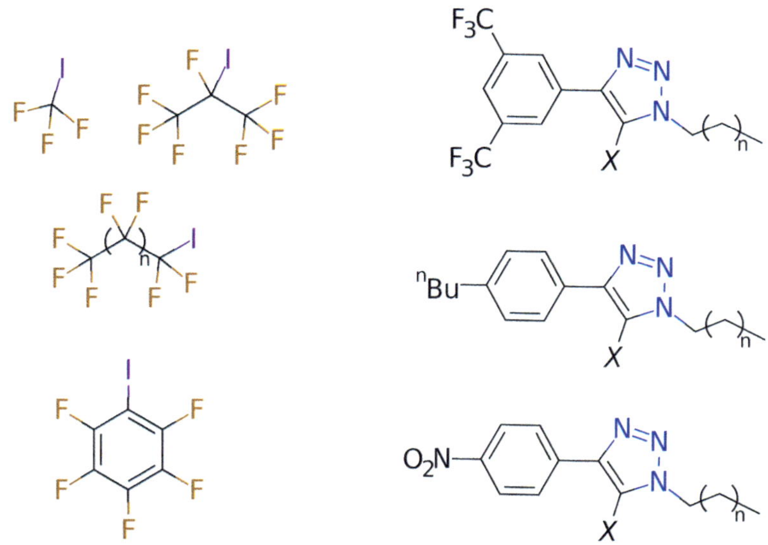

Fig. 1 Example of small synthetic molecules that promote transmembrane anion transport using halogen bonds: perfluoroiodoalkanes and iodobenzene derivatives (left) [3] and 1,2,3-triazole derivatives (right) [4]

achieved with small perfluoroiodoalkanes and iodobenzene derivatives (Fig. 1, left) [3] or 1,2,3-triazole derivatives [4], to name a few (Fig. 1, right).

Halogen bonds [5, 6] are noncovalent interactions between an electrophilic site located at the tip of the halogen and a nucleophilic species, for instance, an anion. This electrophilic site, termed σ-hole [7], corresponds to a positive region on the molecular electrostatic potential (MEP) localized at the tip of the halogen. Before being applied in anion transport, halogen bonds were already a major player in anion recognition [8–10] and drug design [11, 12]. Very recently, the potential role of halogen bonds in mediating membrane–ligand interactions was also demonstrated using molecular dynamics (MD) simulations [13] thus illustrating the large range of applicability of this interaction. Therefore, a correct description of halogen bonds by computational methods is paramount for computer-aided drug design given their prominent role in the field [14].

Modeling halogen bonds is not a trivial task [12, 15]. Although quantum-mechanical (QM) methods can account for the charge anisotropy and provide a good description of σ-holes, and hence, halogen bonds, they are generally too expensive for large biomolecular systems. On the other hand, force field-based methods, such as molecular mechanics/molecular dynamics (MM/MD) simulations, can tackle large-sized systems, also providing a suitable sampling of the conformational space which is particularly relevant in molecular recognition [12]. However, in the majority of classical force fields, single atomic point charges are attributed, and hence, halogens are in most cases described as negatively charged particles which lead to an improper description of the halogen and therefore, halogen bonds. Therefore, to describe the anisotropic charge distribution of halogen atoms, one of the simplest and most common approaches is to add an off-center particle (often called extra-point, EP) along the C–X bond, which was firstly introduced by Ibrahim [16].

In this work, a computational protocol to generate suitable parameters (especially charges) for small synthetic molecules that potentialy promote transmembrane anion transport using halogen bonds, and hence, mimic the function of large transmembrane proteins, will be presented. Taking the small synthetic transporter iodopentafluorobenzene (C_6F_5I) as an example [3] (*see* **Note 1**), and using the strategies used in references [17] and [13], suitable topologies for the **GRO**ningen **MA**chine for **C**hemical **S**imulations molecular dynamics package (GROMACS) [18–20] will be obtained, allowing to perform simulations with the general AMBER force field (GAFF) [21] or the GROMOS 54A7 force field [22]. Notice that the resulting topologies of small transmembrane transporters obtained with this protocol can be further used in membrane simulations following the other protocols provided in other chapters of this book.

2 Materials

1. A Quantum Mechanics (QM) software allowing the generation of QM-based electrostatic potentials. In this protocol, the paid software Gaussian 09 Revision D.01 (http://gaussian.com/) is used.
2. AmberTools16, a collection of freely available programs for biomolecular simulation and analysis, available at https://ambermd.org/AmberTools.php. Among others, it contains programs allowing to extract electrostatic potential files, calculate restrained electrostatic fitting charges (RESP) [23], and create RESP input files. It also contains the Antechamber suite [24] which automatizes the generation of GAFF topologies (including charges) from QM calculations.
3. AnteChamber PYthon Parser interfacE (ACPYPE) [25], a tool that simplifies the automatic generation of topology and parameters in different formats, namely, allowing the conversion of AMBER/GAFF topologies to GROMACS-compatible ones.
4. The Automated Topology Builder (ATB) and Repository [26] whose server is freely available at https://atb.uq.edu.au/. This server provides topologies and parameters for a wide range of molecules that are compatible with the GROMOS family of force fields.

5. A text editor with column support. VScode (https://code.visualstudio.com/), Notepad++ (https://notepad-plus-plus.org/downloads/), and Emacs (https://www.gnu.org/software/emacs/) are just a few examples of editors that can be used.

3 Methods

3.1 Obtaining a GROMACS Topology with an Off-Center Charge (GAFF)

1. Optimize the geometry at the B3LYP/6-311G(d,p) level of theory (or any other trustworthy level of theory) and calculate the frequencies to check if the structure corresponds to a minimum (*see* **Note 2**). A minimal sample input file (`5fluoroiodobenzene-opt.com`) is given in Supporting Information.
2. Taking the optimized geometry, generate a "reference" QM-based electrostatic potential grid at the HF/6-31G* level of theory (6-311G* for iodine). This method/basis is the one recommended for GAFF [21]. A minimal sample input file (`5fluoroiodobenzene-RESP.com`) is given in Supporting Information (*see* **Note 3**).
3. Assign GAFF atom types and calculate RESP charges for the molecule without an off-center charge (*see* **Note 4**). This is accomplished by running *antechamber* from AmberTools, creating a *mol2* file (`5fluoroiodobenzene-RESP.mol2`). The minimal command-line syntax should be:

```
$AMBERHOME/bin/antechamber -i 5fluoroiodobenzene-RESP.log -fi
gout \
  -o 5fluoroiodobenzene-RESP.mol2 -fo mol2 \
  -c resp
```

At this stage, the reader can check the charges and atom types of your molecule by inspecting the *mol2* file. The charge of iodine is 0.140, and the atom type is *i* (*see* **Note 5**).

4. Add an off-center particle to the MEP grid template `ANTECHAMBER`.ESP (*see* **Note 6**). The EP will be placed at an X–EP distance equal to the atomic radius of X derived from the R_{min} Lennard-Jones (LJ) parameter of iodine in GAFF (2.15 Å) as suggested in reference [16], and as used by us in references [13, 27] (*see* **Note 7**). Save the file as `5fluoroiodobenzene-RESP_2.15.ESP` (a sample is provided in Supporting Information).
5. Modify the RESP template `ANTECHAMBER`_RESP1.IN file to reflect the addition of an extra particle and change the total number of atoms. Save it as `ANTECHAMBER`_RESP1_EP.IN.

6. Refit the RESP charges with the added off-center charge (EP). This can be accomplished with the following minimal command-line syntax:

```
$AMBERHOME/bin/resp -O -i ANTECHAMBER_RESP1_EP.IN \
-o 5fluoroiodobenzene-RESP1_2.15.out \
-p 5fluoroiodobenzene-RESP1_2.15.pch \
-t 5fluoroiodobenzene-RESP1_2.15.chg \
-e 5fluoroiodobenzene-RESP_2.15.ESP
```

   The summary of the final fitted charges along with the error of the fit can be found in `5fluoroiodobenzene-RESP1_2.15.pch` (*see* **Note 8**). Iodine is now negatively charged (−0.070), and the EP is positively charged (+0.070).

7. For convenience, build a modified *mol2* file containing the EP and its associated charge (`5fluoroiodobenzene-RESP_2.15.mol2`), changing also the charges of the remaining atoms to the ones found in `5fluoroiodobenzene-RESP1_2.15.pch` (*see* **Note 9**).

8. Check for missing GAFF parameters on your molecule (*see* **Note 10**) by typing:

```
$AMBERHOME/bin/parmchk\
 -i 5fluoroiodobenzene-RESP_2.15.mol2 -f mol2 \
 -o 5fluoroiodobenzene-RESP_2.15.frcmod
```

9. Create an AMBER topology (`5fluoroiodobenzene-RESP_2.15.top`) and coordinates file (`5fluoroiodobenzene-opt-RESP_2.15.crd`) with the command-line program *tleap*, using the template `leap.in` file provided:

```
$AMBERHOME/bin/tleap -s -f leap.in
```

10. Convert the AMBER topology file to a GROMACS-compatible topology using ACPYPE. A minimal syntax is:

```
acpype.py -p 5fluoroiodobenzene-RESP_2.15.top \
 -x 5fluoroiodobenzene-RESP_2.15.crd \
 -b 5fluoroiodobenzene-RESP_2.15_tmp
```

   thus creating `5fluoroiodobenzene-RESP_2.15_tmp_GMX.{top,gro}` files.

11. Add the section "`[ virtual_sites2 ]`" to the GROMACS topology file, erasing all the bonded terms pertaining to the EP (*see* **Note 11**). Save it as `5fluoroiodobenzene-RESP_2.15_GMX.top`. This file, containing an iodopentafluorobenzene (C_6F_5I) topology with an added EP to mimic

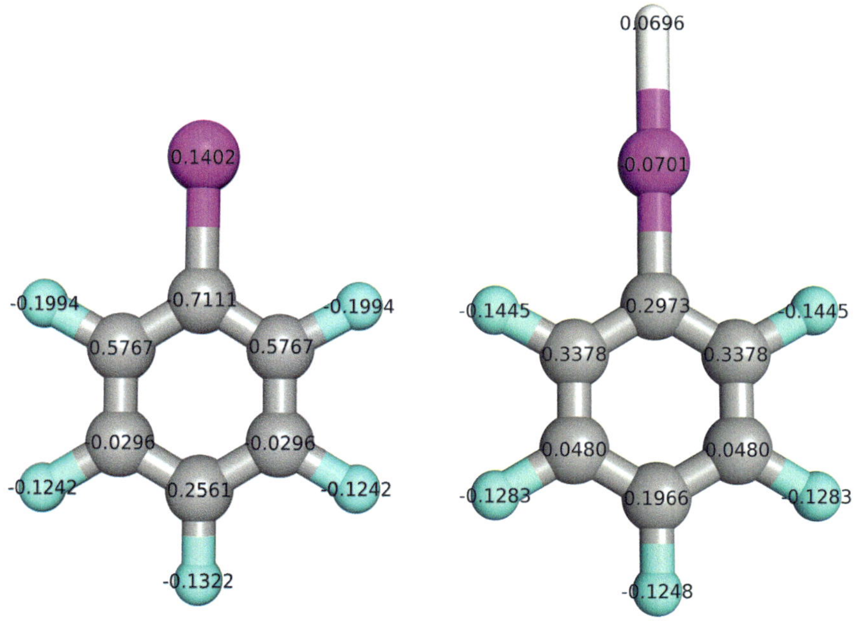

Fig. 2 GAFF-compatible partial atomic charges derived for C_6F_5I parameterized without (left) and with (right) an EP addition

the σ-hole, can now be used in GROMACS in MD simulations.

12. The charges with and without an EP can be compared for analysis purposes (Fig. 2).

3.2 Obtaining a GROMACS Topology with an Off-Center Charge (GROMOS 54A7)

1. Obtain a GROMACS topology (GROMOS 54A7) for iodopentafluorobenzene (C_6F_5I) from the Automated Topology Builder (ATB) server. This file contains the ATB topology without an added off-center charge and will be used as a template (*see* **Note 12**).
2. Optimize the geometry at the B3LYP/6-31G* (6-311G* for iodine) level of theory. The calculations are performed in a water implicit continuum model (*see* **Note 13**). Calculate the frequencies to check if the structure corresponds to a minimum. A minimal sample input file (`5fluoroiodobenzene-water-opt.com`) is given in Supporting Information.
3. Taking the optimized geometry, generate the QM-based electrostatic potential grid at the B3LYP/6-31G* level of theory (6-311G* for iodine) in water, as used in ATB. A sample input file (`5fluoroiodobenzene-water-resp.com`) is provided in Supporting Information (*see* **Note 14**).

4. Generate the necessary template input files for adding an off-center charge (ANTECHAMBER .ESP, ANTECHAMBER _RESP1.IN):

```
$AMBERHOME/bin/antechamber -i 5fluoroiodobenzene-water-resp.
log -fi gout \
  -o 5fluoroiodobenzene-water-resp.mol2 -fo mol2 \
  -c resp
```

   The charges without an added off-center particle can be checked in the 5fluoroiodobenzene-water-resp.mol2 file.

5. Add an off-center particle to the MEP grid template ANTECHAMBER .ESP and save it as 5fluoroiodobenzene-water-resp_2.16.ESP. In this protocol, the EP will be placed at R_{min} (2.16 Å in) which was shown to provide the most efficient sampling of halogen bonds using a GROMOS force field [17] (*see* **Note 15**). Add the EP in the ANTECHAMBER_RESP1.IN file and save it as ANTECHAMBER_RESP1_EP.IN.

6. Fit the RESP charges with an off-center particle (EP):

```
$AMBERHOME/bin/resp -O -i ANTECHAMBER_RESP1_EP.IN \
-o 5fluoroiodobenzene-water-resp_2.16.out \
-p 5fluoroiodobenzene-water-resp_2.16.pch \
-t 5fluoroiodobenzene-water-resp_2.16.chg \
-e 5fluoroiodobenzene-water-resp_2.16.ESP
```

   The summary of the final fitted charges can be found in 5fluoroiodobenzene-water-resp_2.16.pch.

7. As before, a modified *mol2* file containing the EP and its associated charge (5fluoroiodobenzene-water-resp_2.16.mol2), along with the updated charges of remaining elements, can also be built for convenience.

8. Using the template file from ATB (8YUP_GROMACS_G54A7F-F_unitedatom.itp), update the section "[atoms]" from the topology using the newly derived charges and adding a new massless particle. Also, add the corresponding "[virtual_-sites2]" section . Save it as 5fluoroiodobenzene_EP_2.16.itp.

9. The charges with and without an EP can be compared for analysis purposes (Fig. 3).

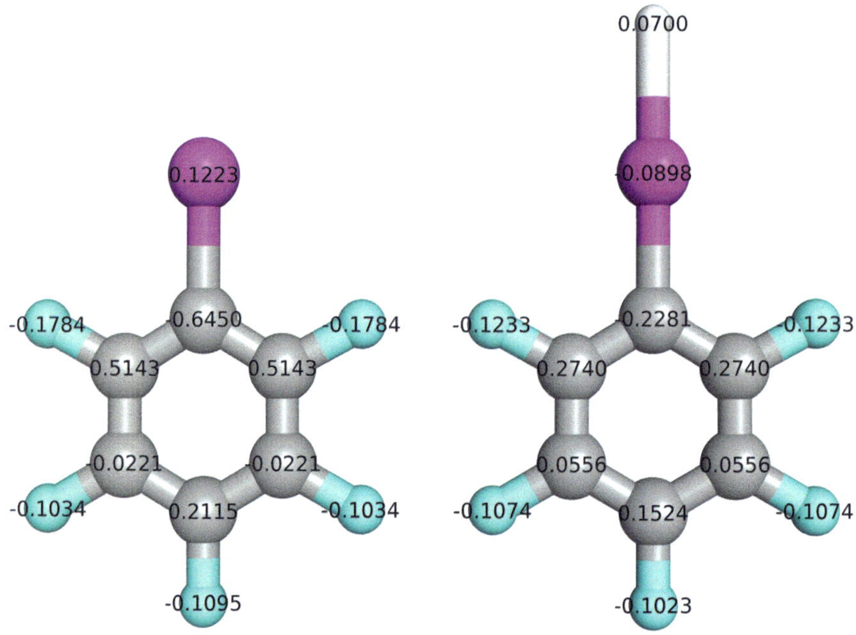

Fig. 3 GROMOS/ATB-compatible partial atomic charges derived for C_6F_5I parameterized without (left) and with (right) an EP addition

4 Notes

1. This protocol uses the small synthetic transporter iodopentafluorobenzene (C_6F_5I) as an example. However, as it can be inferred, the protocol is not limited to this as it can be easily extended to any other halogenated molecule.
2. The 6-311G(d,p) basis set is not directly available for iodine in Gaussian; therefore, it was downloaded from the Basis Set Exchange (https://www.basissetexchange.org/) [28]. For iodopentafluorobenzene, calculating frequencies might seem excessive; however, when applying this protocol to larger and floppier molecules, the frequency calculation is paramount.
3. As 6-31G* is not available for iodine, 6-311G* was used; default Merz-Singh-Kollman (MK) radii are employed for all elements apart from iodine for which there is no default radius. A value of 2.3 Å (used in Gaussian 09 for bromine) was assigned (ReadRadii) although it is noted that this value probably needs optimization [13, 17]. It is also a good policy to read the optimized geometry from the Gaussian checkpoint file.
4. This step is not strictly necessary as you can run the programs called by *antechamber* by yourself (*atomtype*, *bondtype*, *respgen*, *resp*, etc). However, running *antechamber* will not only

automatically assign GAFF atom types and calculate resp charges for a non-EP model, but will also provide the reader with all the necessary template input files for adding an off-center charge. For instance, the MEP grid is placed in `ANTECHAMBER`.ESP whereas the resp input files are in `ANTECHAMBER`_RESP1.IN, `ANTECHAMBER`_RESP2.IN.

5. Notice the positive charge in iodine which is due to the electron-withdrawing nature of the fluorine substituents which massively increases the σ-hole. If one does not use an off-center particle (EP) at iodine, the atomic charge that better reproduces the MEP is positive. However, even with a positive charge at iodine, this type of model is still unable to describe halogen bonds [13].
6. In order to tackle several molecules, the reader is advised to fully automate the addition of the EP. The coordinates of the EP should be added to the `ANTECHAMBER`.ESP file, in atomic units, just before the MEP grid. The coordinates are obtained by extending the C–I vector by an amount equal to the desired X–EP distance. The total number of particles must also be corrected at the top of the `ANTECHAMBER`.ESP file. Compare the `ANTECHAMBER`.ESP and `5fluoroiodobenzene-RESP_2.15.ESP` to see the modifications.
7. Adding the EP at other distances different from R_{min} has also been proposed in the literature. The reader is free to adapt to other X–EP distances, accordingly.
8. In this specific example, we freely fitted all the charges, including the EP. The reader is free to restrain a specific charge if that is needed. Additionally, you will notice that the `ANTECHAMBER`_RESP2.IN template file for the second stage fitting is not actually fitting any charges (negative values on the atoms). All the equivalences were considered in the first stage. Only in very specific cases, the second stage actually does anything. Therefore, in this protocol, only one stage is performed; however, the reader is advised to check if the second stage is required and act accordingly.
9. A bond between the EP and the I atom must also be created and therefore, the number of atoms and bonds must also be corrected. Compare the original `5fluoroiodobenzene-RESP.mol2` with the `5fluoroiodobenzene-RESP_2.15.mol2` provided in Supporting Information.
10. Although GAFF is usually very comprehensive in terms of bonded and nonbonded parameters, sometimes a specific combination is not found. The program *parmchk* checks the molecule indicating with a note "ATTN, need revision" (and

assigning values of zero) the missing or problematic parameters. In this specific case, only parameters pertaining to the EP will be missing. We do not need to fix this issue since the EP will later be transformed to GROMACS type 2 virtual site, defined by the respective C–I bond without additional parameters (*see* **Note 11**).

11. The EP is added as a GROMACS type 2 virtual interaction site. Succinctly, this site is defined as a function of the C–I bond as follows:

```
[ virtual_sites2 ]
13 1 3 1 -1.00996
```

Here, particle 13 (the EP) is defined from atoms 1 and 3 (iodine and carbon, respectively). For this reason, all the other bonded parameters pertaining to the EP must be removed.

12. A sample ATB file (`8YUP_GROMACS_G54A7FF_unitedatom.itp`) is provided. Notice that contrary to the version used in reference [17], the atomic mass of iodine is now correct in the most recent version of ATB.
13. This step can be skipped by using the geometry provided by ATB. However, for large molecules, ATB uses semiempirical methods for geometry optimization, and therefore, the reader might want to optimize the geometry at the DFT level of theory.
14. This protocol uses Gaussian 09 whereas ATB uses GAMESS-US. The sample Gaussian input file tries to emulate, as much as possible, all the default GAMESS parameters (e.g., the C-PCM continuum model with Bondi atomic radii) used in ATB. As before, an MK radius of 2.3 Å was used for iodine. Since it is missing, GAMESS-US assigns a default value of 1.8 Å.
15. The value of 2.16 Å was derived from the C^6 and C^{12} Lennard-Jones parameters for iodine provided by ATB.

Acknowledgments

Fundação para a Ciência e a Tecnologia (FCT), Portugal, is acknowledged for grants UIDB/04046/2020 and UIDP/04046/2020 (BioISI). This work was financed by FCT, Programa Operacional Regional de Lisboa (Lisboa 2020), Portugal 2020, FEDER/FN, and the European Union under project number 28455 (LISBOA-01-0145-FEDER-028455, PTDC/QUI-QFI/28455/2017). D. Vila Viçosa and R. S. Nunes are also acknowledged for their collaboration in this research field.

References

1. Ashcroft FM (2000) Ion channels and disease. Elsevier Science. https://doi.org/10.1016/B978-0-12-065310-2.X5000-0
2. Haynes CJE, Gale PA (2011) Transmembrane anion transport by synthetic systems. Chem Commun 47:8203–8209. https://doi.org/10.1039/C1CC12061A
3. Jentzsch AV, Emery D, Mareda J et al (2012) Transmembrane anion transport mediated by halogen-bond donors. Nat Commun 3:905. https://doi.org/10.1038/ncomms1902
4. Bickerton LE, Sterling AJ, Beer PD et al (2020) Transmembrane anion transport mediated by halogen bonding and hydrogen bonding triazole anionophores. Chem Sci 11:4722–4729. https://doi.org/10.1039/D0SC01467B
5. Costa PJ (2017) The halogen bond: nature and applications. Phys Sci Rev 2:488. https://doi.org/10.1515/psr-2017-0136
6. Desiraju GR, Ho PS, Kloo L et al (2013) Definition of the halogen bond (IUPAC recommendations 2013). Pure Appl Chem 85:1711–1713. https://doi.org/10.1351/PAC-REC-12-05-10
7. Clark T, Hennemann M, Murray JS, Politzer P (2007) Halogen bonding: the sigma-hole. J Mol Model 13:291–296. https://doi.org/10.1007/s00894-006-0130-2
8. Nunes R, Costa PJ (2017) Ion-pair halogen bonds in 2-halo-functionalized Imidazolium chloride receptors: substituent and solvent effects. Chem Asian J 12:586–594. https://doi.org/10.1002/asia.201601690
9. Zapata F, Caballero A, White NG et al (2012) Fluorescent charge-assisted halogen-bonding macrocyclic halo-imidazolium receptors for anion recognition and sensing in aqueous media. J Am Chem Soc 134:11533–11541. https://doi.org/10.1021/ja302213r
10. Caballero A, Zapata F, White NG et al (2012) A halogen-bonding catenane for anion recognition and sensing. Angew Chem Int Ed Engl 51:1876–1880. https://doi.org/10.1002/anie.201108404
11. Wilcken R, Zimmermann MO, Lange A et al (2013) Principles and applications of halogen bonding in medicinal chemistry and chemical biology. J Med Chem 56:1363–1388. https://doi.org/10.1021/jm3012068
12. Costa PJ, Nunes R, Vila-Viçosa D (2019) Halogen bonding in halocarbon-protein complexes and computational tools for rational drug design. Expert Opin Drug Discov 14:805–820. https://doi.org/10.1080/17460441.2019.1619692
13. Nunes RS, Vila-Viçosa D, and Costa PJ (2021) Halogen bonding: an underestimated player in membrane–ligand interactions? J Am Chem Soc (in press). https://doi.org/10.1021/jacs.0c12470
14. Xu Z, Yang Z, Liu Y et al (2014) Halogen bond: its role beyond drug-target binding affinity for drug discovery and development. J Chem Inf Model 54:69–78. https://doi.org/10.1021/ci400539q
15. Costa PJ, Nunes R (2018) Advances in the computational modeling of halogen bonds in biochemical systems. Front Comput Chem 4:144–183. https://doi.org/10.2174/9781681084411118040006
16. Ibrahim MAA (2011) Molecular mechanical study of halogen bonding in drug discovery. J Comput Chem 32:2564–2574. https://doi.org/10.1002/jcc.21836
17. Nunes R, Vila-Viçosa D, Machuqueiro M, Costa PJ (2018) Biomolecular simulations of halogen bonds with a GROMOS force field. J Chem Theory Comput 14:5383–5392. https://doi.org/10.1021/acs.jctc.8b00278
18. Berendsen HJC, van der Spoel D, van Drunen R (1995) GROMACS: a message-passing parallel molecular dynamics implementation. Comput Phys Commun 91:43–56. https://doi.org/10.1016/0010-4655(95)00042-E
19. Van Der Spoel D, Lindahl E, Hess B et al (2005) GROMACS: fast, flexible, and free. J Comput Chem 26:1701–1718. https://doi.org/10.1002/jcc.20291
20. Abraham MJ, Murtola T, Schulz R et al (2015) GROMACS: high performance molecular simulations through multi-level parallelism from laptops to supercomputers. SoftwareX 1-2:19–25. https://doi.org/10.1016/j.softx.2015.06.001
21. Wang J, Wolf RM, Caldwell JW et al (2004) Development and testing of a general amber force field. J Comput Chem 25:1157–1174. https://doi.org/10.1002/jcc.20035
22. Schmid N, Eichenberger AP, Choutko A et al (2011) Definition and testing of the GROMOS force-field versions 54A7 and 54B7. Eur Biophys J 40:843–856. https://doi.org/10.1007/s00249-011-0700-9
23. Bayly CI, Cieplak P, Cornell W, Kollman PA (1993) A well-behaved electrostatic potential based method using charge restraints for deriving atomic charges: the RESP model. J Phys

Chem 97:10269–10280. https://doi.org/10.1021/j100142a004

24. Wang J, Wang W, Kollman PA, Case DA (2006) Automatic atom type and bond type perception in molecular mechanical calculations. J Mol Graph Model 25:247–260. https://doi.org/10.1016/j.jmgm.2005.12.005
25. Sousa da Silva AW, Vranken WF (2012) ACPYPE - AnteChamber PYthon parser interfacE. BMC Res Notes 5:367. https://doi.org/10.1186/1756-0500-5-367
26. Stroet M, Caron B, Visscher KM et al (2018) Automated topology builder version 3.0: prediction of solvation free enthalpies in water and hexane. J Chem Theory Comput 14:5834–5845. https://doi.org/10.1021/acs.jctc.8b00768
27. Nunes R, Vila-Viçosa D, Costa PJ (2019) Tackling halogenated species with PBSA: effect of emulating the σ-hole. J Chem Theory Comput 15:4241–4251. https://doi.org/10.1021/acs.jctc.9b00106
28. Pritchard BP, Altarawy D, Didier B et al (2019) New basis set exchange: an open, up-to-date resource for the molecular sciences community. J Chem Inf Model 59:4814–4820. https://doi.org/10.1021/acs.jcim.9b00725

INDEX

A

B

C

Irina S. Moreira et al. (eds.), *Computational Design of Membrane Proteins*, Methods in Molecular Biology, vol. 2315, https://doi.org/10.1007/978-1-0716-1468-6, © Springer Science+Business Media, LLC, part of Springer Nature 2021

H

I

K

L

M

Printed by Printforce, the Netherlands